Biodiversity and its Significance

Biodiversity and its Significance

Editors
Pramod Tandon
Yash P. Abrol
Suman Kumaria

I.K. International Publishing House Pvt. Ltd.

NEW DELHI • BANGALORE

Published by

I.K. International Publishing House Pvt. Ltd.
S-25, Green Park Extension, Uphaar Cinema Market
New Delhi – 110 016 (India)
E-mail : info@ikinternational.com

ISBN 81-88237-81-7

10 9 8 7 6 5 4 3 2

Published by Krishan Makhijani for I.K. International Publishing House Pvt. Ltd., S-25, Green Park Extension, Uphaar Cinema Market, New Delhi – 110 016 and Printed by Rekha Printers Pvt. Ltd., Okhla Industrial Area, Phase II, New Delhi – 110 020.

Preface

Biological diversity or biodiversity, the study of all organisms, species, populations, and the genetic variations among these, includes the complex assemblage of communities and ecosystems. It provides the basic essentialities, namely, food, fodder, shelter, biomass, etc. for survival of humankind. As one of the resource-ecosystems of environment, it provides basic raw materials, which are primary infrastructures and basic requirement for any economic or other income generating activities. Biodiversity helps to maintain the biophysical systems, chemical balance of the earth and climate stabilization; above all, it is a life supporting system. In other words, biodiversity is fundamental to the existence of life on earth, and the importance of it cannot be underestimated. It creates complex ecosystems that could never be reproduced by humans.

Biodiversity is not only the variety of living organisms on our planet, but also the interdependence of all these living beings, including humans. The ecosystems and species have been negatively affected in a number of ways. In the past few centuries, humans have had an especially negative affect on biodiversity. The use of land is mostly responsible for the contemporary decrease in biodiversity. Around 25% of the earth's surface is covered by farm land; a problem that has not been looked into so far. The most fertile soil containing the rich biodiversity is usually found in the best climates. Besides, the ever-expanding human population, intensive agriculture, the use of fertilizers and pesticides are the growing concern. Many developed countries have become aware and are regulating pollution and toxification. But, the degrees to which this is done vary, and it is still a major concern as many pollutants travel incredibly quickly and cover a broad area.

The knowledge of biodiversity has increased, in recent years, due to its importance in utilitarian values (medicine and agriculture), ecosystem services (continued production of atmospheric oxygen), and moral, ethical and aesthetic values. Plant biodiversity is the single greatest resource that humankind has treasured from nature. More than 2,50,000 species of mosses, ferns, conifers, and flowering plants have been described so far. It has been estimated that there may be more than 50,000 plant species yet to be documented, primarily in the tropical forests of the world. Plants constitute a major remedy in traditional medical system. The traditional knowledge in the use of plants for health care now provides initial leads towards drug development. In the present era of bioprospecting, the study of traditional knowledge about the uses of plants which has hitherto remained unknown to the world at large, gains more significance. The northeastern region of India, inhabited by more than a hundred different tribal populations of various ethnic groups, is a treasure-house of traditional uses of plants. In the backdrop of fast modernization and the considerable headway achieved in bringing the region in the national mainstream, there is an urgent need to record the treasure of indigenous knowledge in the region. Microorganisms occurring virtually in every habitat in the biosphere play a vital role in the functioning of the ecosystem. The value of biodiversity is thus immeasurable, and requires protection. Studies on biodiversity and forest cover dynamics using remote sensing data gains importance in the context of increasing deforestation rates and in studies relating to monitoring, maintenance and conservation of biodiversity.

The loss and extinction of species is a great threat to biodiversity. Extinction is a natural part of evolution, but under normal conditions it is a rare and obscure event wherein about 1-10 species are lost in a year. In contrast, the scientists so far have estimated that extinction rates have accelerated in this century to about 1,000 species per year. These numbers are alarming and have resulted in a global evolutionary upheaval in the diversity and composition of life on Earth. In this event of mass extinction, plants are suffering unprecedented losses. According to a 1997 global analysis of more than 2,40,000 plant species coordinated by the World Conservation Union-IUCN, one out of every eight plants surveyed is potentially at risk of extinction including species already endangered or clearly vulnerable to extinction. Moreover, around 90 percent of these at-risk species are endemic to a single country, that is, found nowhere else in the world. Crop varieties are lost for many reasons viz., extended drought destroys harvests which leads to consumption of the planting seed stocks, voluntarily abandoning of traditional seeds in preference to new varieties, change in agricultural practices or moving out of farming altogether. The significant reasons for the loss of plant diversity are attributed to uncontrolled urbanization, population explosion, industrialization and tremendous pollution, causing very heavy loss to the plant diversity resulting in the environmental degradation at an alarming rate. During the next few decades as many as one quarter of plant species could be lost forever if immediate action is not initiated. Such taxa may remain unexplored for their medicinal, economic and ornamental potentialities for the human welfare. During the next 30-40 years, above 60,000 plant species world over, will face severe threats to their survival or even become extinct.

Broad recognition of the need to safeguard plant resources is largely a twentieth century phenomenon. The dominant approach to conserving plant varieties and species has involved removing them from their native habitat or agricultural setting and protecting them at specialized institutions such as botanical gardns, nurseries, and gene banks. Plant diversity can be securely maintained only by protecting the native habitats and ecosystems where plants have evolved. The conservation of natural resources with some religious beliefs has been in practice in India. Conserving the forest and its ecosystem through traditional practices in the form of sacred groves is worth mentioning. Sacred groves are traditionally protected and managed by the local communities with wide range of management practices. It is now realized by mankind that the preservation of natural heritage is most vital and essential factor for its survival. Botanic Gardens have to play a leading role for the conservation of plant diversity, to face the threat, especially during the 21st century. There are roughly 1500 Botanic Gardens and Arboreta throughout the world and many countries are establishing new botanic gardens for this noble cause. Today botanic gardens play a significant role in the *ex situ* conservation of plants for their acclimatization, multiplication, judicious exploitation and rehabilitation. In modern times botanic gardens should take up database networking of living plant resources that will serve as feedback for accelerating the conservation programmes for saving plant diversity.

Climate of North-East India is its important resource, which has sustained the rich biodiversity of this humid tropical region of India. The summer and winter monsoons control the climate of the region. It is modulated by the peculiar geography of the region. Climate change is a growing concern as well, though it is somewhat debatable as to whether or not humans caused it. In the past, global climate change resulted in major shifts in species ranges and marked reorganization of biological communities, landscapes, and biomes. These changes occurred in a landscape that was not as fragmented as it is today and with little or no pressures from human activities. Current climate change coupled with other human pressures is stressing biodiversity far beyond the levels imposed by the global climatic change that occurred in the recent evolutionary past. The atmospheric concentration of green house gases viz. CO_2, CH_4 and N_2O has grown significantly. These can be attributed largely to human activities like fossil fuel burning, land use change, waste disposal, industrial processes and agriculture. Quantification of carbon fluxes and understanding the human induced global climatic changes forms one of the major studies with respect to International Geosphere Biosphere Programme. Natural changes in weather have had perhaps the greatest affect on biodiversity and ecological systems. The impacts of climate change on biodiversity are therefore of major concern. Many plant and animal species are unlikely to survive climate change. New analyses suggest that 15-37% of a sample of 1,103 land plants and animals would eventually become extinct as a result of climate changes expected by 2050. For some of these species there will no longer be anywhere suitable to live. Others will be unable to reach places where the climate is suitable. A rapid shift to technologies that do not produce greenhouse gases, combined with carbon sequestration, could save 15-20% of species from extinction.

Realizing the importance of biodiversity and its significance, this book is an attempt to cover some aspects of biodiversity in North-East India such as plants, crops, animals, microbes, remote sensing, conservation and sustainable utilization, issues regarding intellectual property rights and climate change. It is envisaged that the book will be of help to the students, researchers, professionals and planners dealing with various aspects of biodiversity.

Pramod Tandon
Y.P. Abrol
Suman Kumaria

E-mail Addresses of the Corresponding Authors

Bhatt, B.P.	bpbhatt@neline.com
Choudhury, H.	hiranjit_c@yahoo.com
Dkhar, M.S.	msdkhar@hotmail.com
Dkhar, Melboreen	melboreen@yahoo.com
Goswami, Sanjiban	sanjiban2000@yahoo.com
Gupta, H.S.	hsgupta@Lycos.com
Hore, D.K.	nbpgrshl@neline.com
Kayang, H.	hkayang@hotmail.com
Kumaria, S.K.	skatri@sancharnet.in
Mahapatra, B.K.	bijoy@icarneh.ren.nic.in
Pattanayak, A.	apat@icarneh.ren.nic.in plantbreeding@neline.com
Sarma, B.K.	bksarma@icarneh.ren.nic.in
Singh, Janak Kumar	janak@icarneh.ren.nic.in
Syiem, D.	donsyiem@hotmail.com
Syiem, M.B.	mayashreesyiem@yahoo.co.in
Tandon, P.	profptandon@yahoo.com tandon1@sancharnet.in
Vinod, K.	vinod@icarneh.ren.nic.in

Editors

Prof. Pramod Tandon, Vice-Chancellor, North-Eastern Hill University (NEHU), was formerly a professor in Jodhpur University. He has worked as a post-doctoral fellow in the Department of Molecular Biology and Biochemistry at the University of California, Irvine, USA during 1979-80. During 1989 to 1995 he was associated in collaborative research with several Universities and National Institute of Agrobiological Resources in Japan, leading to a long-term collaboration on Plant Conservation Biotechnology. He has published a number of research papers and articles in journals of both national and international repute. He has edited three books on biodiversity and biotechnology. Besides being a member of many Academic bodies and National Task Forces, Prof. Tandon is a Fellow of the National Academy of Sciences and the Indian Botanical Society. He is the recipient of several awards and recognitions, notably the B.P. Pal National Environment Fellowship Award on Biodiversity, Ministry of Environment and Forests, Government of India, and Professor Panchanan Maheshwari Medal Award of the Indian Botanical Society.

Prof. Tandon has made a significant contribution in mass micro-propagation, reestablishment and conservation of Indian endemics of NE region. He has been the Coordinator of Bioinformatics Centre, funded by Department of Biotechnology, Government of India.

Professor Yash P Abrol, former Head of the Division of Plant Physiology, Indian Agricultural Research Institute, New Delhi. Prof. Abrol has worked as Emeritus Scientist (CSIR), Senior Scientist (Indian National Science Academy) and is presently Adjunct Professor and Honorary Scientist at Indian National Science Academy. He obtained his Ph.D degree from the University of Chicago (USA) and worked as a Post-doctoral Fellow, at the University of California, Davis. He had been National Fellow (ICAR) from 1978-84. He has profusely published research/review articles in International and National Journals besides editing a number of books/proceedings. His areas of interest include Climate Variability/Change and Agriculture; Land Use/Cover Change, Nitrogen Utilization/Metabolism in crop plants. He is a Fellow of the Indian National Science Academy; Indian Academy of Sciences, National Academy of Sciences and National Academy of Agricultural Science. Prof. Abrol received several awards notably Dr. R.D. Asana Award, Sukumar Basu Award, VASVIK and FICCI Award.

Dr. Suman Kumaria is a Reader in the Department of Botany, North-Eastern Hill University, Shillong. She obtained her Ph.D. degree from North-Eastern Hill University, Shillong. She has been a recipient of merit scholarships during her studies. Dr. Kumaria has an experience of 20 years in Plant Biotechnology research. She has been engaged in postgraduate teaching for the past 10 years. Dr. Kumaria has published research papers and articles in journals/books of National and International repute. She is a Fellow of the International Society for Conservation of Natural Resources, besides being a member of various academic bodies. Her current interests include mass propagation and germplasm conservation of plants, mainly orchids of NE India, and physiology and biochemistry of their regeneration. She has developed improved technologies for large-scale propagation of commercially important orchids of NE India, as part of a research project funded by Department of Biotechnology, Government of India. As a result of this project, a large number of tissue culture–raised hardened orchid plantlets have been supplied to the rural women folk in Meghalaya for their economic upliftment.

Contents

Chapter 1

Nitrogen in Agricultural Ecosystem

M.Z. Abdin[*], R.K. Dwiwedi[*], B. Singh[**] and Yash P. Abrol[***]

Centre for Transgenic Plant Development, Department of Biotechnology[*], Department of Biochemistry[**], and Environmental Botany[***], Hamdard University, New Delhi - 110 062

Nitrogen is a key component of some constituents, such as nucleic acids, vitamins, hormones, proteins and enzymes which regulate various metabolic processes. Consequently, its deficiency in agricultural soils results in impaired plant growth and development that ultimately leads to reduced yield and poor quality of crops. To feed the ever-increasing world population, it is expected that the consumption of nitrogenous fertilizers will be doubled within the next 25 to 30 years. This will not only lead to the drainage of foreign exchange, especially in the case of developing countries like India, but also cause environmental pollution. This problem, however, can be minimised by optimising the N-use efficiency of agricultural crops. The studies conducted so far have revealed that N-use efficiency of plants is not only limited by N-assimilation but also by its uptake and translocation to different plant parts. An understanding of these processes at physiological, biochemical and molecular levels will help to improve their N-use efficiency.

INTRODUCTION

Nitrogen is a key component of some constituents, such as DNA, RNA, vitamins, and hormones, proteins and enzymes, which regulate various metabolic processes. Application of nitrogen influences crops in different ways, like increased nitrogen supply, increased leaf area development and its duration, tiller formation, photosynthetic productivity, etc., which ultimately affect the biomass, grain yield and quality of the produce. Its deficiency in agricultural soils is a world wide problem (Abrol *et al.*, 1999) and it is true of the Indian scenario, as well. More than 63% of the countrywide soils are very poor in nitrogen status (Motsara, 2002). To enhance crop productivity per unit area and agricultural production, so as to meet the requirements of the burgeoning population (which may touch the 8.3 billion mark by 2030), particularly in the developing world, nitrogenous fertilisers will play a crucial role, as there is little possibility of increasing the cultivable land in most of the regions. However, the use efficiency of applied nitrogen is poor. For wheat and rice, the utilisation efficiency of nitrogenous fertilisers under field conditions is around 50 and 25 to 30%, respectively (National Academy of Agricultural Sciences, 1999). This poor efficiency is of great concern for a number of reasons: (i) even if the efficiency of nitrogenous fertilisers remains at the present level, the losses will increase enormously, as their consumption is expected to double within the next 25 to 30 years, (ii) their manufacture involves high cost technology, requiring a whole range of feed stocks, the major one being naphtha, a petroleum product. Its import results in the drainage of foreign exchange reserves of the country (Ramarathnam, 1980), and (iii) this leads to environmental impairment.

In this chapter, various aspects pertaining to nitrogenous fertiliser requirements and attempts to understand/improve their use efficiency at the agronomic, physiological and molecular level are discussed.

NITROGEN FERTILITY STATUS OF SOILS

Nitrogen is the most widely deficient nutrient in agricultural soils, the world over. Its application, therefore, initially results in increased yield. Recently, in India, 63, 26 and 17% soils were classified as heavy low, medium and high fertility, respectively, with respect to soil nitrogen status. The nitrogen status is high in states like Arunachal Pradesh, Meghalaya, Mizoram, Nagaland, Goa, and Himachal Pradesh. The states having medium status of available nitrogen are Assam, Tripura, West Bengal, Gujarat, Madhya Pradesh, Jammu and Kashmir, Punjab and Karnataka. States and union territories which are low in available nitrogen status are Manipur, Orissa, Maharashtra, Haryana, Uttar Pradesh, Andhra Pradesh, Andaman and Nicobar, Islands, Kerala, Pondicherry and Tamil Nadu (Motsara, 2002), Fig 1.1. Those soils, under the influence of arid and semi-arid climate, continue to be typically more sensitive to N-deficiency. High temperature induced disintegration of soil organic matter, the only source of plant available nitrogen, is the leading cause of nitrogen impoverishment (Katyal *et al.*, 1987).

FOOD PRODUCTION AND NITROGEN REQUIREMENT

Food from agriculture supports around six billion people who inhabit this planet today. The population is expected to increase to about 7.2 and 8.3 billion, respectively by the years 2015 and 2030 (Table 1.1). The major increase is being expected in the developing countries. This will be despite the continuous decline in growth rates (Table 1.2). There will be about 77 million added each year to the already burgeoning population. Between 2002 to 2015, the developing countries will absorb 95% of the growth, with two nations, India and China, accounting for 40% of the total increase. India's population will grow by 250 million people, while China will contribute an additional 100 million people to the global landscape (Food and Agricultural Organization, 2002).

1-Jammu & Kashmir; **2**-Himachal Pradesh; **3**-Punjab; **4**-Haryana; **5**-Uttar Pradesh; **6**-Gujarat; **7**-Madhya Pradesh; **8**-Maharashtra; **9**-Andhra Pradesh; **10**-Karnataka; **11**-Goa; **12**-Kerala; **13**-Tamil Nadu; **14**-Pondicherry; **15**-Orissa; **16**-West Bengal; **17**-Meghalaya; **18**-Assam; **19**-Arunachal Pradesh; **20**-Nagaland; **21**-Manipur; **22**-Mizorum; **23**-Tripura; **24**-Andaman Nicobar Islands

Fig. 1.1: Nitrogen fertility status of soils of Indian states (Source- Motsara, 2002)

Table 1.1: World population projection (millions)

Year	*1979-81*	*1997-99*	*2015*	*2030*	*2050*
World	4430	5900	7207	8270	9322
Developing countries	3259	4595	5858	6910	7987
Industrial countries	789	892	951	979	985
Transition countries	382	413	398	381	349

Source: FAO, 2002

Table 1.2: Population growth (% per annum)

Year	*1979-99*	*1989-99*	*1979-99 to 2015*	*2015-2030*	*2030-2050*
World	1.6	1.5	1.2	0.9	0.6
Developing countries	1.9	1.7	1.4	1.1	0.7
Industrial countries	0.7	0.7	0.4	0.2	0
Transition countries	0.5	0.1	-0.2	-0.3	-0.3

Source: FAO, 2002

The declining growth rate in population will reduce the growth in the demand for agricultural production, though it will still be in the range of 1.7 to 2.2% in developing and 0.4 to 0.7% in developed countries. Cereals will remain the mainstay of human food supply (Table 1.3). This is primarily to meet the nutritional requirements of the millions, who are undernourished. This scenario is not, however, uniform. In China, the average food consumption reached to 3,040 kcal (1997-99), as compared to India with an average figure around 2,500 kcal and a population increase of 1% a year over the next 20 years. Most of the undernourished people are in the developing world, with the highest population being in Asia and Sub-Saharan regions (Table 1.4). The future economy will rise by 1.9% per capita, between 2000 and 2050, which will be higher than the 1.2% seen in the 1990s. Closely related to population and income changes, is the transformation of demographic patterns. The most vital demographic characteristic, particularly for projecting the kind and quantity of future food needs, is the fast-growing rate of urbanization, besides fast-growing economy in the developing countries. In 2000, approximately, 44% of the population of developing countries was residing in urban areas, up from 22% in 1960 and 30% in 1980. This had a significant effect on food demand structure. Urbanisation has accelerated dietary transition from basic staples, such as sorghum, millet, maize and root crops to fine cereals, fruits, vegetables, livestock products and processed foods. Production of cereals is expected to grow by one million tons by the year 2030. In developing countries, dietary changes have already taken place during the period 1964-66 to 1997-99 and they are expected to alter still further, by the year 2030 (FAO, 2002). Most growth will come from yield increase, and yield growth rates will be faster in developing countries, especially for wheat and rice. Expansion of area under cultivation will not be a significant source of future growth in cereal production as a whole. To meet the requirements of the urban population, importance should be given to the quality of food which is giving importance.

Table 1.3: Projection on world's cereals demand (million tons)

Year	*1974/76*	*1984/86*	*1997/99*	*2015*	*2030*
World	612	810	1003	1227	1406
Developing countries	438	618	790	1007	1185
Industrial countries	103	119	142	150	155
Transition countries	70	73	72	70	66

Source: FAO, 2002

Table 1.4: Undernourished population projection of the world

	Undernourished							
	Million people				**Percentage of Population**			
Year	*1990-92*	*1997-99*	*2015*	*2030*	*1990-92*	*1997-99*	*2015*	*2030*
World		815					14	
Developing countries	816	777	610	443	20	17	11	6
Industrial countries		11				1		
Transition countries		27				6		

Source: FAO, 2002

Demand for livestock products will increase by almost 60% in the developing countries, from 16.5 kg/year/capita by the year 2020, while this increase will be only 5.5% for the developed countries, where the present average per capita meat consumption of 80 kg/year is nearly four times higher than that of developing countries.

According to the recent FAO projection, meat consumption at the global level will increase by about two-fold. The consumption will be in the range of 55.0 to 66.0 in East Asia and Latin America. The present consumption in South Asia, particularly India is not high, and it is not expected to reach even the present level of consumption in the other regions, due to the dietary preference for less meat than in other societies (Table 1.5, FAO, 2000).

Table 1.5: Changes in the commodity consumption of food

	Meat, carcass weight (kg/person/year)					
	1964/66	*1974/76*	*1984/86*	*1995/97*	*2015*	*2030*
World	24.1	27.4	30.7	34.7	40.2	43.8
South Asia	3.9	3.9	4.3	5.5	8.2	11.8
East Asia	8.7	10.0	17.0	33.3	47.2	55.0
Sub-Sharan Africa	9.9	9.5	10.2	9.7	11.6	13.6
North-East/North Africa	11.9	13.7	20.5	20.0	26.6	32.0
Latin America/Carribean	31.7	35.6	39.7	28.5	57.8	66.0

India's population, estimated at 880 million in 1993, has tripled since Independence in 1947 and doubled since 1980. The growth rate is slowing down from 2.1% in the 1980s to 1.8% in the 1990s, and is expected to slow even further in the next century. Still, the population is projected to reach 1.3 billion by the year 2020 (Table 1.6). To keep pace with the rising food demand for 1,329.1 million people, the cereal demand in 2020 for direct human consumption will be 246.1 million tons, a 67% increase over the 1993 demand of 147.1 million tons (Table 1.7). The baseline projection for the total cereal demand of 296.2 million tons includes 50.1 million tons (or 17%) for feed and 246.1 million tons (83%) for direct human consumption. This would mean a doubling of total cereal demand after 1993, which is the kind of increase that India has experienced over the past 30 years (Table 1.7).

Table 1.6: Population projection (in millions) for the year 2020, India

Year	*Rural*	*Urban*	*Total*
1993	648.7	231.6	880.3
2020	863.5	465.6	1329.1

Source: Government of India, 1996

Table 1.7: Food, demand projection for the year 2020, India

Projected total cereal requirements for the year 2020 in comparison with other countries

Country/year/source	*Food*	*Feed*	*Total*	*kg/capita/day*
		(million metric tons)		
India				
1993 (actual)	047.121	3.712	150.833	0.47
2020 with per capita				
Income growth [a] of:				
2 percent	231.510	25.749	257.259	0.53
3.7 percent	246.080	50.111	296.191	0.61
6 percent	267.210	107.522	374.732	0.77
2020 (IMPACT)[b]	223.6	13.3	237.3	0.51
2020 (Kumar)[c]	237.6	16.9	254.5	0.55
China				
1993 (actual)	251.243	72.640	323.883	0.73
2020 (IMPACT)[b]	293.441	177.989	471.430	0.90
Pakistan				
1993 (actual)	20.361	1.313	21.674	0.47
2020 (IMPACT)[b]	42.406	2.826	45.232	0.49
United States				
1993 (actual)	34.822	158.200	193.022	2.03
2020 (IMPACT)[b]	42.390	202.101	244.491	2.08

[a] IFPRI, 1999; [b] IMPACT model baseline results (Rosegrant *et al.*, 1995); [c] Kumar's projections (Kumar, 1998).

This scenario requires increased food production that will certainly have to come through intensification of cultivation, as there are limited land resources. Consequently, there will be more pressure on soil and water resources, and that is of particular concern, as in many regions, there are already "exhaustion symptoms" because of unsustained use of natural resources (Abrol *et al.*, 2002, Kumar *et al.*, 2002). For their sustained use, management of soil fertility with efficient utilisation of available water resources will be a prerequisite. Thus, additional production will be highly input sensitive, and mineral fertilizers, particularly nitrogen will play a key role in realising the potential of future crop cultivars evolved by wide breeding and transgenesis, besides conventional breeding.

The International Fertiliser Development Centre (IFDC) has made detailed projections of future fertiliser demands. On this basis, global consumption of nitrogen fertiliser is projected to rise from 83.5 million tons in 2000 to 115 million tons in 2020, with an increase of about 40% (Table 1.8). Fertiliser-N use in India has risen dramatically since independence, with a modest rise of 0.058 million tons in 1951-52; its consumption galloped to 11.3 million tons in the year 2002. Currently, India ranks second in the world in the production, as well consumption of fertiliser-N.

Table 1.8: Fertiliser-Nitrogen Demand (in million tons) and projection of world and India

Year	*Developed Countries*	*Developing Countries*	*India**
2000	34.5	49	11.6
2020	41.8	72.6	30-35

Source: Bumb and Baanante (1996), *NAAS (1995)

It is evident that food grain production and fertiliser-N are closely interrelated (FAI, 2002; Fig. 1.2 and Fig. 1.3). With the rising food demand for 1329.1 million people, India has to add 30 to 35 million tons of fertiliser-N by the year 2020 (Table 1.8). The demand for fertiliser-N will stretch by almost 15 million tons

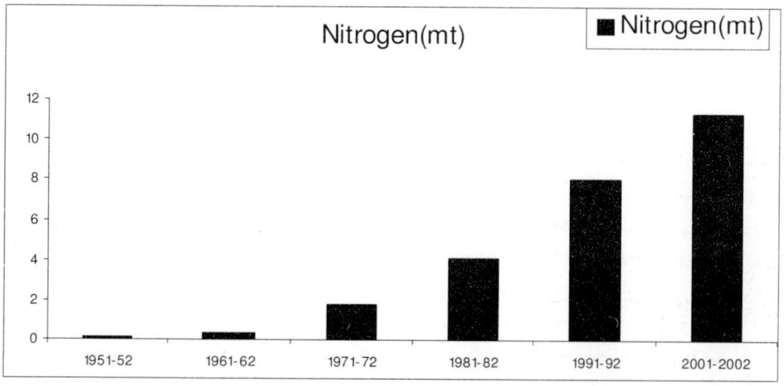

Fig. 1.2. Trends in fertilizer-nitrogen application in India. (Source: FAI, 2002)

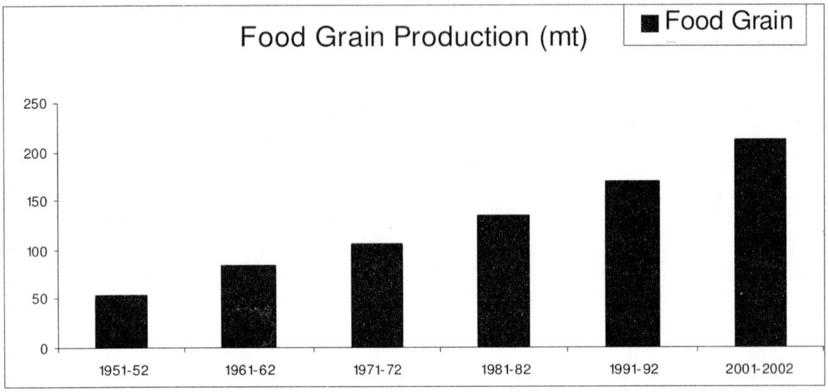

Fig 1.3. Trends in food grain production in India (Source: FAI, 2002)

more, if the requirements for horticulture, vegetables and plantation crops, sugarcane, potato, cotton and oilseeds are included (NAAS, 1995).

FERTILISER NITROGEN USE EFFICIENCY

Not all fertiliser-N nutrients applied to soil are absorbed by growing crops. A part is always left unutilised. This unused part can stay in the soil or be lost through erosion (surface runoff, leaching, volatilisation or nitrification-denitrification). Research done so far, shows that the aggregate loss of nitrogen via these pathways accounts for 50% or more of the fertilizer-N (Table 1.9). Ostensibly, more than 50% of the applied nutrients are lost. This implies only partial benefits from the applied fertiliser nutrients. The overall situation is alarming when viewed from the point of view of loss in production and energy and risks to the environment. Of all the fertiliser nutrients, nitrogen is implicated in environmental degradation more often than other nutrients, because: (i) irretrievable loss of nitrogen in quantitative terms is the highest, and (ii) its role in biosphere (atmosphere + hydrosphere) pollution is maximum. Impact of nitrogen loss from rice paddies is cited as an example of a moderate rate of 40% nitrogen loss (nitrogen loss up to 60% is not uncommon from rice paddies), 1.6 million tons of nitrogen is lost out of approximately 4 million tons of nitrogen applied to rice annually (rice is estimated to consume 40% of the total fertilizer-N consumed in India). It will take 4 fertiliser factories (of approximately 1000 tpd nitrogen production capacity) to replenish this loss in one year. This nitrogen loss also translates into significant drain of energy, nearly 16 million barrels of gasoline (1 tons nitrogen = 10 barrels) (NAAS, 1999). If the loss of nitrogen is diminished by 10% points, it could mean 4 million tons of additional paddy rice production (1 ton nitrogen equals almost 10 tons of paddy rice). Similarly, yield increases with wheat can be cited (Katyal, 1997) by reducing fertilizer-N loss with appropriate application techniques. The gain from saving fertiliser-N can be as high as with rice. In a country like India, which is compelled to produce 3.5 to 4 million tons additional food grains annually, every possible effort has to be made to minimise fertiliser losses.

Table 1.9: Current status of nutrient use efficiency *

Nutrient	*Efficiency (%)*
Nitrogen (N)	30-50
Phosphorus (P)	10-20
Potassium (K)	<80**
Zinc (Zn)	2-5
Iron (Fe)	1-2
Copper (Cu)	1-2
Mangnese (Mn)	1-2

*Use efficiency (%) refers to utilisation of a nutrient by a crop of that applied through fertilisers.
** By different methods

Source: NAAS, 1999

Apart from economics, nitrogen loss has serious social implications. For example, nitrogen in the form of gaseous nitrous oxide (N_2O) has a role in the destruction of the protective ozone layer, that shields the earth against harmful ultraviolet radiation. Surface runoff takes nitrogen to water bodies, where it accelerates the speed of eutrophication, a process primarily promoted by excess phosphorus. The leached-N pollutes the groundwater, making it unsafe for drinking. Already underground water, in high intensity areas of the country, has shown signs of excessive nitrate loading. Since it is the unutilised portion of fertiliser-N, which is the cause of poor nitrogen use efficiency and damage to the environment, minimisation of nitrogen loss is a basic step to rectify the situation.

Agronomic, Physiological and Molecular Approaches for Improving Nitrogen Use Efficiency

A number of agronomic strategies, to improve the efficiency of the utilisation of nitrogenous fertilisers include the application of different types of fertilisers, their mode of application avoiding runoff, mitigation of losses from soil and plants, use of slow-release fertilisers, nitrification, green manuring, use of legumes in cropping systems, correction in their imbalanced use and integrated nutrient management (Abrol *et al.*, 1984; Abrol *et al.*, 1999; Kanwar and Katyal, 1999; Dinnes *et al.*, 2002).

During the recent past, investigations at the physiological and molecular levels, which relate to acquisition and utilisation of nitrogen, were conducted. These are related to understanding the regulation of uptake and the assimilatory processes, redistribution within the cell and balance between storage and current use at the cellular and whole plant levels. It is expected that an understanding of these processes will make it possible to apply the tools of molecular biology to genetically manipulate the plant to improve its Nitrogen Use Efficiency (NUE) and thus, contribute to precision management (Gadgil *et al.*, 1999).

The predominant form of nitrogenous fertilisers applied to the soil is urea. Under the semi-aerobic conditions prevailing in the field, it is converted to nitrate. The nitrate is taken up by the plant and assimilated by a well-documented assimilation pathway involving the enzymes, nitrate reductase and nitrite reductase. The end product, NH_4^+ is incorporated into amino acids via the GS-GOGAT pathway (Crawford, 1995). During senescence, the proteins are degraded and mobilised to the sequentially formed foliage and finally to the grains, where the storage proteins serve as a source of human consumption (Abrol *et al.*, 1984).

Nitrate Uptake

Nitrogen is predominantly available in soils in the form of nitrate, under semi-aerobic/aerobic conditions. It is the most preferred form of nitrogen taken up by plant roots from the soil (Glass *et al.*, 2001; Williams and Miller, 2001). In agricultural soils, concentration of NO_3^- varies from 1 to 10 mM (Crawford and Forde, 2002). Its concentration in Indian soils is reported to vary from 0.5 to 20 mM during the cropping season (Abrol *et al.*, 1984). Nitrogenous fertilisers are generally given at the early stages of crop growth. At this stage, the concentration of nitrogen in the soil goes up to even 100 mM, while it declines subsequently, because of its utilisation and losses due to leaching and other processes. The concentration becomes as low as 10 µM in the flag leaf, during its emergence, flowering and grain development (Redinbaugh and Campbell, 1991; Abrol *et al.*, 1999; Ahmad *et al.*, 2003).

The net nitrate uptake rate is the result of two opposite nitrate fluxes, viz. influx and efflux, mediated by nitrate transporters and nitrate channels, respectively (Forde, 2000). Transporters exist in the plasma membrane of the root epidermal and cortical cells. Entry of NO_3^- through plasmalemma requires active transport systems, which are protons, coupled and dependent on a proton motive force generated by the H^+-ATPase (Glass *et al.*, 1992). The transport systems responsible for NO_3^- uptake include: Low Affinity Transport System (LATS), Inducible High Affinity Transport System (iHATS) and Constitutive High Affinity Transport System (cHATS). Thus, it is likely that while both the low and high affinity transporters are functional during the early stages of growth, the high affinity transport systems are operative at the later stages too, when the soil nitrogen concentration is low. These systems offer great potential in future studies on the mechanisms of transport functions, particularly when combined with site-directed mutagenesis.

Two types of variations in nitrate uptake have been observed in higher plants:

(a) Temporal variations in response to changing environmental conditions, such as light intensity, temperature or stress, and
(b) Variations that occur during ontogeny (Imsande and Touraine, 1994).

In the first case, due to the discrepancy between internal nitrogen supply and growth rate, the nitrate uptake varies, so that the amount of nitrogen in the different internal pools remains relatively constant, while in the latter, the rate of nitrate uptake varies dramatically during ontogeny of the plant due to variations in the level of NO_3^- in the soil solution. Though environmental factors may not be controlled under field conditions, the NO_3^- level in the soil solutions and, hence, NO_3^- uptake can be optimised by the application of nitrogenous fertilisers in split doses at appropriate phenological stages of the crop, during its ontogeny (Abrol *et al.*, 1999). Molecular approaches can also be used to optimise NO_3^- uptake by introducing iHATS in the crop plants. Some understanding and recent studies on nitrate uptake are being given below.

Low-affinity transport system (LATS) is active for nitrate uptake at concentrations above 250 µM NO_3^-. Constitutive high-affinity transport systems (cHATS) are characterised by low K_m and low V_{max} (typically 6-20 µM and 0.3-0.82 µ mole g^{-1} h^{-1}, respectively), while induced high affinity transporters (iHATS) with high K_m and high V_{max} values (typically 20-100 µM and 3-8 µ mole g^{-1} h^{-1}, respectively) are induced within hours to days of exposure to NO_3^- (Crawford and Glass, 1998). These high affinity transport systems are upregulated in response to nitrate.

According to the nomenclature of transporter genes, proposed by the ISPMB, (Caboche *et al.*, 1994) based on their sequence analysis, the nitrate transporter genes have been classified into two families: Nrt1 and Nrt2 (Crawford and Glass, 1998; Forde, 2000). The Nrt2 and Nrt1 genes encode the NNP (Nitrate-Nitrite Porter) and the PTR (Peptide Transporter) families, respectively. The Nrt1 and Nrt2 proteins share the same structural topology with 12 transmembrane domains, distributed in two sets of six helices connected by a cytosolic loop. Besides their common structure, no primary sequence homology is found between the Nrt1 and Nrt2 classes, Nrt1 and Nrt2 codes for LATS and HATS, respectively. Some reports, however, suggest that Nrt1 codes for both the systems (Wang *et al.*, 1998). There is plenty of evidence to suggest that Nrt2 and Nrt3 code for HATS in chlamydomonas (Navarro *et al.*, 2000). Enough evidence is available on nitrate uptake by plant roots depending not only on external factors, but also on the nitrogen status of the whole plant (Glass and Siddiqi, 1995). The Nrt2 genes in *N. plumbaginifolia* and *Arabidopsis* were shown to be induced by very low levels of nitrate (10-50 µM) (Krapp *et al.*, 1998; Filleur and Danial, 1999). In barley, however, the accumulation of Nrt2 mRNAs was the highest, when external nitrate concentration was maintained at 50 µM under quasi-steady conditions (Vidmar *et al.*, 2000). The Nrt2 genes are nitrate inducible in various plants, but that does not mean that they are components of iHATS. Cerezo *et al.*, (2001) have shown that, when none induced wild type plants and AtNrt 2-A mutants supplied nitrate, the wild-type plants showed a marked induction of AtNrt2.1 expression and HATS–mediated influx, while the mutants exhibited only a very limited response. This shows that AtNrt2.1 and/or AtNrt2.2 are strictly required for the activity of iHATS. Thus, the control of nitrate inducibility of iHATS occurs at the transcriptional level.

Approximately, one hour after the addition of nitrate, the induction by nitrate is the highest and, thereafter, Nrt2 mRNA steady state levels decrease (Krapp *et al.*, 1998; Filleur and Daniel, 1999; Zhou *et al.*, 1999). Nitrate influx shows the same transient induction, followed by a downregulation when plants are starved of nitrogen (Lejay *et al.*, 1999). These regulations are thought to be due to the repression of nitrate transport, by reduced nitrogen metabolites accumulating in root tissues when plants are given enough nitrogen nutrition.

A few approaches to improve NUE, however, may be worthy of investigation: (i) it is likely that over-expression of the high affinity transporter operative at the later stages of growth will be beneficial to the plant, when soil nitrogen concentrations are low and the potential nitrate assimilatory activity is not fully utilised due to the lack of availability of nitrate. Nitrogen acquired at this stage seems to help in improving the grain protein content and the number of fully developed grains (Abrol *et al.*, 1999), (ii) constitutive expression of the nitrate transporter in the root cells of transgenic plants will circumvent the problem of repression and may result in continuous uptake. Expression of the low affinity transporter gene in *Arabidopsis* and rice resulted in enhanced nitrate uptake (Lin *et al.*, 1997). Such an enhancement can be coupled with an increase in xylem loading of nitrate by constitutively expressing another gene involved in nitrate translocation (Kuo *et al.*, 1997), and (iii) a considerable amount of variation in C_{min} (lowest concentration at which roots can extract ions from the soil solution) at species level has been reported. Screening of cultivars for low C_{min} may be helpful in identifying the ones which have a higher influx of nitrate, as compared to efflux (Marschner, 1995). At the whole plant level, the rate of nitrate uptake matches the growth of the plant. Its demand increases when there is an internal deficiency. It seems to be regulated by specific amino acids (Ismande and Touraine, 1994). When plants are adequately supplied with nitrogen, transport processes are downregulated so that only a small fraction of the potential for transport is expressed. Over-expression of the transporter genes for increased nitrate transport may be feasible, if (i) the feedback inhibition of the transport system by amino acids is relaxed, and (ii) adequate storage of nitrate in organs like the shoot is possible. It appears doubtful whether over-expression for increased transport will be a feasible approach. However, manipulating the storage capacity of organs may be possible. In wheat and barley, nitrate, whose concentration is high in the soil, is taken up by the plant and stored in the internodes and subsequently, assimilated by the upper laminae which show sub-optimal activity (Table 1.10; Ramraj *et al.*, 1999). Similar observations have been reported in maize (Schroeder, 1978).

Table 1.10: Nitrate concentration (μM g^{-1} dry wt.) in the laminae at various growth stages in high (cv. Shera) and low (cv. Pusa Lerma) nitrate reductase (NR) activity wheat Cultivars (cv)

Variety	*16 days after sowing*	*23 days after sowing*	*30 days after sowing*
High NR	267.7	151.5	159.7
Low NR	380.3	218.2	245.8

Source: Ramraj *et al.*, 1999

Nitrate Utilisation

Prior to its incorporation into amino acids, it is reduced to ammonium. This reduction takes place in two steps. Nitrate reductase (NR, E. C.1.6.6.1) catalyses the first step i.e. reduction of nitrate to nitrite which requires two electrons, NADH or NADPH which serve as the electron donors for this reduction, which takes place in the cytoplasm (Solomonson and Barber 1990; Wang *et al.*, 2000). The second step occurs in the chloroplasts or plastids and is catalysed by nitrite reductase (NiR), where nitrite is reduced to ammonium. This reduction requires six electrons donated by ferredoxin in the chloroplast, and probably by a ferredoxin- like protein in the plastids (Suzuki *et al.*, 1985; Wang *et al.*, 2000). The resulting ammonium is incorporated into amino acids by the glutamine synthetase (GS) and glutamate synthetase (GOGAT) pathways. (Lea *et al.*, 1990). As the flow of inorganic nitrogen into organic nitrogen in plants is regulated by nitrate reductase, which is an unstable enzyme and has poor affinity for nitrate and is a very slow enzyme, hence, nitrogen utilisation efficiency is primarily dependent on its activity. Consequently, an understanding of the biochemical and molecular regulation of nitrate reductase may help in enhancing the nitrogen utilisation efficiency of plants.

Nitrate Reductase

Nitrate reductase is the key enzyme involved in nitrate assimilation. It is a complex enzyme that is regulated at the transcriptional and post transcription levels. Its activity is regulated by both exogenous and endogenous factors. Exogenous factors include environmental factors, e.g., light, substrate availability, moisture stress, while endogenous factors include carbohydrate supply, phytohormones, phytochrome and covalent modifications (Abrol *et al.*, 1999).

Nitrate triggers the transcription of inducible genes for NR (Crawford *et al.*, 1986; Li and Oaks, 1993). Employing NR specific antibodies and NR cDNA probes, it has been shown that NO_3^- induced increase in NR activity and NR protein is due to the enhanced steady state level of NR-mRNA in several plant species viz, barley, squash, *Arabidopsis* and rice (Melzer *et al.*, 1989; Sueyoshi *et al.*, 1995). It was further confirmed by other experiments, where nitrogen-starved plants rapidly accumulated NR-mRNA upon nitrate feeding (Chang *et al.*, 1991; Aslam *et al.*, 1993; Tischner *et al.*, 1993; Krapp *et al.*, 1998; Filleur and Daniel, 1999; Zhou *et al.*, 1999; Kawachi *et al.*, 2002). Nitrate induced increases in the steady state level of NR-mRNA have also been reported in the leaves of wheat barley (Abdin *et al.*, 1992; Kumar *et al.*, 1993).

In roots of maize and barley, the supply of μM levels of nitrate led to a rapid increase in the accumulation of NR-mRNA (Long *et al.*, 1992; Ohlen *et al.*, 1995). Nitrate deprivation resulted in a rapid decrease in NR-mRNA levels without changes in the tissue nitrate concentration in barley roots (Sueyoshi *et al.*, 1995). Subsequently, it was shown that cytosolic nitrate concentration was maintained in barley roots for at least 24 hours after the removal of external nitrate (Vander Leij *et al.*, 1998). These results indicated that nitrate the acts as an external factor in inducing NR gene expression in roots. In contrast to medium roots, mechanisms of NR gene regulation by nitrate in leaves have not been well-documented, due to the complexity of the processes involved in xylem nitrate transport and leaf nitrate absorption. Nitrate absorbed by root cells can be reduced by NR in the cytoplasm or stored in the vacuoles, or transported to the xylem for translocation to the leaves (Crawford and Glass 1998). This complexity makes it difficult to define the role of nitrate in the regulation of NR gene expression in leaves. Shaner and Boyer (1976) suggested that NR activity in maize leaves was influenced by nitrate flux to the leaves from the roots. Subsequently, Sueyoshi *et al.* (1995) reported the existence of a close relationship between the levels of NR-mRNA accumulation in leaves and the nitrate concentration in the xylem sap of barley seedlings, suggesting that xylem nitrate plays an important role in regulating the expression of the NR gene in leaves. However, the exact concentration of the xylem sap triggering NR gene expression has not been determined, yet. Furthermore, Kawachi *et al.* (2002) tried to show that a higher level of nitrate (>mM) in the xylem sap was required for maintaining adequate NR gene expression in barley leaves.

Three different isoforms of NR have been found in eukaryotes. The most common form in higher plants is NADH-specific NR (EC 1.6.6.1). A NAD(P)H-bispecific NR (EC 1.6.6.2) has been identified in several plant species. It occurs either as a second isoform, along with the NADH-specific NR, as in maize, barley, rice, mustard and soybean (Kleinhofs and Warner, 1990; Ahmad and Abdin, 1999) or it is the sole isoform, as found in *Betula pendula* (Friemamin *et al.*, 1991), *Psophocarpus tetragonolobus* and *Erythring senegalensis* (Kleinhofs and Warner, 1990). The third isoform is NADPH-specific (E.C. 1.6.6.3.). It has not been found in higher plants, but is present in fungi and mosses (Padidam *et al.*, 1991). The predominant form, inducible NADH: NR (EC1661), in higher plants is mostly present as a homodimer, with molecular weight ranging from 200 to 230 kDa (Abrol *et al.*, 1999). The amino acid sequences of monomeric units are now available from several species. There are about 881 (ban) to 926 (spinach) amino acid residues in each unit (Hoff *et al.*, 1994). Acidic amino acids are more abundant than the basic ones (Redinbaugh and Campbell, 1991). The C terminal sequences are more conserved among different species than the N-terminal, indicating that the N-terminal part was not involved in the catalytic action of the enzyme. Cloning and sequencing of higher plant NRs show that amino acid sequences present at the active site are well conserved. Two cysteine residues, corresponding to Cys 191 and Cys 245 in Arabidopsis, are found conserved in higher plant NRs. The residues are proposed to be involved in forming an interchain disulphide bridge during subunit association.

The prosthetic groups, FAD, heme, and molybdenum cofactor (MoCo) occur in 1:1:1: stoichiometry. The molecular weights of these prosthetic groups are 23, 10 and 90 kDa, respectively (Abrol *et al.*, 1999). The N-terminal of NR binds MoCo, the central part binds heme, and the C-terminal region binds FAD. These functional domains are linked by a protease sensitive hinge region (Fido, 1991). The prosthetic groups comprise individual redox centres and are used for the catalystic transfer of electrons from NAD(P)H to nitrate. The electron pathway is from NAD(P)H-FAD-heme-MoCo to nitrate (Hoff *et al.*, 1994).

Histidine residues act as ligands for binding heme prosthetic groups of higher plant NRs. The Cys-731 and Cys-889 residues are found to be conserved in all higher plant NRs. These amino acids may be essential for NADH binding. However, six amino acids, namely gly-pro-pro-pro-met-ile, upstream of Cys-889, determine their relative affinity for either NADH or NADPH. In birch NR, two proline residues are replaced by alanine and serine that perhaps, allow it to bind either NADH or NADPH, equally well (Schondorf and Hachtel, 1995). Arginine residues are also implicated in the binding of NADH (Baijal and Sane, 1988). On the other hand, histidine residues may be involved in electron transfer from the reduced FAD to the heme-iron protein of the cytochrome domain of NR (Su *et al.*, 1996; Pattanayak and Chatterjee, 1998). Furthermore, ser-543 present at hinge 1 region, receding the MoCo domain, appears to be involved in phosphorylation-dephosphorylation (Mackintosh *et al.*, 1995). Site-directed mutation of ser-543 to asp in *Arabidopsis* NR prevents phosphorylation, modification, and inactivation of NR (Pattnayak and Chatterjee, 1998). In brief, various roles of the different amino acids in the NR molecule have been elucidated so that site-directed mutagenesis could be used to modify the activity of the enzyme at will.

NR is a cytosolic enzyme encoded by a single *Nia* gene per haploid genome in *Nicotiana* species, thus, leading to the existence of two homologous genes, *Nia1* and *Nia2* (Vaucheret *et al.*, 1989). It is conserved among plants and is present in multiple copies per haploid genome. The stimulatory effect of nitrate on NR gene transcription is mediated by a constitutively expressed nitrate-sensor protein. This protein either binds to the nitrate-responsive promoter of the *Nia* gene in higher plants or activates *Nit-2*-like transacting DNA-binding protein (TIF) in the presence of nitrate and allows enhanced transcription of NR mRNA. Light induction of *Nia* mRNA is also mediated via phytochrome. Over-expression of NR is found in transgenic tobacco plants transformed by oat *phyA* gene (Pattnayak and Chatterjee, 1998).

A relationship between two activity of enzymes, NR and grain yield was suggested by Hageman and Below (1990). It was postulated that since this enzyme was rate limiting, any increase in its content/activity would be beneficial and thus, improve nitrogen utilisation efficiency of the plant. In some studies, functional complementation of apoenzyme nitrate reductase-deficient mutant has been achieved by the introduction of the wild-type alleles of either the tobacco *Nia1* or *Nia2* gene into tobacco mutants (Vaucheret *et al.*, 1990), and the *Arabidopsis* mutant (Wilkinson and Crawford, 1991), or the tomato *Nia* gene into *N. plumbaginifolia* mutants (Dorbe *et al.*, 1992). However, the expression of introduced genes never reaches the level of expression of the host genes in wild-type plants. For example, when the *Nia30* tobacco mutant was transformed by genomic clones, carrying either the tobacco *Nia1* or *Nia2* gene, 85% of the transformants did not express detectable NR activity and only a low level of activity was found in the remaining transformants, ranging from 0.1 to 7% of the activity found in wild-type plants (Vaucheret *et al.*, 1990). Some transgenic plants, expressing NR activity at a level five to six times that of the wild type obtained by transformation of the E23 *Nia* mutant with tobacco *Nia* 2 cDNA under the control of 35S promoter, showed no differences in chlorophyll, protein or amino acids accumulation. Plants do not seem to benefit from excess NR activity. It needs to be investigated if *in vitro* NR activity reflects the *in vivo* situation (Hoff *et al.*, 1994).

An interesting approach, that can improve the NR enzyme potential and thus possibly improve nitrogen use efficiency, was postulated by Abrol (1999). According to this, the reconstitution of the chloroplasts of higher plants with cyanobacterial type NR would make the organelle self-containing and independent in terms of nitrogen metabolism. Cyanobacterial-NR derives its energy from reduced ferradoxin through photosynthetic electron transport. The enzyme is small (80 kDa) and the gene encoding for it was isolated and characterised from *Synechococcus* (Andriesse *et al.*, 1994). There are two possible approaches to introduce *Synechococcus* NR in higher plant chloroplasts: (i) engineering the chloroplast genome by inserting the NR gene into it and, (ii) synthesising the enzyme in cytosol and transporting it to the organelle with the aid of transit peptide sequences.

Another approach for the efficient use of nitrogenous fertilisers can be made by identifying the genotypes and managing various caltivars. This involves complete analysis of nitrogen uptake, assimilation of nitrate, storage and mobilisation of nitrogen throughout growth and development (Abrol *et al.*, 1984; Abdin *et al.*, 1994). Following the prevalent management practices, nitrate assimilation is high in the first-formed laminae and declines in the subsequently formed ones (Fig. 1.4). This suggests, either a lack of availability of substrate at the later stages of growth (Abrol *et al.*, 1984) or reduction in the capacity to take up and/or assimilate nitrate. Measurements of soil nitrate concentrations throughout the growth period and feeding of nitrate to the excised laminae from field-grown plants showed a lack of availability of the substrate viz. nitrate at the later stages (Abrol *et al.*, 1999).

At the early stages of growth, however, the laminae are saturated and one

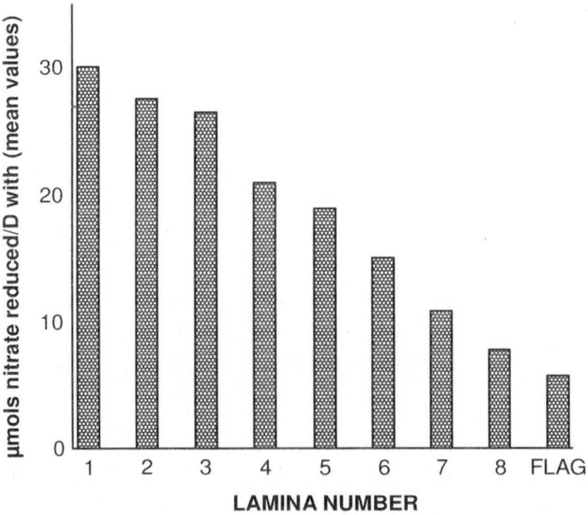

Fig 1.4. Shows the mean *in vitro* NR activity of the laminae on the main shoot. Each value is a mean of 3 values at full expansion (Abrol *et al.*, 1999)

can say that the nitrate assimilatory pathway, as per the present management practices, is rate limiting. In fact, it was observed that at the optimal level, as determined by agronomic experiments, there is reduction in NR activity compared to the activity at sub-optimal levels. This may possibly be related to toxicity (Abrol *et al.*, 1999).

In wheat and barley, while the laminae contribute to approximately 65 to 70% of the total nitrate assimilated, roots, stems (leaf sheaths + internodes), and reproductive parts reduce the rest of the amount equally. Maximum nitrate reduction takes place during the flag laminae emergence. During the post-flowering phase, ear components make significant contributions (Abrol *et al.*, 1999; Fig 1.5). Amongst the laminae, the upper ones, despite their low NR activity, reduce the major amount of nitrate taken up by the plant. This supports the observation mentioned earlier that the nitrate acquired at the earlier stages is stored in the internodes.

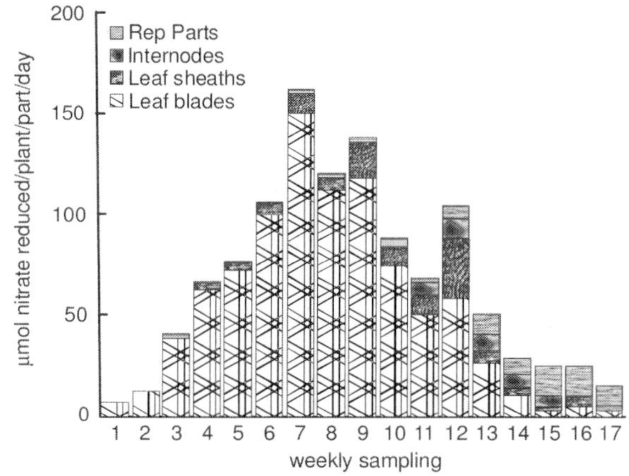

Fig. 1.5. Nitrate Reductase activity (*in vivo*) in individual plant parts expressed as μ mole.

Response of the wheat cultivars, which differed two- to three-fold in NR activity, revealed that: (i) there was more than 100% increase in activity in the upper laminae of the high NR cv., when they were placed in Hoagland's solution containing nitrate. The magnitude of increase was much less in the upper laminae of low NR cultivars, (ii) a major amount of the nitrate was reduced by the upper laminae, which is because of their larger size, as compared to the lower ones, and (iii) at optimal doses of nitrogenous fertiliser application, as recommended by agronomic studies, high as well as low NR cultivars, reduce much less nitrate as compared to their total reduction potential (Abrol *et al.*, 1984). Application of the same amount of nitrogen, in more than two splits under field conditions, increases the nitrogen availability at the later stages of growth so that the suboptimal activity of the upper laminae can be exploited. It was observed that there was a significant improvement in the nitrate assimilatory activity of the upper laminae (Abrol *et al.*, 1999, Fig. 1.6) and enhancement in the total nitrogen harvest and grain protein content (Table 1.11). The magnitude of enhancement was higher in the high NR cv., compared to the low NR ones. Application of additional nitrogen at the later stages of growth was also useful, as has been demonstrated by a number of studies.

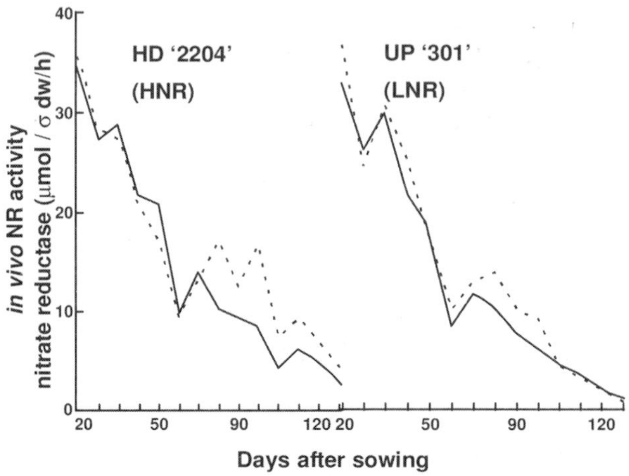

Fig. 1.6. *In vivo* nitrate reductase activity (μ mole g^{-1} dw h^{-1}) in the fully expanded laminae (pooled), at different intervals during the ontogeny of the field grown wheat (high NR (Hd 2204) and low NR (UP 301) cv.).

The nitrogen was applied in two (solid lines) splits, i.e. 60 kg N each, at 0 and 28 days after sowing, and four (dotted lines) splits, viz. 30 kg N each at 0, 28, 54 and 76 days after sowing (Abrol *et al.*, 1999).

It needs to be mentioned that high NR cv. shows a better response than the low NR cv., at low soil nitrogen levels as well (Abrol *et al.*, 1984).

Table 1.11: Additional nitrogen accumulated as a result of increasing the splitting frequency of nitrogen application

	mg additional N in		
Cultivar	*Whole plant*	*Grains*	*% additional N in grains*
Low NR			
Pusa Lerma	32.70	29.70	90.80
UP 301	17.80	15.10	84.80
High NR			
HD 2177	57.40	42.40	73.80
HD 2204	43.80	40.00	91.30

Source: Abrol *et al.*, 1999

NITROGEN, SULPHUR AND N-USE EFFICIENCY (NUE)

Both nitrogen and sulphur are required for the synthesis of S-containing amino acids. Because of the central role of sulphur and nitrogen in providing functional integrity to proteins, the requirement of one would depend upon the supply of the other. Nitrogen nutrition, therefore, has a strong regulatory influence on sulphur uptake and assimilation and *vice-versa* (Duke and Reisenaure, 1986;

Hawkesford *et al.*, 1995). Consequently, plants maintain a relatively constant ratio of organic nitrogen to organic sulphur, particularly in their vegetative tissues. In general, nitrogen and sulphur concentration in plants is in the ratio of 20:1 (Crame, 1990). When nitrogen is excess in relation to sulphur, accumulation of non-protein compounds such as amides occurs, resulting in nitrogen to sulphur ratio being greater than 15:1. The ratio of total nitrogen to total sulphur and protein-S, determines the degree of deficiency of nitrogen in protein, and nitrogen and sulphur ratio is often used as a diagnostic criteria for sulphur deficiency (Spencer and Frency, 1980). Higher yields in wheat were obtained when sulphur concentration and N/S ratio in the flag leaf was 0.2% and 18%, respectively (Reneau *et al.*, 1986). Similar observations were made by Mahler and Maples (1987). Dev and Kumar (1982) observed maximum response in terms of grain yield in maize, mustard, groundnut and wheat with N/S ratio of 15.6, 3.1, 14.8 and 7.1, respectively. N/S ratio not only involved in higher seed production, but also in the quality of oil in oilseed crops. Ahmad and Abdin (2000) found a positive correlation between the seed N/S ratio and the quality of rapeseed oil. Schung and Haneklaus (2000) argued that the use of N/S ratio may not be a criterion for the diagnosis of S and N deficiency, because the same ratio of N/S can be obtained at totally different concentrations of nitrogen and sulphur in tissues. Surplus of one element may, therefore, be falsely interpreted as deficiency in the other. The activities of ATP sulphurylase (ATPS) and adenosine 5'-phosphosulphate (APS) reductase (APR), which has been shown to be identical to APS-sulphotransferase (Suter *et al.*, 2000), decreased under nitrogen-deficient conditions in *Lemna minor* L. and cultured tobacco cells (Reuveny *et al.*, 1980). At the same time, the addition of nitrate or ammonia to the medium quickly restored the activity of these two enzymes. Further, a marked depression in the ability of cereal plants to take up nitrate and ammonium was observed when plants were starved of sulphur, which was accompanied with the augmented capacity of sulphur uptake (Clarkson *et al.*, 1989). The activities of enzymes involved in sulphur and nitrogen assimilation and their respective assimilatory pathways are greatly influenced by S and N nutrition (Reuveny *et al.*, 1980; Barney and Bush, 1985; Bell *et al.*, 1995; Ahmad *et al.*, 1999a). Plants starved of sulphur, accumulate arginine and asparagine with low levels of sulphur amino acid i.e. methionine and cysteine (Gilbert *et al.*, 1997). Also, the reduced protein synthesis is accompanied by the accumulation of organic and inorganic nitrogenous compounds. Sulphur uptake and assimilation has been shown to be dependent upon a constant supply of the precursor of cysteine, o-acetylserine, which in turn, is dependent upon adequate nitrogen availability. Excess cysteine or other reduced S-compounds represses the uptake and assimilation of sulphur, when either sulphur is in excess, or nitrogen is limiting (Zhou *et al.*, 1999). Cystein may be involved in the repression of the synthesis of other amino acids, ultimately repressing N-uptake (Fig 1.7). The regulatory interaction between assimilatory sulphate and nitrate reduction is believed to occur at the transcriptional level (Prosser *et al.*, 2001).

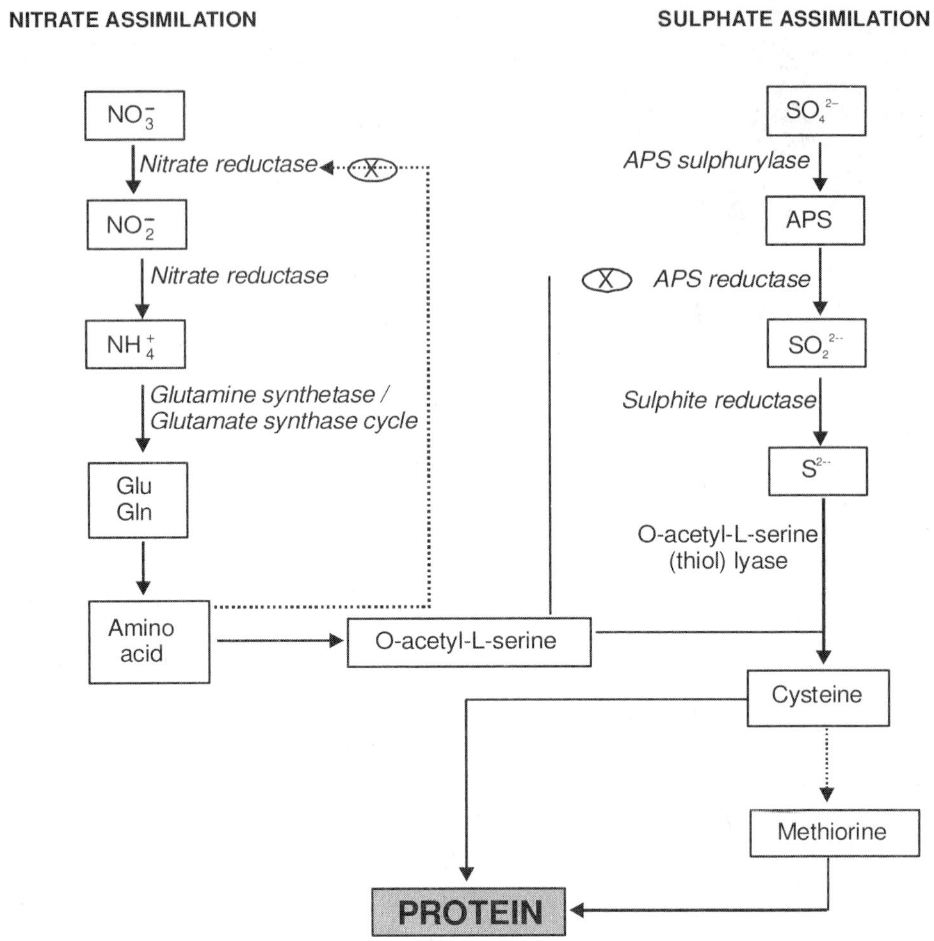

Fig. 1.7. Metabolic coupling between sulphur and nitrogen assimilatory pathways in plants Ⓧ = postulated points of regulation (Abdin *et al.*, 2003)

NITROGEN, SULPHUR AND PHOTOSYNTHETIC N-USE EFFICIENCY (PNUE)

Nitrogen nutrition plays an important role in determining plant photosynthetic capacity in both natural and agricultural environments (Abrol *et al.*, 1999). Appropriate nitrogen supply increases the leaf area development (LAI) and its duration, tillering and the survival and photosynthetic productivity of a crop canopy. Leaf nitrogen plays a key role in determining a crop's photosynthetic capacity. In C_3 plants, Net photosynthesis and nitrogen content in leaf have been co-related in numerous studies (Nakano *et al.*, 1997; Kumar *et al.*, 2002). Leaf N affects the size and morphology of chloroplasts. Ample nitrogen increases the number of chloroplasts per mesophyll cell and their cross-sectional area, as well as the length, compared to N-deficient chloroplasts, which have slightly more thylakoid membranes, but lower stromal volume. The density of protein (predominantly rubisco) in the stroma is greater with high nitrogen supply (Kutik *et al.*, 1995). Rubisco is one of the rate-limiting enzymes of photosynthesis. The dependence of the rate of photosynthesis on Rubisco activity is in turn dependent on the level of nitrogen supplied during plant growth.

The relationship between nitrogen supply and Rubisco content is complex, and it depends on the species and habitat in trees, the proportion of nitrogen allocated to Rubisco can be independent of nitrogen supply (Brown *et al.*, 1996). In C_4 leaves, N-deficiency decreases photosynthesis with a selective reduction, not only in rubisco, but also in the activities of phosphoenalpyruvate corboxylase (PEP) and pyruvate orthophosphate dikinase (Khamis *et al.*, 1990). Intensive study of C_3-crops has shown that Rubisco content increases linearly (but not proportionately) with nitrogen uptake and leaf nitrogen content (Evans, 1993; Makino *et al.*, 1997; Theobald *et al.*, 1998). In these plants, photosynthetic rate increases curvilinearly with respect to the amount of Rubisco (Lawlor *et al.*, 1989; Ahmad and Abdin, 2000a; Kumar *et al.*, 2002).

The increased amount of leaf nitrogen invested in Rubisco at high leaf nitrogen in crops (Theobald *et al.*, 1998) may mitigate the lower CO_2 concentration, due to the increased internal resistance. This notion is consistent with changes in chloroplast morphology (Kutik *et al.*, 1995). When nitrogen is not limiting, an over-investment in Rubisco not only acts as a reserve of nitrogen, but also increases the leaf's ability to exploit short periods of intense illumination (Millard and Catt, 1988). A high carboxylation capacity also contributes to increased water use efficiency. These considerations may explain that with high nitrogen supply, rubisco often appears to be in excess of requirements for the growth environment (Lawlor *et al.*, 1989; Theobald *et al.*, 1998).

While nitrogen directly affects the photosynthesis efficiency of plants, sulphur affects the photosynthesis efficiency indirectly by improving the nitrogen utilisation efficiency of the plants, as is evident from the relationship between nitrogen content and photosynthetic rate in the leaves of +S and –S treated rapeseed-mustard plants. In –S treated rapeseed-mustard plants, photosynthesis is linearly related to leaf-N content, only up to 1.5g m^{-2}. At higher leaf-N content, however, the relationship becomes curvi-linear (Abdin *et al.*, 2003; Fig. 1.8). This observation was also supported by Lawlor *et al.*, (1989), who reported a curvilinear relationship between photosynthetic rate and leaf-N content, beyond 2.0 g m^{-2} in case of wheat. It is, however, very interesting to note that in +S treated rapeseed-mustard plants, the relationship between photosynthetic rate and leaf-N was linear, even beyond 1.5 g m^{-2}. It has been suggested that under S deficiency (-S treatment), the extra leaf-N (beyond 1.5 g m^{-2}) remains non-utilised in the form of non-protein N in the leaves of –S treated plants (Eppendorfer, 1971; Abdin *et al.*, 2003). Consequently, adequate sulphur supply increases the N-utilisation into protein (Rubisco) synthesis, thereby enhancing the photosynthetic rate which ultimately leads to a higher crop yield (McGrath and Zhao, 1996; Ahmad *et al.*, 1998; Fismes *et al.*, 2000). McGrath and Zhao (1996) observed an

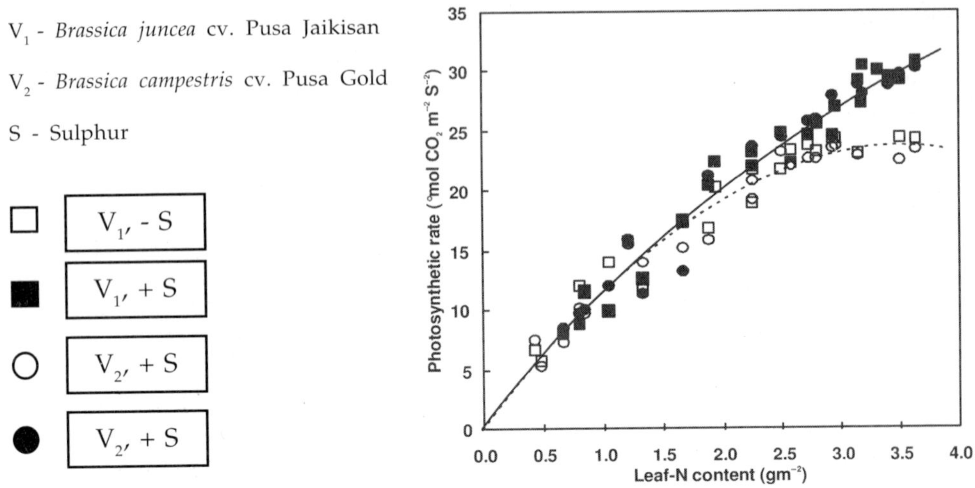

Fig. 1.8. Relationship between Leaf-N and photosynthetic rate of *Brassica* genotypes grown under +S and –S conditions (Source: Ahmad and Abdin, 2000a)

increase of 42-67% in the seed yield of *Brassica napus* with the application of 40kg S ha^{-1} and 180 and 230 kg N ha^{-1}. The seed yield declined when nitrogen was supplied at the rate of 180-230 kg ha^{-1} without Sulphur. Similar observations were reported by Abdin and his group (Ahmad *et al.*, 1998, 1999a, 1999b) in rapeseed-mustard. In wheat, the yield increased linearly with increased nitrogen application along with sulphur (Reneau *et al.*, 1986). S-deficiency in plants grown on S-limiting soils, suppresses the development of the reproductive organs and even leads to pod abortion in rapeseed (Fismes *et al.*, 2000). Reproductive growth and the proportion of the reproductive tissues in total dry matter were significantly increased by sulphur application during pod development (McGrath and Zhao, 1996). Under S-deficient conditions, excess of unassimilated NO_3^-—N and/or free amino acids accumulates in leaves (Hue *et al.*, 1991; Abdin *et al.*, 2001), and protein is degraded within the chloroplasts (Dannehl *et al.*, 1995; Ahmad and Abdin, 2000a; Abdin *et al.*, 2003). Sunrapi and Anderson (1997) have shown that high levels of nitrogen inhibit the proteolysis process in soybean and hence, the export of sulphur and nitrogen from mature leaves to developing leaves or developing grains is impaired.

Nitrogen utilisation efficiency also varies with plant species. For example, plants with C_4 photosynthesis have greater nitrogen use efficiency (g.d.w. gain mg^{-1} N utilised) than C_3 plants. This could be attributed to a higher supply of carbon for the synthesis of carbohydrates and proteins and the availability of reductant and ATP for nitrogen assimilation in C_4 plants. The low N-use efficiency of C_3 plants is reported to be mainly due to carbon deficiency caused by the inhibition of mitochondrial PDC by monocovalent cations, in particular by NH_4^+ produced during photorespiration. On the other hand, NH_4^+ cannot interfere with the carbon flow required for NO_3^- assimilation in C_4 plants, as NH_4^+ production is localised in the bundle sheaths cells, and NO_3^- assimilation occurs in mesophyll cells (Oaks, 1994).

ROLE OF NITROGEN IN GRAIN QUALITY

Agricultural crops act as the major food supplier for feeding a growing population. Today, however, there is also a great demand of increasing the nutritional quality of food. Improved nutritional quality may solve problems in developing countries encountered in cases where plant foods are the major or sole source of food which is consumed as food as well as plant feed for livestock. The demand for these sectors focusses particularly on improving the levels of the essential amino acids, minerals and basic vitamins to allow the healthy growth of humans and livestock.

Nitrogen fertilisation in a crop often results into a higher grain yield and to a lesser extent, grain protein concentration. More commonly, a negative correlation exists between grain yield and grain protein concentration in cereals, as well as in oilseed crops. Low nitrogen in soil results in a low content of proteins in wheat (Terman *et al.*, 1969), whereas nitrogen fertilisation increases wheat protein content (Robinson *et al.*, 1979; Knowles *et al.*, 1991, Ortiz-Monasterio *et al.*, 1997). Low rates of N-fertiliser, however, can decrease grain protein content, if grain yield is increased more by nitrogen fertilisation than nitrogen uptake (Ramon Fernandez and Laird, 1959, Bole and Dubetz, 1986). In prevalent agronomic practices of wheat cultivation, nitrogen is often applied in two splits: half of the total nitrogen at sowing and the other half at the time of tillering. But, the N-utilisation efficiency of the crop never exceeds 50% as it has been found that the activity of nitrate reductase, a substrate inducible enzyme regulating the inflow of reduced nitrogen in plants, is limited by reductant (NAD(P)H) supply during the early stages (Abdin *et al.*, 1992). Moreover, the upper leaf blades of field-grown wheat are starved of nitrate during the crucial reproductive stage and there is a scope for enhancing nitrate assimilation potential (Abrol, 1990), as well as grain protein yield by the application of nitrogen in more than two splits. Wheat genotypes with high nitrate reduction potential ('HNR' genotypes) have been observed to accumulate 10.8% more proteins in the grains, as compared to the genotypes with low nitrate reducing potential ('LNR' genotypes), with the application of nitrogen in three splits (Abrol, 1990; Abdin *et al.*,1992, 1997). Applications of N including anthesis, later in the season, are more effective than the earlier applications in winter wheat in increasing grain protein content than (Kelley, 1995). Similar observations were also made for rice, where fertiliser nitrogen application upto the panical initiation stage resulted into the increased protein content and possible increased grain yield (Patrick and Hoskins, 1974). Foliar application of urea, triazines or substituted ureas at the heading stage also increased the grain protein content of rice (De Datta *et al.*, 1972, Nishizawa *et al.*, 1977). Nitrogen application at the flowering stage provides an option to improve milling and the nutritional quality of rice grain. High protein content improves the whole grain or head rice content (Jongkaewwattana *et al.*, 1993) and milled-rice translucency. Protein distribution within the grain becomes even more, as protein content increases because of the increase in the number of protein bodies (Juliano *et al.*, 1973). The increase in protein content in milled rice is reflected mainly in the storage of proteins, glutelin, and prolamin (Nishizawa *et al.*, 1977) located in protein, bodies (Juliano *et al.*, 1973). An increase in milled rice protein by less than 10 percent, reduces the lysine content slightly, but a drastic reduction in lysine content results with the increase in grain protein by more than 10 percent (Nishizawa *et al.*, 1977). This has been confirmed by actual nitrogen balance studies of milled rice, differing in protein content in rats (Eggum and Juliano, 1975, Nishizawa *et al.*, 1977) and in human (Juliano, 1993).

An increase in the rice seed protein concentration of two soybean cultivars, NC 107 and N 87-984-16, in response to supra-optimal N (30mM), was associated with relatively increased amounts of the B-subunit of 7S protein and decreased amount of the basic polypeptide of 11S protein.

Molecular Approaches to Improve Grain Quality

Altered seed traits, that result in modified oil, protein, carbohydrate, and vitamins content, as well as the composition, can bring enhanced value to farmers, producers, and consumers. Such traits, referred to as seed quality or output traits, alter the nutritional or functional properties of the harvested plant material for use in foods, animal feeds, or industrial products. Developing plants with improved grain quality traits involves overcoming a variety of technical challenges inherent in metabolic engineering programmes. The new genetic and genomic approaches that are promising for the genetic introgression of foreign genes into cultivated species as well as the identification of novel genes regulating plant quality traits of food crops have been discussed here.

Lysine and methionine are the most important essential amino acids, because they are present in limiting levels in the major feed and food crops. Cereal grains generally contain low levels of lysine, while legumes are generally deficient in methionine (Duke, 1981). Due to the vital nutritional significance of lysine and methionine, most research efforts have focussed on enhancing the levels of these two essential amino acids in the free or protein-bound forms.

The essential amino acids, lysine, methionine, isoleucine, and threonine are synthesised from aspartate by several different branches of the aspartate-family pathway (Fig. 1.9, Galili, 1995). Methionine receives its sulphur moiety from cysteine (Fig. 1.9, Ranavel *et al.*, 1998). Lysine regulates its own synthesis, mostly by feedback inhibition of the activity of dihydrodipicicolinate synthetase (DHPS); threonine synthesis is primarily regulated by the sensitivity of aspartate kinase, to feedback inhibition by lysine and threonine (Galili, 1995); methionine synthesis is subjected to a more complex control, but it is the major point of regulation of methionine production which occurs through the competition between cystathionine β-synthase and threonine synthase for their common substrate, phosphohomoserine (Fig. 1.9, Ravanel *et al.*, 1998).

Approach to Enhance Lysine Content in Grain Crops

The synthesis of lysine can be enhanced by reducing the sensitivity of DHPS to feedback inhibition by lysine (Fig. 1.9). This has been shown in several studies where recombinant genes, encoding bacterial feedback-insensitive DHPS enzymes, were expressed in transgenic dicot and monocot plants (Galili, 1995; Brinch-Pedersen *et al.*, 1996; Mazur *et al.*, 1999). Constitutive expression of genes for bacterial enzymes in plants resulted in amino acid overproduction, but in many cases this expression was also associated with abnormal phenotypes and partial sterility (Shaul and Galili, 1993). To overcome these problems, the bacterial enzymes were produced in a seed-specific manner, using seed storage protein promoters to control the expression of the genes. Seed-specific expression of the bacterial DHPS gene was first performed in tobacco plants, using the bean phaseolin promoters (Karchi *et al.*, 1994). Lysine synthesis was enhanced specifically in the developing seeds of these transgenic plants, but its level in mature seeds was not higher than the non-transgenic plants. Seed-specific expression of the bacterial DHPS gene was correlated with a significant elevation in the activity of lysine ketoglutarate reductase (LKR), the first enzyme in the α-aminoadipic acid pathway, which catabolises lysine in to glutamate, α-aminoadipic acid, and acetyl CoA (Galili *et al.*, 2002). These results were the first indirect and unexpected evidence that lysine catabolism may be an important factor in regulating free lysine accumulation in seeds. Moreover, it was also suggested that lysine autoregulates its own catabolism, at least in seeds, by stimulating LKR activity (Karchi *et al.*, 1994; Galili *et al.*, 2002). In subsequent studies, bacterial DHPSs have been expressed in a seed-specific manner in a number of transgenic crop plants, including soybean, rapseed, maize, and narbon beans (Falco *et al.*, 1995; Mazur *et al.*, 1999). In contrast to tobacco, all of these transgenic crops showed significant elevation of free lysine levels, and nearly doubled total seed lysine levels in soybean and rapeseed. Lysine overproduction in these plants was also associated with the increased levels of various catabolic products, showing that most seeds possess an active process of lysine catabolism. The negative effects of lysine catabolism are twofold: first it reduces the extent of free lysine accumulation; second, some of the catabolic products of lysine, such as glutamate and the products of α-aminobutyric acids and α-aminoadipic acid, act as neurotransmitters in animals and may be toxic at high levels (Karlsen *et al.*, 1982; Welinder *et al.*, 1982; Bonaventure *et al.*, 1985; Reichenbach and Wohlrab, 1985). Therefore, the reduction of lysine catabolism may be an important trait to be considered in the breeding of high lysine crops. Lysine catabolism can be reduced by antisense, co-suppression, or knockout of gene encoding enzymes in this part of the pathway.

The use of seed storage protein gene promoters for the expression of the bacterial DHPS gene is based on the assumption that amino acid synthesis and storage protein production are subjected to the concerted spatial and temporal regulation of expression during seed development. This is apparently true for dicot plants in which seed storage protein genes are expressed in developing embryos, and utilisation of such promoters to express the bacterial DHPS results in lysine overproduction. However, whether this is also true for monocot plants, in which storage protein synthesis occurs mainly in endosperms, is still debatable. Mazur *et al.* (1999) have expressed a bacterial feedback-insensitive DHPS gene in transgenic maize, using either endosperms or embryo-specific promoters. Increased free lysine levels were detected only when DHPS production was controlled by embryo-specific promoters. This study raises an important fundamental issue. If amino acid biosynthesis in monocot seeds occurs mostly in the developing embryo, a mechanism should exist for rapid delivery of the amino acids into the endosperms to support the massive process of storage protein synthesis. If amino acids are synthesised in the endosperm tissues, then lysine level may be regulated by additional factors, such as lysine catabolism.

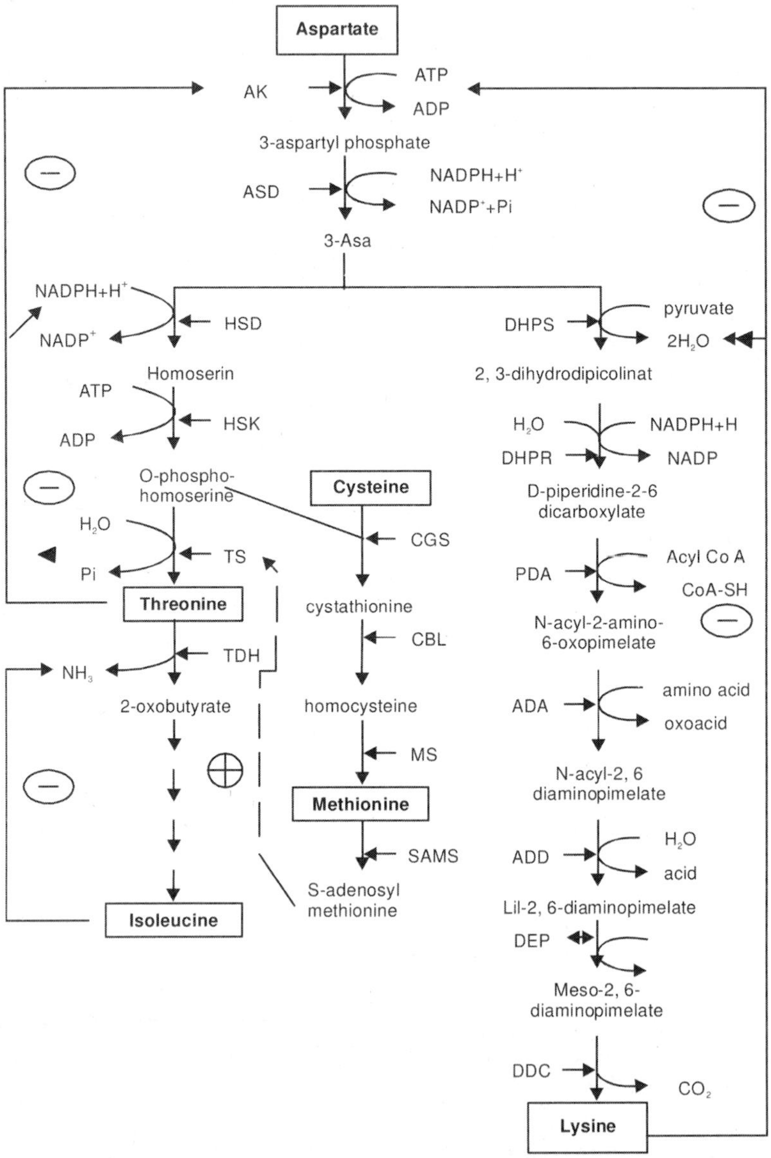

Fig 1.9. The biosynthetic pathway of the aspartate-family, essential amino acids, lysine, threonine, methionine, and isoleucine.

Curved arrows with a (-) sign represent major feedback inhibition loops by the end product amino acids. Dashed arrows with a (+) sign represent enzyme activation. Enzyme abbreviations: AK, aspartate kinase; ASD, aspartic semialdehyde dehydrogenase; HSD, homoserin dehydrogenase; HSK, homoserine kinase; TS, threonine synthase; TDH, threonine dehydratase; THPS, dihydrodipicolinate synthase; DHPR, dihydrodipicolinate reductase; PDA, Ä′ piperidine dicarboxilate acylase; ADA, acyldiaminopimelate; aminotranferase; ADD, acyldiaminopimelate deacylase; DEP, Diaminopimelate epimerase; DDC, diaminopimelate decarboxilase; C2GS, Cystathionine β-synthase; CBL, cystationine b-lyase; MS; methionine synthse; SAMS; *S*-adenosylmethionine synthase. (Galili *et al.*, 2002)

Another strategy for the production of high lysine plants is to introduce genes encoding lysine rich proteins. These proteins will serve as a lysine sink and may reduce the problem of lysine catabolism. In maize, a variety of genes for neutral, modified and synthetic proteins including the most successful ones, encoding hodrothionine (HT12) and barley high lysine protein 8 (BHL8), containing 28 and 24% lysine, respectively were tested. These proteins accumulated between 2 to 6% of total grain proteins, and when introduced together with a bacterial DHPS they resulted in a marked elevation of total lysine to over 0.7% of the dry seed weight (Jung and Falco, 2000), compared with around 0.2% in wild type maize. If these dramatic elevations of lysine levels dose do not interfere with yield and other yield quality factors, then the commercial application of transgenic maize, expressing these high lysine proteins (either alone or together with a bacterial feedback-insensitive DHPS) for feeding human and non-ruminant livestock, looks very promising.

LOSS OF NITROGEN

Nitrogen fertiliser consumption has risen spectacularly increasing 54-fold in India and about 7-fold globally, between 1961-62 to the year 2000-01 (IFIA, 2002). This increased use of nitrogen fertiliser has attracted increasing concern about nitrogen losses to the environment. The losses are due to nitrate leaching (NO_3^-), ammonia (NH_3) volatilisation and denitrification associated with the production of nitrous oxide (N_2O), nitric oxide (NO) and dinitrogen (N_2). Nitrate (NO_2^-) is highly water-soluble and is not strongly adsorbed with the soil surface due to negative charge. These chemical characteristics of nitrate make it susceptible to leaching down through the soil into groundwater. Since groundwater is a major source of drinking water, the health effects of ingesting nitrate through drinking water are a concern, especially when nitrate levels are high. Ingestion of nitrate in food and water may cause serious health problems, like methemoglobinemia, termed as blue baby syndrome and cancer. The world Health Organization and U.S. Environmental Agency set the maximum permitted NO_3^- - nitrogen level in drinking water to be 10mg L^{-1}. High nitrate content has been reported in the groundwater of developed countries like USA and European Union (50mg NO_3^- - N L^{-1}; Pathak *et al.*, 2002). Increased agricultural and industrial activities to feed the increasing population have caused an increasing incidence of nitrate pollution of groundwater in India also. In agriculturally advanced states like Punjab, Karnataka, Maharashtra, Delhi, Andhra Pradesh and U. P., a higher nitrate content in groundwater has been reported (Gupta *et al.*, 2000; Kulshrestha and Singh, 2001; Sharma *et al.*, 2002; Pathak *et al.*, 2002). The high NO_3^- concentration has also been reported in groundwater from states like Rajasthan, where cropping intensity is less, but NO_3^- leaching is due to light textured soil. Hence, there is an alarming trend of groundwater pollution by nitrates. Application of higher amounts of N-fertiliser not only results in groundwater, pollution but also creates excess accumulation of NO_3^- in plants (Rao and Puttanna, 2000). In this respect, Rao and Puttanna, (2000) have studied the dietary intake of nitrates (Table 12).

Table 1.12: Nitrate Content in Different Crops

Crop	*Nitrate Content (mg kg^{-1})**
Cereals	20-76
Pulses	39-114
Leafy vegetable	30-270
Root and tubers	31-2043
Condiments & spices	145-4680

World agriculture is currently responsible for more than half of the atmospheric increase of nitrous oxide (N_2O) and two thirds of the global ammonia inputs into the atmosphere. These compounds play an important role in atmospheric chemistry, ozone depletion, aerosol formation and greenhouse warming. Biological denitrification and nitrification are the major processes involved in the production of NO and N_2O. Denitrification is the reduction of nitrate (NO_3^-) or nitrite (NO_2^-) to gaseous nitrogen oxide and molecular N_2 by aerobic bacteria, whereas nitrification involves the biological oxidation of ammonium (NH_4^+) to NO_2^- or NO_3^- under aerobic conditions. Under oxygen limited conditions, nitrifiers can use NO_3^- as a terminal electron acceptor to avoid accumulation of the toxic NO_2^-, whereby N_2O and NO are produced. It has been estimated that global emissions of N_2O–N and NO–N are about 3.5 mt and 2.0 mt, respectively, from cropped fields. Out of this, 34% global annual N_2O–N emission is from the developed countries, while 66% is from the developing countries. The annual emission of NO–N from fertilised crops is about 1.5 mt. Out of this, 44% is contributed by the developed, whereas 56% is contributed by the developing countries. The total fertiliser induced N_2O–N and NO–N emissions are about 0.9 mt (0.8% of total N application) and 0.6 mt (0.5% of total nitrogen application), respectively. The emission rate of N_2O–N for urea and other straight nitrogen fertiliser was recorded the highest, but the highest NO–N loss was recorded with the component NK–N and other straight fertilizers, as presented in the Table 1.13 (FAO, 2002).

Table 1.13: Global N use and fertiliser induced N_2O and NO emission estimate by fertiliser type

Fertiliser Type	*Global N use*	*N_2O-N*	*NO-N loss*
	mt/year	**Loss (%)**	**(%)**
Ammonium sulphate	2.4	1.0	0.6
Urea	34.4	1.1	0.6
Ammonium Nitrates	7.5	0.9	0.5
Calcium ammonium nitrate	7.5	0.9	0.5
Ammonia direct application	4.6	0.8	0.5
Nitrogen solution	4.0	0.9	0.6
Other straight fertiliser	10.1	1.0	0.8
Ammonium phosphates	4.1	0.8	0.6
Other compound NP–N	1.7	0.8	0.5
Compound NK–N	0.0	0.8	0.8
Compound NPK–N	0.1	0.7	0.5
Total mineral fertiliser	78.5	1.0	0.6
Total organic fertiliser	33.0	0.6	0.4
Total mineral and organic Fertiliser	111.6	0.8	0.5

Source: FAO, 2002

Very limited studies on N_2O emission have been performed in India. Some measurements of N_2O emission from grasslands and wastelands in India have been reviewed by Pathak (1999). Studies showed that the emission of N_2O from rice fields, fertilised with urea ranged from 0.06 to 0.8% of the applied fertilisers. Soils amended with urea emitted the highest amount of N_2O, followed by ammonium sulphate and potassium nitrate amended soils of wheat crop. However, in paddy soils, potassium nitrate applied plots emitted the maximum amount of N_2O, compared to ammonium sulphate. In India, the total emission of N_2O from soil is estimated to be 5–15 $x10^{12}g\ N\ yr^{-1}$ (Pathak, 1999).

Ammonia is constantly formed in soils because of the biological degradation of organic compounds and NH_4^+ yielding mineral and organic fertilisers. As it is a gas, any NH_3 present in soil, water or fertilisers can volatilise to the atmosphere. Global ammonia losses reported from mineral fertilisers are 11 million tons, 14% of the mineral fertilizer nitrogen used (Pathak, 1999). Nearly 2.3 million tons N/year is lost through ammonia volatilisation from the 11.8 mt of nitrogen-fertilizer in wet land rice cultivation, which generally occurs in developing countries (97%). About 61.7 million tons of nitrogen-fertilizers are used globally in upland crops and 14% of that is lost through NH_3 volatilisation. Developing countries have got a major share (18%) in NH_3 volatilisation as compared to the developed countries (8%). It is referred to as a pollutant, because of its corrosive nature due to the formation of ammonium salts.

Strategies to Reduce Nitrogen Losses

With proper management of agricultural systems, nitrogen losses could be reduced and pollution problems can be minimised. In this respect, Pathak (1999) has reviewed the strategies to mitigate nitrogen-losses from agricultural system.

Table 1.14: Strategies to mitigate N_2O from Agricultural System

Practice Followed	*Estimated decrease in emission (Tg N yr⁻¹)*
1. Match N supply with crop demand:	0.24
a. Use soil/plant testing to determine fertiliser needs	
b. Minimise fallow periods to limit mineral nitrogen accumulation	
c. Optimise split application schemes	
d. Match nitrogen application to reduce production goals in regions of crop overproduction	
2. Tighten nitrogen flow cycles:	0.14
a. Integrate animal and crop production system in terms of manure. Reuse in crop production.	
b. Maintain plant residue nitrogen on the production site	
3. Use advanced fertilisation techniques:	0.15
a. Controlled release fertilisers	
b. Place Fertilisers below the soil surface.	
c. Foliar application of fertilisers	
d. Use Nitrification inhibitors	
e. Match fertiliser type to seasonal precipitation	
4. Optimise tillage, irrigation and drainage	0.15
Total	0.68

Source : Pathak (1999)

CONCLUSION AND FUTURE PROSPECTS

Nitrogenous fertilisers are costly agricultural inputs. They are essential for higher yields and better quality of crops. With the increased food demand of the ever increasing world population, the consumption of nitrogenous fertilisers is expected to double within the next 25 to 30 years. This will result, not only in destabilisation of the economy of developing countries like India due to the drainage of foreign exchange, but also cause environmental pollution that varies from 30 to 50 percent, as a result of the low N-use efficiency of crops under field conditions. Intensive studies, therefore, are being conducted, world over by scientists and agrotechnologists, to optimise the N-use efficiency of field crops, employing various strategies that include physiological, agronomic and molecular approaches. The following conclusions may be drawn and future research strategies developed from these studies:

1. NO_3^- uptake is an active process facilitated by NO_3^- transporters located in the plasma membrane of root epidermal and cortical cells. There are three kinds of NO_3^- transporters, viz. Low Affinity Transport system (LATS), Inducible High Affinity Transport system (iHATS) and Constitutive High Affinity Transport system (cHATS). The genes for these transporters have already been identified and cloned. Out of these, iHATS is the most promising transport system that can operate optimally, both at the early and late stages of plant growth and development with very high and very low NO_3^- concentrations in the soil. The genes of iHATS can be mobilised

to various agricultural crops using a suitable transformation and expression system so as to optimise their N-uptake potential. This will help in minimising the leaching of NO_3^- from the root zone and its escape into the environment.

2. NO_3^--N is assimilated in the organic-N in plant organs, mainly in leaf cells, via the nitrogen assimilatory pathway operating in the chloroplast, except in the first step i.e., NO_3^- reduction operating in the cytosol. The first step of this pathway, however, is found to be rate-limiting and is catalysed by nitrate reductase, an unstable and sluggish enzyme. This enzyme, therefore, regulates the flow of inorganic-nitrogen into reduced-nitrogen, the usable form of nitrogen in the plants. Consequently, efforts should be made to optimise the catalytic efficiency of this enzyme through site-directed mutagenesis, its overexpression at the cellular level and relocalisation into chloroplasts/plastids. This will help to optimise the N-assimilation potential of crops and subsequently, their N-use efficiency, as the NO_3^- taken up by these crops will be converted into proteins and other organic compounds that constitute the structural and functional components of the plant cells and their metabolic machinery.
3. The N-use efficiency of crops is also limited by the losses of applied nitrogen through leaching and volatilisation. These losses can be minimised by using crop cultivars with high nitrate reduction potential, as they have the potential to assimilate higher amounts of nitrogenous fertilizers, even at the later stages of their growth and development, as compared to the cultivars with low nitrate reductase potential.
4. NO_3^- levels in the soil vary throughout the growth period of crops, high at the early stages when there is a low demand, and low at the later stages when there is a high demand. To ensure the adequate availability of substrate i.e., NO_3^- for the synthesis of organic nitrogenous compounds during the entire growth period for better growth and development, nitrogenous fertilisers need to be applied in split doses at appropriate phenological stages. This will not only help in maintaining the adequate NO_3^- level in the soil during the entire growth period of the crop, but also improve N-use efficiency of the crop due to reduction in its losses through leaching and volatilisation. The nitrogen losses through leaching and volatilisation via microbial biotransformations of nitrogenous fertilisers can also be minimised with the use of slow release N-fertilisers.

Future investigations aiming at optimising N-use efficiency of agricultural crops, will include genetic modification and mobilisation of genes encoding inducible high affinity nitrate transporters (iHATS) and nitrate reductase enzymes, so that N-acquisition and assimilation potential of these crops could be further augmented. Another important future research aspect in N-assimilation could be the relocalisation of nitrate reductase (NR) from cytosol to the chloroplast/plastid. This will improve NO_3^- reduction, as the reducing equivalent may not be a limiting factor in this organelle and the energy in the transport of NO_2^- from cytosol to chloroplast will be saved. Also, the understanding of processes related to the storage and mobilisation of nitrogen from the vegetative parts to the reproductive and biochemical and molecular levels is also essential. Finally, the application of sulphur, under S-deficient situations is essential to optimise N-use efficiency of the crops.

REFERENCES

Abdin M.Z., Kumar P.A. and Abrol Y.P. (1992). Biochemical basis of variability in nitrate reductase activity in wheat (T. aestivum L.) genotypes. Plant Cell Physiol 33: 951-956.

Abdin M.Z., Lakkineni K.C. and Kumar P.A. (1993). Molecular aspects of nitrate reductase regulation in higher plants. Proc Indian Natn Sci Acad B59: 219-226.

Abdin M.Z., Kaim M.S., Bansal, K.C. and Abrol Y.P. (1994). Nitrates assimilation and nitrogen harvest in Uniculm 'Gigas' wheat (Triticum aestivum L). J Agronomy Crop Science 173: 225-229.

Abdin M.Z., Ahmad A., Khan I., Qureshi M.I. and Abrol Y.P. (2001). Effect of S and N Nutrition on N- accumulation and N-Harvest in rapeseed-mustard (Brassica Juncea L and Brassica campestris L). In: Plant Nutrition-Food Security and Sustainability of agro-ecosystem through basic applied research (Horst WJ ed), Kluwer Academic Publishers, The Netherlands, p 816-817.

Abdin M.Z., Ahmad A., Khan N., Khan I., Jamal A. and Iqbal M. (2003). Sulphur interaction with other nutrients. In: Sulphur in Plants (Abrol YP and Ahmad A eds), Kluwer Academic Publishers, The Netherlands, p 359-374.

Abrol Y.P. (1990). Pattern of nitrate assimilation and grain nitrogen yield in field grown wheat (Triticum astivum L). In: Plant Nutrition-Physiology and Application. (Van Beusichem ML ed), Kluwer Academic Publishers, The Netherlands, p 773-778.

Abrol Y.P., Kumar P.A. and Nair T.V.R. (1984). Nitrate uptake and grain nitrogen accumulation. Adv Cereal Sci Tech 6: 1-48.

Abrol Y.P., Chatterjee S.R., Kumar, P.A., and Jain V. (1999). Improvement in nitrogen use efficiency: Physiological and Molecular approaches. Current Science 76: 1357-1364.

Abrol Y.P., Sangwan S. and Tiwari M.K. (eds) (2002). Land Use–Historical Perspective–Focus on Indo-Gangatic Plains, Allied Publishers, New Delhi.

Ahmad A., Abraham G., Gandotra N., Abrol Y.P. and Abdin M.Z. (1998). Interactive effect of nitrogen and sulphur on growth and yield of rapeseed-mustard (Brassica juncea L Czern and Coss and Brassica campestris L) genotypes. J Agron Crop Sci 181: 193-199.

Ahmad A. and Abdin M.Z. (1999). NADH: Nitrate reductase and NAD (P) H: Nitrate reductase activity in mustard seedlings. Plant Science, 143: 1-8.

Ahmad A., Abraham G. and Abdin M.Z. (1999a). Physiological investigation on the impact of nitrogen and sulphur applications on seeds and oil yield of rapeseed (Brassica campestris L) and mustard (Brassica juncea L Czern and Coss) genotypes. J Agron Crop Sci, 183: 19-25.

Ahmad A., Abrol Y.P. and Abdin M.Z. (1999b). Effect of split application of sulphur and nitrogen on growth and yield attributes of Brassica genotypes differing in their time of flowering. Can J Plant Sci, 79: 175-180.

Ahmad A. and Abdin M.Z. (2000). Photosynthesis and its related physiological variables in the leaves of Brassica genotypes as influenced by sulphur fertilisation. Physiol Plant 110: 144-149.

Ahmad A., Khan I., Abrol Y.P. and Abdin M.Z. (2003). Relationship between soil nitrate content and activities of NADH: and NAD(P)H: nitrate reductase in Indian Mustard. Biologia Plantarum 46(2): 295-296.

Andriesie X., Bakker H. and Weisbeck P. (1994). In: Inorganic nitrogen in Plants and Microorganisms (Ullrich WR, Rigand C, Fuggi A and Aparicio PJ eds), Springer- Verlag, Berlin, p 303-307.

Aslam M., Travis R. and Huttaker R.C. (1993). Comparative induction of nitrate and nitrite uptake and reduction systems by ambient nitrate and nitrite in intact roots of barley seedlings. Plant Physiol 102: 811-819.

Baijal M. and Sane P.V. (1988) Arginine residue (s) at the active site (s) of the nitrate reductase complex from Amaranthus. Phytochemistry, 27: 1969-1972.

Barney Jr. P.E. and Bush L.P. (1985). Interaction of nitrate and sulphate reduction in tobacco. Influence of the availability of nitrate and sulphate. J Plant Nutr, 8: 507-515.

Bell C.I., Clarkson D.T. and Cram W.J. (1995). Partitioning and redistribution of sulphur during S-stress in Macroptilium atropurpureum cv Siratro. J Exp Bot 46: 73-81.

Bole J.B. and Dubetz S. (1986). Effect of irrigation and N fertilisers on the yield and protein content of soft white spring wheat. Can J Plant Sci 66: 281-289.

Bonaventure N., Wioland N. and Roussel G. (1985). Stereospecific effects of the alpha-aminoadipic acid on the retina: a morphological and electrophysiological study. Doc Ophthalmol 61: 71-77.

Brinch-Pedersen H., Galili G., Kmudsen S., Holm P.B. (1996). Engineering of the aspertate family biosynthetic pathway in barley (Hordeum vulgare L) by transformation with heterolgogous genes encoding feedback-insensitive, aspertate kinase and dihydrodipicolinate synthase. Plant Mol Biol 32: 611-620.

Brown K.R., Thompson W.A., Camm E.L., Hawkins B.J. and Guy R.D. (1996). Effects of N addition rates on the productivity of Picea sitchensis, Thuja plicarta and Tsuga heterophylla seedlings II photosynthesis, ^{13}C discrimination and N Partitioning in foliage. Trees 10: 198-205.

Caboche M., Campell W., Crawford N.M., Fernandez E., Kleinhorf A., Ida S. (1994). Genes involved in nitrate assimilation. Plant Mol Biol, Res 12: 45-49.

Cerezo M., Tillard P., Filleur S., Munas S., Daniel-Vedele R. and Gojon A. (2001). Alterations of the regulation of root NO_3- uptake are associated with the mutation of Nrt 2.1 and N2t2.2 genes in Arabidopsis. Plant Physiol, 127: 262-271.

Chang C.L., Acedo G.N., Dewdney J., Goodman H.M., and Cankling M.A. (1991). Differential expression of the two Arabidopsis nitrate reductase genes. Plant Physiol, 96: 275-279.

Clarkson D.T., Sarker L.R. and Purves J.V. (1989). Depression of nitrate and ammonium transport in barley plants with diminished sulphate status. Evidence of co-regulation of nitrogen and sulphate intake. J Exp Bot, 40: 953-963.

Crame W.J. (1990). Uptake and transport of sulphate. In: Sulphur Nutrition and Sulphur Assimilation in Higher Plants (Renennberg H, Brunold C, De Kok LJ and Stulen I eds), SPB Academic Publishers, p 3-11.

Crawford N.M., Campbell W.H. and Davis R.W. (1986). Nitrate reductase from squash: cDNA Cloning and nitrate regulation. Proc Natl Actl Acad Sci USA, 83: 8073-8076.

Crawford N.M. (1995). Nitrate: Nutrients and signals for plant growth. Plant Cell 7: 859-868.

Crawford N.M. and Glass A.D.M.C. (1998). Molecular and Physiological aspects of nitrate uptake in plants. Trends plant Sci, 3: 389-395.

Crawford N.M. and Forde B.G. (2002). Molecular and Developmental Biology of Inorganic Nitrogen Nutrition. American Society of Plant Biologists. The Arabidopsis Book.

Dannehl H.A., Herbik A. and Godde D. (1995). Stress-induced degradation of the photosynthetic apparatus is accompanied by changes in thylakoid protein turnover and phosphorylation. Plant Physiol, 93: 179-186.

De Datta S.K., Obcemea W.N. and Jana R.K. (1972). Protein content of rice grain as affected by nitrogen fertiliser and some triazines and substituted ureas. Agron J 64: 785-788.

Dev G. and Kumar V. (1982). Secondary nutrients. In: Review of Soil Research in India. Vol. I, Indian Society of Soil Science, New Delhi, p 342-360.

Dinnes D.L., Karlen P.L., Jaynes D.B., Kaspar T.C., Halfreld J.L., Colvin T.S. and Combardella C.A. (2002). Nitrogen management strategies to reduce nitrate leaching in tile-drained Midwestern soils. Agron J 94: 153-171.

Dorbe M.F., Caboche M. and Denial-Vedele F. (1992). The tomato Nia gene complements a Nicotiana plumbaginifolia nitrate reductase-deficient mutant and is properly regulated. Plant Mol Bio, 18: 363-375.

Duke J.A. (1981). Handbook of legumes of economic importance. Plenum Press, New York.

Duke S.H. and Ressenauer H.M. (1986). Roles and requirements of sulphur in plant Nutrition. In: Sulphur in Agriculture, Agronomy, Monograph no 27 (Tabatabai MA ed), American Society of Agronomy, Crop Science Society of America, Soil Science Society of America, Madison, Winsconsin, p 123-168.

Eggum B.O. and Juliano B.O. (1975). Higher protein content from nitrogen fertiliser application and nutritive value of milled-rice protein. J Sci food Agric 26: 325-427.

Eppendorfer W.H. (1971). Effect of sulphur, nitrogen and phosphorous on amino acid composition of field bean (Vicia faba) and responses of the biological value of the seed protein and sulphur-amino acid content. J Sci Food Agric, 22: 501-505.

Evans J.R. (1993). Photosynthesis acclimation and nitrogen partitioning within a lucerne canopy 2. Stability through time and comparison with a theoritical optimum. Aust J plant physiol, 20: 69-82.

FAI (2002). Fertiliser statistics, 2001-2002. FAI, New Delhi, India. p I-85 to IV-94.

Falco S.C., Guida T., Locke M., Mauvais J., Sandress C., Ward R.T. and Webber P. (1995). Transgenic canola and soybean seeds with increased lysine. Biol Technol 13: 577-582.

FAO (2000). Agriculture towards 2015/2030. Technical Interim Report, FAO, Rome.

FAO (2002). World agriculture towards 2015/2030. Long term perspectives – The outlook of agriculture, FAO, Rome.

Fido R.J. (1991). Isolation and Partial amino acid sequence of domains of nitrate reductase from spinch. Phytochemistry, 30: 3519-3523.

Filleur S. and Daniel V.F. (1999). Expression analysis of high affinity nitrate transporters isolated from Arabidopsis thaliana by differential display. Planta, 207: 461-469.

Fismes J., Vong P.C., Guckert A. and Frossard E. (2000). Influence of sulphur on apparent N-use efficiency, yield and quality of oilseed rape (Brassica napus L) grown on a calcareous soil. Euro J Agron, 12: 127-141.

Forde B.G. (2000). Nitrate transporters in plants: structure, function and regulation. Biochem Biophys Acta 1465: 219-235.

Friemamin A., Brinkmann K. and Hachtel (1991). Seqences of a cDNA encoding bispecific NAD(P)H-nitrate reducatse from the tree Betula pendula and identification of conserved protein regions. Mol Gen Genet 227: 97-105.

Gadgil S., Abrol Y.P. and Seshgiri A. (1999). On growth and fluctuation of Indian foodgrain production. Current Science, 76: 548-556.

Galili G. (1995). Regulation of lysine and threonine synthesis. Plant Cell 7: 899-906.

Galili G, Galili S, Lewinsohn E, and Tadmor Y (2002). Genetic, molecular, and genomic approaches to improve the value of plant foods and feeds. Critical Reviews in Plant Sciences 21(3): 167-204.

Gilbert S.M., Clarkson D.T., Cambridge M., Lambers H. and Hawkesford M.J. (1997). SO_4^{2-} deprivation has an early effect on the content of rubulose-1, 5-bisphosphate carboxylase/oxygenase and photosynthesis in young leaves of wheat. Plant Physiol 115: 1231-1239.

Glass A.D.M., Shaff J.E. and Kochian L.V. (1992). Studies of the uptake of nitrate in barley. IV Electrophysiology. Plant Physiol, 93: 281-289.

Glass A.D.M. and Siddiqi M.Y. (1995). Nitrogen absortion by plant roots and nitrogen nutrition in higher plants (Srivantava HS and Singh RP eds), Associated Publishing Co, p 21-56.

Glass A.D.M., Brito D.T., Kaisar B.N., Krozucker H.J., Kumar, A., Okamoto M., Rawat S.R., Siddique M.Y., Silim S.M., Vidmar J.J. and Zhuo D. (2001). Nitrogen transport in plants, with an emphesis on the regulation of fluxes to match plant demand. J Plant Nutr Soil Sci, 164: 199-207.

Government of India (1996). Population projections for Indian states, 1996-2016. Report of Technical group on population projections, New Delhi Registered General.

Gupta S.K., Gupta R.C., Seth A.K., Gupta A.B., Bassin J.K., Gupta A. (2000). Methaemoglobinamia in areas with high nitrate concentration in drinking water. Natl Medl J India, 13(2): 58-61.

Hageman R.H. and Below F.E. (1990). In : Nitrogen in higher plants (Abrol YP ed), JWiley and Sons, Ny, p 313-334.

Hawkesford M.J., Schnader A., Belcher A.R. and Clarkson D.T. (1995). Regulation of enzymes involved in the sulphur assimilation pathway. Z Pflanzenernahr, Bodenk, 158: 55-57.

Hoff T., Truon H.M., and Caboche M. (1994). The use of mutants and transgenic plants to study nitrate assimilation. Plant Cell and Environment 17: 489-506.

Hue H., Spark D. and Evan J.J. (1991). Sulphur deficiency influences vegetative growth, chlorophyll and element concentrations and amino acids of pea. J Am Soc Hotri Sci, 116: 974-980.

IFPRI (1999). Prospect for India's cereal, supply and demand by 2020. Food, Agriculture; and the Environment. Discussion paper 29. Washington, DC.

International rice research institute (1964). Annual report for 1963, p 149-161. The institute: los Banos, Laguna, Philippines.

International Fertiliser Industry Association On-line database, internet reference: http://www.fertilizer.org/STATSIND.International Food Policy Research Institute

Ismande J. and Touraine B. (1994). Nitrogen demand and regulation of nitrate uptake. Plant Physiol, 105: 3-7.

Jongkaewwattana S., Geng S., Brandon D.M. and Hill J.E. (1993). Effect of nitrogen and harvest grain moisture on head rice yield. Agron J 85: 1143-1146.

Juliano B.O., Antonio A.A. and Esmama B.V. (1973). Effect of protein content on the distribution and properties of rice protein. J Sci Food Agric 24: 295-306.

Juliano B.O. (1993). Rice in Human Nutrition FAO. Food Nutr Ser No26. International rice Research Institute, Manila, Philippines.

Kanwar J.S., and Katyal J.C. (1999). Plant nutrient needs, supply, efficiency and policy issue: 2000-2025. National Academy of Agricultural Sciences, New Delhi, p 329.

Karchi H., Shaul O. and Galily G. (1994). Lysine synthesis and catabolism are coordinately regulated during tobacco seed development. Proc Natl Acad Sci, USA 91: 305-311.

Karlsen R.L., Pedersen O., Schousboe A. and Langeland A. (1982). Toxic effects of DL-alpha-aminoadipic acid on muller cells from rats in vivo and cultured cerebral astrocytes. Exp Eye Res 35: 305-311.

Katyal J.C. (1997). Sustainability: Relevance of integrated concept. In the Hindu Survey of Indian Agriculture, The Hindu, Chennai, Tamil Nadu, India, p 25-32.

Katyal J.C., Singh B. and Buresh R.J. (1987). Efficient nitrogen use as affected by urea application and irrigation sequence: Soil Sci Soc Am J, 51: 366-370.

Kawachi Y., Shoji Y., Sugimoto T., Oji Y., Kleinhofs A., Warner R.L., Ohtake N., Ohyama T. and Sueyoshi K. (2002). Role of xylem sap nitrate in the regulation of nitratereductase gene expression in the leaves of barley (Hordeum Vulgare L) Seedlings. Soil Sci Plant Nutr, 48(1): 79-85.

Kelley K.W. (1995). Rate and time of N application for wheat following different Crops. J Prod Agric 8: 339-345.

Khamis S., Lamaze T., Lemoine Y. and Foyer C. (1990). Adaptation of the Photosynthesis apparatus in maize leaves as a result of nitrogen limitation. Plant Physiol 94: 1436-1443.

Kleinhofs A. and Warner R.L. (1990). Advances in nitrate assimilation. In: The biochemistry of plants (Miflin BJ and Lea PJ eds), Academic press, San Diego, p 89-120.

Knowles T.C., Doerge T.A. and Ottman M.J. (1991). Improved N management in irrigated durumwheat using stem nitrate analysis: II Interpretationon nitrate-n contents. Agron J 83: 353-356.

Krapp A., Fraisier V., Scheible W.R., Quesada A., Gojon A., Stitt M. (1998). Expression studies of Nrt2: its role in nitrate uptake. Plant J 14: 723-731.

Kulshrestha S. and Singh R.V. (2001). Water pollution due to rising nitrate level in some parts of Jaipur and its remedies. J Env Polln, 8(4): 339-343.

Kumar P.A. (1998). Food demand and supply projections for India. In: Agrucultural Economics Policy Paper 98-01. Indian Agricultural Research Institute, New Delhi.

Kumar P.A., Krusse E., Andriesse X., Weeisbeek P. and Kloppstech (1993). Intergation of a cynobacterial protein involved in nitrate reduction (narB) into isolated synechococcus, but not into pea thaylakoid membranes. Eur J Biochem, 214: 533-537.

Kumar P.A., Parry A.J.M., Mitchell R.A.C., Ahmad A. and Abrol Y.P. (2002). Photosynthetic nitrogen assimilation and associated carbon and respiratory metabolism In: Advances in Photosynthesis, 12 (Foyer CH and Noctor G eds) Kluwer Academic Publishers p 23-34.

Kuo H.F., Liang Shih-yi and Tsay Y.F. (1997). 5th International Congress on Plant Mol Biol, Singapore, Abst 91.

Kutik J., Natr I., Demmers-Derks H.H. and Lawlor D.W. (1995). Chloroplast structure of Sugarbeet (Beat Vulgaris, L) cultivated in normal and elevated CO_2 concentration with two contrasted nitrogen supplies. J Exp Bot 46: 1797-1802.

Lawlor D.W., Kontturi M. and Young A.T. (1989). Photosynthesis by flag leaves of wheat in relation to protein, ribulose bisphosphate carboxylase activity and nitrogen supply. J Exp Bot 40: 43-52.

Lea P.J., Robinson S.A. and Stewart G.R. (1990). The enzymology and metabolism of glutamine glutamate and asparagine. In: The biochemistry of Plants (Miflin BJ and Lea PJ eds), Vol 16, Intermediary Nitrogen Metabolism, Academic press, New York, p 147-152.

Lejay L., Tillard P., Lepetit M., Olive F.D., Filleur S., Daniel-Vedele F. and Gojon A. (1999). Molecular and functional regulation of two NO_3^- uptake systems by N- and C- status of Arabidopsis plants. Plant J 18: 509-519.

Li X.Z. and Oaks A. (1993). Induction and turnover of maize nitrate reductase. Influence of NO_3^-. Plant Physiol 102: 1251-1257.

Lin M., Lee H.T., Tsai Y.L., YU S.M. and Tsay Y.F. (1997). 5th International Congress on Plant Mol Biol Singapore, Abst 311.

Long D.M., Oaks A. and Rothstein S.J. (1992). Regulation of maize root nitrate reductase mRNA levels. Physiol Plant 85: 561-566.

Mackintosh C., Douglas P. and Lillo C. (1995). Identification of a protein that inhibits the phosphorylated form of nitrate reductase from spinach (Spinach oleracea) leaves. Plant Physiol, 107: 451-457.

Mahler R.J. and Maples R.L. (1987). Sulphur fertilisation of wheat grown in Arkansas. Bulletin of Arkansas Agricultural Experiment Station No 906, p 23.

Makino A., Sato T., Nakano H. and Mae T. (1997). Leaf Photosynthesis, Plant Growth and Nitrogen allocation in Rice under Different Irradiances. Planta, 203: 390-398.

Marschner H. (1995). In: Mineral Nutrition in Higher plants, Academic Press, London, p 889.

Mazur B., Krebbers E. and Tingey S. (1999). Gene discovery and product development for grain quality traits. Science 285: 372-375.

McGrath S.P. and Zhao F.J. (1996). Sulphur uptake, yield response and the interactions between N and S in winter oilseed rape (Brassica napus). J Agric Sci (Cambridge), 126: 53-62.

Melzer J.M., Kleinhofs A. and Warner R.L. (1989). Nitrate reductase regulation: effects of nitrate and light on nitrate and reductase mRNA accumulation. Mol Gen Genet, 217: 341-346.

Millard P. and Catt J.W. (1988). The influence of nitrogen supply on the use of nitrate and ribulose-1, 5-bisphosphate carboxylase oxygenase as leaf nitrogen stores for growth of potato-tubers (Solanaum tuberosum L). J Exp Bot 39: 1-11.

Motsara M.R. (2002). Available nitrogen, phosphorus and potassium status of Indian soils, as depicted by soil fertility maps. Fertiliser News, 47(8): 15-21.

NAAS (1995). Agricultural Scientists' perceptions on plant nutrient needs, supply, efficiency and policy issues: 2000-2025, New Delhi, India.

NAAS (1999). Agricultural Scientists' perception on Indian Agriculture: Scene, Scenario and Vision, New Delhi, India.

Nakano H., Makino A. and Mae T. (1997). The effects of elevated partial pressures of CO_2 on the relationship between photosynthetic capacity and nitrogen content in rice leaves. Plant Physiol, 115: 191-198.

Navarro M.T., Guerra E., Fernandez E. and Galvan A. (2000). Nitrate reductase mutants as an approach to understanding nitrite assimilation in Chlamydomonas reinhardtii. Plant Physiol 122: 283-289.

Nishizawa N., Kitahara I., Noguchi T., Hareyama S. and Honjyo K. (1977). Protein quality of high protein rice obtained by spraying urea on leaves before harvest. Agric Biol Chem 41: 477-485.

Oaks A. (1994). Efficiency of nitrogen utilisation in C_3 and C_4 cereals. Plant Physiol, 106: 407-414.

Ohlen E., Ingemarsson B., Cambell W.H. and Larsson C.M. (1995). Relationship between external nitrate availability, nitrate uptake and expression of nitrate reductase in roots of barley grown in N-limited split root cultures. Planta, 196: 485-491.

Ortiz-Monasterio I.J., Pena R.J., Sayre K.D. and Rajaram S. (1997). Cimmyt's genetic progress in wheat grain quality under four N rates . Crop Sci 37: 892-898.

Padidam M., Venkatesvarly K., Johri M.M. (1991). Ammonius represses NADPH-nitrate reductase in the mosses Funaria hygrometrica. Plant Sci 75: 184-194.

Pathak H. (1999). Emission of nitrous oxide from soil. Current Science, 77(3): 359-369.

Pathak H., Biswas D.R. and Singh R. (2002). Fertiliser use and environmental quality. Fertiliser News, 47(11): 13-20.

Patrick R.M. and Hoskins F.H. (1974). Protein and amino acid content of rice as affected by application of nitrogen fertiliser. Cereal Chem 51: 84-95.

Pattanayak D. and Chatterjee S.R. (1998). Light mediated regulation of nitrate reductase in higher plants. J Plant Biochem and Biotech, 7(1): 73-78.

Prosser I.M., Purves J.V., Saker L.R. and Clarkson D.T. (2001). Rapid disruption of nitrogen metabolism and nitrate transport in spinach plants deprived of sulphate. J Exp Bot, 52: 113-121.

Ramarathnam K.V. (1980). Fertiliser demand and supply situation. India Fertiliser News 25: 25-29.

Ramon Fernandez G. and Laird R.J. (1995). Yield and protein content of wheat in Central Mexico as affected by available soil moisture and N fertilisation. Agron J 51: 33-36.

Ramraj V.M., Guru S.K. and Abrol Y.P. (1999). Nitrate content in wheat leaf blades. Current Science 76: 29-30.

Ranavel S., Gakiere B., Job D. and Douce R. (1998). The specific features of methionine biosynthesis and metabolism in plants. Proc Natl Acad Sci USA 95: 7805-7812.

Rao Prakasa E.V.S. and Puttanna K. (2000). Nitrate, agriculture and environment. Current Science, 79(9): 1163-1168.

Redinbaugh M.G. and Campbell W.H. (1991). Higher Plant responses to environment nitrate. Physiol Plant 82: 640-650.

Reichenbach A. and Wohlrab F. (1985). Effects of alpha-aminoadipic acid on the glutamate-isolated P III of the rabbit electroretinogram. Doc Ophthalmol 59: 359-364.

Reneau R.B. Jr., Bran D.E. and Donohue S.J. (1986). Effect of sulphur on winter wheat grown in the coastal plain of Virginia. Commu Soil Sci Plant Anal 17: 149-158.

Reuveny Z., Dougall D.K. and Trinity P.M. (1980). Regulatory coupling of nitrate and sulphate assimilation pathways in cultured tobacco cells. Proc Natl Acad Sci USA 77: 6670-6672.

Robinson F.E., Cudney D.W. and Lehman W.F. (1979). Nitrogen fertiliser timing, irrigation, protein and yellow berry in durum wheat. Agron J 71: 304-308.

Schnug E. and Hankelaus S. (2000). Significance of interactions between sulphur and nitrogen supply for growth and quality of crop plants. In: Sulphur Nutrition and Sulphur assimilation in Higher Plants: Molecular, Biochemical and Physiological Aspects (Brunold C, Rennenberg H, De Kok LJ, Stulen I and Davidian JC eds), p 345-347.

Schondorf T. and Hachtel W. (1995). Choice of reducing substrate is altered by the replacement of an alanine by a proline in the FAD domain of bispecific NADPH-nitrate reductase from birch. Plant Physiol 108: 203-210.

Schroeder L.M. (1978). On nitrogen in the environment (Nielsen DR and Mac Donald JE eds), Academic Press, NY, vol 2, p 101-141.

Shaner D.L. and Boyer J.S. (1976). Nitrate reductase activity in maize leaves. I. Regulation by nitrate flux. Plant Physiol 58: 949-952.

Sharma S.K., Tiwari A.N. and Nawale V.P. (2002). Nitrate pollution in ground water of Nagpur city area. In: Proc Natl Conf Polln Prev Contl India, IAEM, Nagpur, p 173-176.

Shaul O. and Galili G. (1993). Concerted regulation of lysine and threonine synthesis in tobacco plants expressing bacterial feedback-insensitive aspertate kinase and dihydrodipicolinate synthase. Plant Mol Biol 23: 759-768.

Solomonson L.P. and Barber M.J. (1990). Assimilatory nitrate reductase functional properties and regulation. Annu Rev Plant Physiol Plant Mol Biol 41: 225-253.

Spencer K. and Freney J.R. (1980). Assessing of sulphur status of field grown wheat by plant analysis. Agron J 72: 469-472.

Su W., Huber H.C. and Crawford N.M. (1996). Identification in vitro of a post translational regulatory site in hinge1 region Arobidopsis nitrate reductase. Plant cell 8: 519-527.

Sueyoshi K., Mitsuyama T., Sugimoto T., Kleinhofe A., Warner R.L., Sunarpi and Anderson J.W. (1997). Effect of nitrogen nutrition on the export of sulphur from leaves in soybean. Plant Soil, 188: 177-187.

Sunarpi and Anderson J.W. (1997). Effect of nitrogen nutrition on the export of sulphur from leaves in soybean. Plant Soil 188: 177-187.

Suter M., von Ballmoos P., Kopriva S., Op den Camp R., Schaller J., Kuhlemeier C., Schurmann, P. and Brunold C. (2000). Adenosine 5'-phosphosulphate sulphotransferase and adenosine 5'-phasphosulphate reductase are identical enzymes. J Biol Chem, 275: 930-936.

Suzuki A., Oaks A., Jaequot J.P. Vidal and Gadal P. (1985). An electronic transport system in maize roots for reactions of glutamate synthase and nitrite reductase. Plant Physiol 78: 374-378.

Terman G.L., Ramig R.E., Dreier A.F. and Olson R.A. (1969). Yield-protein relationships in wheat grain, as affected by nitrogen and water. Agron J 61: 755-75.

Theobald J.C., Mitchell R.A.C., Parry M.A.J. and Lawlor D.W. (1998). Estimation of excess investment in ribulose-1, 5 bisphosphate carboxylase/oxygenase in leaves of spring wheat grown under elevated CO_2. Plant physiol 118: 945-955.

Tischner, R., Waldeck, B., Goyal, S., and Rains W.D. (1993). Effect of nitrate pulses on the nitrate- uptake rate, synthesis of mRNA coding for nitrate reductase, and nitrate reductase activity in the roots of barley seedlings. Planta 189: 533-537.

Vander L.M., Smith S.J. and Miller A.J. (1998). Remobilisation of vacuolar stored nitrate in barley root cells. Planta 205: 64-72.

Vaucheret H., Kronenberger J., Rouze P. and Caboche M. (1989). Complete nucleotide sequence of the two homologus tobacoo nitrate reductase gene. Plant Mol Biol 12: 597-600.

Vaucheret H., Chabaud M., Kronenberger J. and Caboche M. (1990). Functional complementation of tobacco and Nicotiana plumbaginifolia nitrate reductase deficient mutants by transformation with the wild type allele of the tobacco structural genes. Mol Gen Genet 220: 468-474.

Vidmar John J., Zhuo D., Siddiqi M.Y. and Glass A.D.M. (2000). Isolationnand Characterisation of HVNRT2.3 and HVNRT2.4, cDNAs Encoding high affinity nitrate transporters from roots of barley. Plant Physiol 122: 783-792.

Wang R., Liu B., and Crawford N.M. (1998). The Arabidopsis CHL1, protein plays a major role in high-affinity nitrate uptake. Proc Natl Acad Sci USA, 95: 15134-15139.

Wang R., Guegler K., Labrie S.T. and Craford N.M. (2000). Genomic analysis of a nutrient response in Arabidopsis reveals diverse expression patterns and novel metabolic and potential regulatory genes that are induced by nitrate. Plant Cell 12: 1491-1510.

Welinder E., Textorius O. and Nilsson S.E. (1982). Effects of entravitreally injected $_{D}$L-alpha-aminoadipic acid on the c-waves of the D C-recorded electroretinogram in albino rabbits. Invest Ophthalmol Vis Sci 23: 240-245.

Wilkinson J.Q. and Crawford N.M. (1991). Identification of the Arabidopsis CHL3 gene as the nitrate reductase structural gene NIA2. Plant Cell 3: 461-471.

Williams L.E. and Miller A.J. (2001). Transporters responsible for the uptake and partitioning of nitrogenous solutes. Annu Rev Plant Physiol Plant Mol Biol 52: 659-688.

Zhou F.J., Hawkesford M.J. and McGrath S.P. (1999). Sulphur assimilation and effects of yield and quality of wheat. J Cereal Sci, 30: 1-17.

Chapter 2

Intellectual Property Rights and Community Rights

Suman Sahai
Gene Campaign, J-235/A Sainik Farms
Khanpur, New Delhi - 110 062

In order to understand the framework within the new demand for IPRs on biological materials, one needs to understand the recent developments and global trends in the latest and most dominant technology of our times, biotechnology. Biotechnology, which was till now confined to the laboratory, has entered the market. The raw materials of biotechnology are genes, cells and living organisms like bacteria, plants and animals and its tools are genetic engineering. In almost all cases the technological strength of germplasm owning countries is weak but there are a few exceptions of which India is perhaps the most outstanding. India is one of those germplasm-owning countries that also has a good indigenous technology stand. Combination of technology with germplasm, both its own, puts India in a unique position to take advantage of the field of biotechnology and emerge as a global player in this field. Local communities own the bio-resources in their region since it is they who maintain them and it is they who possess the knowledge of their properties and their use. Wild plants growing in the forests have been used over generations by communities of men and women to breed races of several food and cash crops. That is the reason why the claim to place Farmers Rights at par with Breeders Rights is so natural. Farmers have a greater and more innovative share in the creation of new plant varieties than scientists. The role of indigenous knowledge in the realm of medicinal plants is even more obvious than in the case of crop varieties.

INTRODUCTION

In today's market language, biotechnology spells mega-bucks. It is the most dominant technology of the times. According to the projection of several reputed institutions, biotechnology is slated to account for almost 60 to 70 percent of the global economy, for at least the next two to three decades. What is of special relevance is that biotechnology covers a wide span of economic sectors, which is unprecedented. It will play a role in fields as diverse as mining, feedstock, chemicals, energy, pharmaceuticals, and enzyme-mediated processes and of course, agriculture which is the best developed branch of biotechnology and also its most lucrative prize.

Although biotechnology has been used by human civilizations since long, its impact on the global economy is more recent. In agriculture, biotechnology has made possible those gene conferring advantageous traits which can be brought into food and cash crops from any source. In conventional plant breeding, genes (traits) could only be transferred within related species. It was not possible, for example, to take the disease resistance gene from tomato and put it into wheat. With biotechnology this is possible, and being done. This approach can be used with effect in breeding new varieties. This new technology is being used to a great advantage in the pharmaceutical industry, where drug production has been revolutionised for those drugs that were originally obtained from animal and human tissues like insulin. So is the case with drugs like antibiotics derived from microorganisms.

The latest in this area of biotechnology is the news that vaccines can be produced in plants because it is possible to introduce antibody-producing genes into potatoes and peaches. It seems entirely likely that soon it will be possible to harvest antibodies for vaccines from a field. The implications are enormous for lowering the price of vaccines and opening new vistas in medical care. The profits to be earned are mind-boggling. The growing influence of biotechnology can be seen in almost all sectors that relied on chemical processes to produce products. Instead of a chemical reaction that brings molecules together, or "synthesizes" a new product, from two sources in biotechnology, the gene synthesises the product directly. The result is that there are no toxic chemical wastes, and biotechnology is much more kind on the environment than the chemical industry.

In the field of mining, where strongly polluting industries create acid lakes that are hard to clean up, new bacterial strains are being discovered for biomining. Bacterial strains are already being tried for gold, copper and iron, and they are being discovered for other metals too. These bacteria break down the ore to release the metal by 'feeding' on the ore. In biomining, large volumes of bacteria in a culture solution are simply poured over the ore. After sometime, the released metal can be collected and the digested material washed away. Suddenly, with biotechnology it has become possible to exploit low-grade ores, which are economically not feasible with the existing methods. This in itself is of special importance to mineral-rich developing countries like India. These countries do not have efficient state of the art technologies to exploit low-grade ores and are therefore, unable to exploit them.

Unconventional applications in various economic sectors have made biotechnology a formidable money-spinner. It is the control over this technology that is sought for, by the corporate giants of the world. The issue of gene and seed patents has exploded the scene, simply because the battle for the political and economic control of the genetic resources of the world has begun. This battle cannot be fought in laboratories, between scientists. Therefore, it has been brought out onto trade for a place, where corporate gladiators can set the rules, where trade sanctions can be used to force nations to grant access and where a country's export can be blocked till it agrees to the conditions demanded by western nations with respect to its biological wealth.

This battle is between the developed and a developing country, quite simply because the genetic wealth of the world is concentrated almost exclusively, in the tropical countries, that is, in today's developing countries. The major biodiversity of this planet is found in the tropical rain forests, in the thousands of crop varieties in the fields of farmers of all tropical countries and in regions as disparate as the Andes mountains of South America, the jungles of Borneo and Costa Rica and the fields and forests of India and Nepal. The developed world has the technological tools needed to convert genes to products and then to money. It does not have the raw materials. Its forests are large monocultural tracts and the diversity in its fields and forests is limited. Its agriculture is based on the germplasm of the South. So conscious is the North of the limitations of its access to genetic resources, that it has been a subject of strategic planning for a long time.

JP Kendrick of the University of California, at Davis, the Mecca of Agricultural research in the USA, had warned many years ago that "if we all had only to rely on the genetic resources now available in the US, for the genes and recombinants needed to minimise genetic vulnerability of all crops in the future, we would soon experience losses equal to, or greater than those caused by the southern Leaf Blight several years ago at a rapidly accelerating rate across the entire crop spectrum". In order to overcome its limitations of raw materials and to emerge as leaders of the commercial exploitation of genetic material, the corporate interests of the North need guaranteed and continued access to the genetic wealth located in the countries of the South. In order to achieve this, the governments of the developed countries have dreamed up stratagems like GATT/TRIPS and Biodiversity Convention. Undoubtedly, others are being thought of to coerce the countries of the South by using their economic vulnerability as a lever against them.

The position on the one hand is that there are the developed countries which have the cutting edge technology in the field of genetics and genetic engineering, but have little or no germplasm. On the other hand, are the developing countries, which are the repositories of the genetic wealth of this world. In addition to this, they are also the repositories of indigenous knowledge about how to use this wealth. In almost all the cases, the technological strength of germplasm-owning countries is weak, but there are a few exceptions, of which India is perhaps the most outstanding. India is one of those germplasm-owning countries that also have a good indigenous technology stand. This is because the enormous investments made in agricultural research, especially during the days of the Green Revolution, created a strong scientific cadre from scientists to technical assistants. Therefore, India is suddenly emerging as a global player in the field of biotechnology. There are obvious reasons for this:

1. India is a germplasm-owning country and in South-South collaborations with other germplasm-owning countries, it has access to a large range of genetic resources.

2. It has a high technology stand available within the country.
3. Most of all, it has a large repertoire of skilled manpower which makes available comparative skills at half the cost. A technician in India getting Rs. 4000 per month is about 15 times cheaper than a technician in Germany, who would get at least DM 2500 per month, the equivalent of 62,500 rupees.
4. The great thing about biotechnology, from India's point of view, is the cost of the technology itself. Biotechnology, unlike every other major technology to have developed in recent times, is not capital but labour intensive. This is a tailor-made situation for a country strapped for cash, but rich in manpower.

This combination of fortuitous circumstances is the reason why Gene Campaign fights for the protection of the country's genetic resources. The Campaign's sustained position continues to be against patents and privatisation in this field. This is one modern technology where we can be highly competitive and should not allow ourselves to be tied up by patent laws that are not of our own choosing. For the first time since independence, India is in a position to be, not just the consumer of a new technology, but one of its important producers.

AREAS OF CONCERN IS GATT/TRIPS. WHAT POSITION SHOULD INDIA TAKE?

There are 3 areas of concern pertaining to the section on Intellectual Property Rights (IPR), as is relevant to biological materials, which is the crux of the IPR question for India:

1. Geographical indication

The protection based on Geographical Indication is to be found in Section 3 of Trade Related Intellectual Property Rights (TRIPs).

Article 22, as also 23 and 14, deal with the protection of goods that are geographically indicated.

Basmati rice which is one of our special export products needs to be protected in the international market, both for its "brand name", as also for its genes. It is possible to do this under one of the clauses of the Intellectual Property Rights section that has so far not received any attention in the public debates on General Agreement on Tariffs and Trade/World Trade Organization (GATT/WTO).

This clause called, "Geographical Indication" is of almost as much importance to our agriculture as patents on seeds. Our government, as also our rice exporters, should understand its ramifications clearly. They should exploit this clause and use it to monopolise select sections of the agricultural commodities market.

The need to protect the Basmati name and the Basmati germplasm is highlighted by the fact that the Americans have hijacked this special rice variety. The Americans have recently begun to market a new product called Texbasmati or Texmati. This long grain, fragrant rice is the basmati rice that is grown in

Texas and other American states. Texbasmati is an infringement of the geographically indicated rights of India and Pakistan, the two traditional Basmati growers and owners of Basmati germplasm.

Basmati rice is associated with India and Pakistan, as a product special to these regions. It is the same kind of association that France has with Champagne wine and Scotland with Scotch whisky. The Intellectual Property regime in GATT / WTO acknowledges the claim of a region over products that are associated in a special and specific way only with that region.

The TRIPs chapter includes a clause named, "Geographical Indication" which offers proprietary rights to a specific geographical area over the products associated with it. Americans growing and selling Basmati violate the Geographical Indication of Basmati, just as they would violate the indicated rights of France and Scotland if they were to sell American wine and whisky as "Champagne" and "Scotch".

Actually there are two important points that arise with respect to the Americans growing Basmati rice and marketing it as such. The first is the fact that by growing and selling Basmati rice without an agreement with India and Pakistan, the Americans are violating the convention on Biological Diversity, which grants countries ownership rights over the germplasm found in their territories.

The other matter with respect to Texbasmati is that of Geographical Indication. We should be vigilant on this score, since we grow and market special products, which are associated with our region like Basmati rice and Darjeeling tea. Other products could be Alfonso and Dasheri mangoes or the Shahi Lichi.

What to do?

i. India and Pakistan must join together and take the Basmati case to the Dispute Settlement body constituted in GATT/WTO. This is the body where Mexico took its Dolphin-Tuna case against the Americans.
ii. In addition to the WTO, India and Pakistan must also approach the Convention on Biological Diversity. Here, they must establish their ownership rights over Basmati germplasm and claim compensation for the unauthorised use of Basmati varieties by the Americans.

2. Patents on microorganisms

Patents on microorganisms like bacteria, algae, fungi and viruses will have far reaching consequences. Agriculture will be affected if our ability to develop biofertilisers and biopesticides, both based on microorganisms, are hindered by foreign patents.

Bacterial strains like those which act on soil phosphates can make a tremendous difference to our agriculture. These bacteria break down inert soil phosphates to a form that plants can use as nutrients. Such bacterial use could potentially slash our phosphate fertiliser imports dramatically. We need to keep our avenues of research open.

Similarly, strains of nitrogen-fixing bacteria could significantly improve nitrogen uptake of plants and improve the protein content of our foods. This can be of immense significane in enhancing the quality of nutrition for the poorer sections of society.

The role of microorganisms in other areas like pharmaceuticals, mining, energy, etc. is already explained above.

What to do?

India should get out of the patent on microorganisms, already accepted by the previous government, by invoking the clauses of ODRE PUBLIC and offence to the prevailing norms of MORALITY. The GATT/TRIPs section that details the requirements of the IPRs for member nations has a section on exclusions from patentability. In this section, among other exclusions, it is also mentioned that nations may exclude from patentability of those products or categories, the patenting of which would offend the sense of morality of the people or society and that would militate against the order prevailing in those societies.

India should take the position, that when the government attempted to obtain a consensus in Parliament (or wherever else) on legislating for patents on microorganisms, as accepted by the earlier regime, a storm of protest broke out over the possibility of granting ownership over life.

This point should elaborate that according to the religions being practised in India and the sensibilities of the Indian people, it would be considered a moral affront to grant even limited ownership of life, as required by a patent, to individuals or corporations. According to the multitude of Indian beliefs, such a right is vested only in God. Any departure would outrage the sense of morality of the Indian people.

The Green Party and other conscience objectors in Europe who oppose the patents granted by the European Patent Office have successfully used this clause. The use of the Odre Public clause led to the striking down of patents on living organisms, like the famed OncoMouse. A detailed study of the European cases invoking the odre public/morality clauses against patents on life forms would provide our government with the technical details of how to use this clause.

In addition to our own case, we should refer to the European cases, as precedence and case law examples where such conditions have been accepted against life form patents.

3. Sui Generis System for Plant Varieties

The GATT/WTO requires every member country to provide either a patent or an effective sui generis system to protect newly developed plant varieties. Nowhere does it state that WTO members have to follow the UPOV model. The model is not in India's interest for the following reasons:

The UPOV system is far too expensive. Calculating at today's exchange rates, the costs of testing, approval and acquiring an UPOV authorised Breeders Right certificate will amount to something like Rs. 2 to 3 lakhs, at least. This could go up to 8 to 10 lakhs. These rates will effectively preclude the participation of all, but the largest seed companies. There certainly will be no space in such a system for small companies, farmers co-operatives or farmer/breeders.

Farmers play a significant role as breeders of new varieties. They often release very successful varieties by crossing and selecting from their fields. These varieties are released for use as such. In addition, in almost all cases, agriculture universities as breeding material for producing other varieties take up these varieties. Such farmers/breeders would not be able to participate in an expensive system like UPOV.

Their material along with their labour and innovation would be misappropriated by those with money to translate such valuable germplasm into money-spinning varieties, registered with the UPOV. Poor farmers, unable to pay the costs of getting an UPOV certificate, would tend to sell their varieties for small sums to larger seed companies. This will be the ultimate irony, creating an institution that will snatch away from the farmer his material and his opportunities.

Once we are in the system, we shall be forced to go in the direction that UPOV goes. If not today, then tomorrow. The writing on the wall in UPOV is clear. It is a system headed towards outright patents. Starting with its first amendment in 1978, when limited restrictions were placed on protected seeds, the 1991 amendment brought in very strong protection for the plant breeder. In this version, breeders are not exempt from royalty payments for breeding work, and the exemption for farmers to save seeds has become provisional.

UPOV now also permits dual protection of varieties, which means in the UPOV system, the same variety can be protected by Plant Breeders Right (PBR) and patents. It would seem obvious that UPOV is ultimately headed towards patent protection of plant varieties. It would be wise for India to stay out of a system, which has plant patents as its goal since that is neither our goal nor our interest.

The UPOV system is wrong for us because it embodies the philosophy of the industrialised nations where it was developed and where the primary goal is to protect the interests of powerful seed companies who are the breeders. In the UPOV system, rights are granted only to the breeder, but there is nothing for the farmer. In India the position is very different. We do not have big seed companies in the essential seed sectors and our major seed producers are farmers and farmers' cooperatives. Logically, our law will have to concentrate on protecting the interests of the farmer in his role as the producer, as well as the consumer of seeds.

UPOV laws are formulated by countries, which are industrial, not agricultural economies. In these countries, the farming community is by and large rich, constituting 2 to 7% of the population. These countries do not have such a large number of small and marginal farmers like we do.

UPOV laws are framed in countries with a completely different agricultural profile than ours. These are countries where subsidy to agriculture is of a very high order, unlike India. Because they produce a massive food surplus, farmers in industrialised countries get paid for leaving their fields fallow. The UPOV system does not have to protect the farming community of Europe in the way that our seed law will have to protect ours.

In Europe, agriculture is a purely commercial activity. For the majority of Indian farmers, however, it is a livelihood. These farmers are the very people who have nurtured and conserved the genetic resources, the very same genetic resources that breeders want to corner under Breeders Rights. We must protect the rights of our farmers and these rights must be stated unambiguously in our sui generis legislation.

Almost all agricultural research and plant breeding in India is financed with the taxpayers' money. It is conducted in public institutions like agricultural universities and institutions, like the Indian Council of Agricultural Research (ICAR). This research belongs to the public.

On the other hand, the laws of UPOV are formulated by societies where seed research is conducted more in the private domain than in public institutions; where big money is put into breeding using recombinant DNA technology, which is expensive. Because they invest in expensive breeding methods and need to secure returns on their investments, seed companies in Europe seek market control through strong IPRs. These conditions do not apply in India.

What to do?

1. We should devise our own sui generis system. It is possible to build in greater flexibility into, what is defined as, "Sui generis" in TRIPs, as there would be in accepting a precast UPOV model. As a matter of strategy, it makes sense for us to draft a wide-bodied law now, without accepting the restrictions of UPOV. In this way, when the TRIPs come up for review in 1999 and the pressure increases to tighten our Intellectual Property regime further, we will have more space to negotiate.
2. India's seed law must emphasise that it is a germplasm owning country. Its position should be such that seed companies wanting protection for plant varieties through Plant Breeders Right will first have to pay for the genetic material used by them. Genetic resources, like other natural resources, are a source of revenue and must be paid for. The economic value of genes can be assessed by case studies in order to help fix a price for their use.

   Our law should require the use of germplasm and indigenous knowledge about the germplasm to be paid for. The farmers must continue to have complete freedom to use, reuse, sell and modify seeds. Scientists and breeders should have unrestricted access to germplasms for breeding new varieties.

3. India's sui generis system must recognise the Farmers Right, which is absolutely different from the farmer being exempted from saving and selling seeds under the older UPOV rules. Even now, there is confusion among officials on this score. Farmers Rights refer to the rights of the farming community of the Third World in creating and maintaining the genetic resources of the world. It is in acknowledgement of this stupendous contribution that the FAO passed the Farmers Rights bill in 1998. This bill acknowledges that the contribution of farmers is at par with those of the breeder and if one is to be rewarded by a Breeders Rights, the other must be rewarded by a Farmers Right.

   In June 1996, the Food and Agricultural Organisation (FAO) held a meeting in Leipzig, in which the principle of Farmers Rights was strengthened. There is also a commitment that new and additional funds will have to be created to honour Farmers Rights. India should build on this progress by being the first country to incorporate Farmers Rights into its Plant Variety Act.
4. Our law should be distinctly supportive of sustainable agriculture and conservation of genetic resources. It should encourage genetic diversity in the field by promoting the release of several varieties. Instead of a few large seed companies pushing their few successful varieties, a decentralised seed industry should be established in rural areas. Such regional seed-producing units will be able to produce region-specific seeds, far more effectively. Our sui generis system will finally have to ensure the country's food security and the livelihood of farmers.
5. India should seize this initiative to do some creative thinking and craft a creative and sensitive sui generis legislation to present the viewpoint of the South, emphasising the strengths of the South. If India succeeds in doing this, it will provide an alternative sui generis model for all developing countries to follow. In time, with successful lobbying, such a legislation should succeed in ending the era of gene-exploitation by demanding the true rights of countries that have been the developers and custodians of the genetic resources of the world.

INDIGENOUS KNOWLEDGE AND BENEFIT SHARING

Local communities own the bioresources in their region since they maintain them and possess the knowledge of their properties and their use. In other words, they possess the technology about bioresources. This indigenous knowledge automatically confers on them certain rights, including the right to share the benefits derived from the commercial exploitation of bioresources.

The importance of indigenous knowledge cannot be understood when one realises that there are no rice or wheat plants, nor cotton or mustard found lying around in the forest. What are found in the forest are wild plants, out of which communities of men and women have bred races of several food and cash crops over generations. These communities have bred, out of the wild plants of the forests, the thousand of land races, which are the basis of the world's agriculture. The land races bred by farming communities are the foundation material of modern plant breeding and global food security. These land races are the self same varieties that plant breeders use to breed other varieties and for which they seek special and exclusive privileges like Plant Breeders Rights.

It needs to be remembered that farm women and men have not only created several thousand races of food and cash crops, but they have also identified valuable genes and traits in these crops and maintained them over generations, through highly sophisticated systems of crossing and selection. Communities have not only developed complex systems of pest management and biological control, but they have identified and managed a series of genes conferring valuable traits for commercial and domestic needs. So it is, that genes for traits as diverse as disease resistance, high salt tolerance, resistance to water logging and drought tolerance have been maintained in the repertoire of communities. Along with these commercial traits, characteristics like cooking time, taste, digestibility, milling and husking and how much grain breaks during milling operations are recognised and maintained. Women, who have been the traditional custodians of the seed and responsible for its selection, are the repositories of this knowledge and in the true sense, owners of this complex seed technology and its know-how.

This work of genetic selection, maintenance and cross breeding is the result of innovative and creative scientific experimentation in the field. Scientists in the experimental plots of agricultural research stations conduct this work in no way less than in scientific experimentation. We need to overcome the bias that most of us suffer from, that of acknowledging the research conducted by scientists in white coats, working in laboratories of universities, as 'Science' and dismissing the complex knowledge systems contained in rustic, rural communities as something infinitely less and not worthy of being acknowledged.

The fact is that there would be no plant breeders in long white coats working on experimental farms, if it were not for the prior knowledge gained from communities. Indigenous knowledge is not only the foundation of modern science in this and many other fields; it is also what could be described as the reference and referral centre for modern plant breeding.

Today, faced with the threat of global warming and climate changes across agricultural zones, scientists are on the look out for crop varieties that are more heat tolerant. The scientists do not acquire information about the location of heat resistant wheat or millet varieties by sitting in their expensively appointed laboratories and mediating for guidance. They acquire this information by going to deserts and hot regions and asking local farming communities about the varieties that grow in that region and those that can withstand extreme heat. Armed with the benefit of indigenous knowledge, these scientists return to their experimental farms and engage in a breeding and selection program that results in a combination of traits, which they seek to achieve in the new variety that is to be designed for post global warming agriculture.

If credit had to be apportioned for the breeding of a new crop variety, then it could be shared perhaps as 80:20 or at least, 70:30 between the farming and scientific communities. One could quite easily say that if the breeding of a crop variety entailed 100 steps, then indigenous knowledge contributed the first 80 or 70 steps and laboratory science contributed the next 20 or 30 steps. It stands to reason, therefore, that credit, reward and recognition for a new variety should be similarly shared. That is the reason why the claim to place Farmers Rights on par with Breeders Rights is so natural. Farmers have a greater and more innovative share in the creation of new plant varieties than scientists. Their contribution must be recognised with at least the same degree of enthusiasm, if not more than that accorded to scientists.

The role of indigenous knowledge in the realm of medicinal plants is even more obvious than in the case of crop varieties. Knowledge about the characteristics of a particular plant and its properties as a healing substance, or stated differently, the technology of its use, is what gives medicinal plants their social and economic value. This technology of their use has been acquired through a few thousand years of experience, trial and error, and incremental refinement. As a result of this, communities have developed the knowledge of the plant, animal and mineral world to a mature and scientifically sound technology level, which exists in several forms, the best known of which are perhaps, Ayurveda and Siddha. Apart from these, knowledge of the healing properties of plants is found in communities that live around the forest or close to nature. Tribal communities, island communities and others have developed their own knowledge base about the flora, fauna and mineral wealth of their region and use this knowledge to their socio-economic advantage.

It stands to reason, that the technology pertaining to the medicinal uses of plants and animals belongs to indigenous communities and must be considered their property. It must be considered to be their property, in the same way that technology for making high-grade chrome steel is considered to be the property of the Japanese company that developed it. It stands equally to reason that when someone wants to use indigenous technology to produce medicines from medicinal plants, they must first ask for permission and then agree on terms of payment for the use of this technology. When a company like Dabur or Baidyanath commercialises community knowledge and benefits financially from it, it should certainly pay royalty or make an arrangement for profit-sharing. This would be even more the case, if foreign companies wanted to use this technology.

If a pharmaceutical giant like Merck were to show interest in the production of medicines based on Ayurvedic or tribal knowledge, it would explore the forest wealth of regions like India and Costa Rica via the local vaids or shamams. The scientist from Merck cannot make head or tail of medicinal flora when he is bioprospecting, if he does not have information from the vaid or the tribal ojha. Merck will begin to look for a cure for stomach ulcers in plants that local vaids use for stomach ailments, and not in plants that they use for ailments of the eye. The scientist from Merck could not enter a tropical forest on his own and choose random plants, say for example, that the creeper with blue flowers could yield drugs against cancer or the shrub with yellow flowers could contain ingredients effective against diabetes. The plants in the forest or in the field, or for that matter along the banks of rivers or on the roadside, have value only because people have special knowledge about their characteristics and the range of their utility. It is clear that Merck must pay for this technological know-how.

It stands unequivocally established that biological resources have a value only when accompanied by the technology of their use. It is equally clear that bioprospecting, as an activity is only possible when indigenous technology is made available to those seeking accesses to biological resources for commercial use. Now that bioresources are becoming a highly sought after raw material in the era of biotechnology, we must create a framework for the just and proper use, and appropriate payment, of indigenous technology. At present, the awareness and acceptance with respect to the recognition of indigenous knowledge as valuable technology with a high price is inadequate in India.

We need to address ourselves to a few important aspects in order to lay down a comprehensive national policy in this regard. These can be listed as follows.

DOCUMENTATION OF THE LOCATION OF BIOLOGICAL RESOURCES AT THE REGIONAL AND NATIONAL LEVEL

If we want to use bioresources as the foundation for national growth, we must at least know where we have what and whether previously recorded populations of plants, animals and insects are healthy, threatened, on the verge of extinction or already extinct.

Along with documentation of the bioresources, we must document the existing local/community knowledge about the various uses of these resources. This documentation, which should be compiled as a National Bioresource Register, will serve several functions.

The first is that of a data bank for people seeking access to information. This access should be made available for a fee, accompanied by the conditions governing the use of this information. This is the normal practice with data banks everywhere. The fee for bioprospecting must be levied and paid to a Community Gene/Technology Fund for the dispensation of which, representatives of communities will have a say. It would be advisable to have a basic fee for the right to prospect, irrespective of whether a viable product emerges from this exploration. A profit-sharing formula should be worked out, in addition, if a commercial product is developed, to pay for the use of an Indian raw material and Indian/Indigenous technology.

The documentation can be used to stake the claim of communities or individuals for royalty payments for the transfer of indigenous technology. This database can also be used to identify communities, which should be included in the National Authority, that will govern the use of bioresources and implement conditions of the Convention on Biological Diversity (CBD), like prior Informed Consent and Material/Information Transfer Agreements.

Finally, this data bank will serve the important function of establishing community knowledge firmly in the public domain. This will provide the technical basis for rejecting patent claims deriving from indigenous knowledge. The recent furore over the American patent granted on the wound healing properties of Haldi, is a case in point. As is also, the patent granted to WR Grace for a Neem Based pesticide. *Phyllanthus nirurii,* called Bhoomi Amla in India is known in several Asian countries for its efficacy in treating liver ailments. A liver medication extracted from *Phyllanthus* obviously derived from the traditional knowledge of Indian/Asian communities, has been patented by Bloomberg (USA). All these patents should have been challenged on the grounds that they are derived from knowledge/technology owned by indigenous communities. Furthermore, since this knowledge was used unlawfully, without either Prior Informed Consent or Material/Information Transfer Agreements, therefore, a strong case exists for annulling these patents.

If India is to protect its interests, and the interests of its indigenous communities, it will have to be aggressive and proactive in laying down the guidelines governing the use of bioresources. It will then have to act resolutely to challenge any infringements. Apart from challenging the grant of patents in other countries, it would be advisable for Indian legislation to include clauses barring the grant of patents on any products or processes derived from indigenous knowledge. The knowledge of communities must remain in communities and not be privatised.

We must ensure that the information that is documented is banked in a government owned repository and is legally admissible in a court of law as evidence of prior knowledge. In order to strengthen the claim of indigenous communities over their knowledge base, our laws must admit Oral Tradition as documentation of use. This will be of importance when dealing with knowledge other than that documented in Ayurveda, like tribal knowledge or the knowledge existing in far islands or hill populations.

Drafting of national legislation. New laws should be drafted quickly to deal with all aspects of bioresources and policies governing their use. These laws are to be drafted primarily in the context of two international treaties, the CBD signed in Rio in 1992 and the GATT/TRIPs which was finalised in 1994 and led to the formation of its successor, the World Trade Organisation (WTO).

The new law to be formulated, regarding the conditions of the CBD, will have to take into account three principal areas. These are:

To establish ownership right over the biological resources found in the sovereign territory of India. This is necessary to overturn the earlier principle of the Common Heritage of Mankind, according to which all genetic resources were considered to be the property of everybody, with no particular ownership.

To formulate the guidelines and structures for Prior Informed Consent, according to which users will have to seek the permission of some kind of National Authority authorised to grant or refuse access to genetic resources. Representatives of communities that have been responsible for maintaining genetic resources over generations and are the repositories of indigenous knowledge should be members of such a National Authority.

The conditions for Material and Information Transfer Agreements will have to be laid down so that the use of biological resources is just, equitable and sustainable. This law would seek to ensure that indigenous communities are not denied their share of the profits that accrue from the commercial exploitation of the genetic resources that they have conserved.

The question of Intellectual Property Rights will have to be addressed in the CBD, although indirectly. Our position should be that India should not grant IPR protection over products and processes derived from indigenous knowledge. The rationale for this is that the knowledge belonging to communities should not be privatised. Whereas, this knowledge can and should be used for commercial exploitation and the betterment of communities, it should not be monopolised.

With respect to GATT/TRIPs or now WTO/TRIPs, a sui generis legislation will have to be brought in to determine the use of genetic resources in the breeding of new crop varieties and the kind of Intellectual Property Rights that would be granted for the development of new varieties. Our sui generis legislation should place Farmers Rights at Par with Breeders Rights and acknowledge and reward the contribution of farm men and women to the development of land races and therefore, to the development of new varieties. The indigenous knowledge involved in the location of favourable genes, needed for successful breeding work, should be paid for. The recipient can be the Community Gene/Technology Fund, as mentioned earlier.

Another prospective legislation emanating from the WTO/TRIPs, will have to decide on an IPR regime for microorganisms. Microorganisms do not feature as much in the mainstream of community knowledge as plants and animals. Even then, there are pockets of specialised knowledge, regarding their use in cases like human and veterinary care and the processing and preservation of many foods. Care will have to be taken that any IPR regime, that India finally accepts, will have provisions for acknowledging and compensating the rights accruing from indigenous knowledge.

BIORESOURCE AND BIOTECHNOLOGY POLICY FOR THE ASIAN REGION: RECOMMENDATIONS FROM AN INTERNATIONAL SEMINAR

Gene Campaign had organised a seminar on Bioresource and Biotechnology Policy for the Asian Region in New Delhi on May 10 and 11, 1997. The purpose of the meeting was to try and reach a consensus on Bioresource use and to devise a regional approach to deal with the new and important issues like patents, community rights, farmer's rights, etc. These issues have become important international issues and are dealt within the Trade Related Intellectual Property Rights section of the GATT/WTO, as also in the Convention on Biological Diversity. This meeting brought together experts and policy makers from India, Bangladesh, Nepal, Sri Lanka, Malaysia and the Philippines.

At the seminar, the participants recognized

1. Since most of the world's biological resources are located in the developing countries and it is their communities that have protected these resources over generations, these countries have a major stake in how the bioresource base of the world is to be used and how benefits of this use are shared. These biodiversity rich countries should formulate articulate views on the rights of communities, so that justice is done to the custodians of the world's bioresources. It is equally important that these countries reflect carefully about what kind of the IPR regimes they should accept.
2. That the South and South-East Asian region is gene rich and has three major centres of diversity of important crops, including cereals, pulses, oilseeds, vegetables, fruits and flowers. This region has contributed in a very significant way to the germplasm pool of the international centres and the countries of the north.
3. That Agriculture which is the mainstay of the economy and the livelihood base of this region is supported by research and development through public funding. Conversely, in the countries of the north, where the percentage of people working on farms does not exceed 5% of the population, the private sector plays the most crucial role in agriculture. Therefore, what is useful and acceptable in the north, may not be good for the south.
4. For the countries of the Asian region, where agriculture is the backbone of the economy and where bioresources still form the socio-economic foundation of tribal and rural communities, retaining control over these resources is a compulsion for survival. As the foundation material of one of the world's most dominant technologies, biotechnology, bioresources also offer the promise of a vehicle for self-reliant growth to the countries in our region.
5. That biotechnology has emerged as a powerful tool for enhanced and improved productivity in various areas, such as food, medicine and industrial products. Crops with better nutritional quality, increased resistance to biotic and abiotic stress, as also improved capacity for post-harvest processing, can be produced through biotechnology. Given its crucial role, biotechnology should become an integral part of the development of bioresources. This technology should be taken to the villages to increase food and agricultural production, conserve genetic diversity, and enhance rural incomes.
6. Genetic resources recognise natural, not political boundaries, so similar biological resources are found in more than one country. Therefore, it is of utmost importance that the countries of a particular region work together to formulate a regional policy. This is needed to strengthen their position as a gene rich centre and derive the maximum possible benefits from their biological wealth, as also to ensure that no one country is able to undermine the larger interests of the region.

7. These issues are complex and need to be handled by technical experts. Bureaucrats, because of their poor knowledge base should not be allowed to be principal negotiators in this highly technical field. It is important to maintain continuity and coherence in the teams that negotiate at international treaties and follow-ups. This should be done by keeping a core-negotiating group of technically strong people, assisted by the bureaucrats of the day.
8. Nations in the region should set up autonomous Bioresource and Biotechnology Authorities to deal with all aspects of biological resources. Members of this high level authority should have a strong representation from bioresource stakeholders and experts.
9. That NGOs have a special role to play since they are more flexible and can move where governments cannot. The NGO community must take the lead in formulating and changing policies and execute policies without waiting for governments to do so.
10. The IPR regime on bioresources, which is being demanded by the North, is not desirable for the Asian region. They should devise IPR regimes appropriate for their particular situation.

The Recommendations of the seminar were as follows:

1. IPR awareness and patent literacy must be improved at all levels. Patent offices, industry, the scientific community and those in government departments need to be trained, discerning and competent in the field of IPRs for biological materials. In addition, appropriate steps must be taken to encourage social and political awareness amongst scientists, and promote scientific literacy amongst social scientists and politicians.
2. The Region should move quickly to enact legislation pertaining to the Convention on Biological Diversity, so as to protect its bioresources and indigenous knowledge base. GATT derived legislations on IPRs must be made subservient to the CBD. It is possible to do this by following Article 22 of the CBD, which places the obligation to protect biological diversity above the all other requirements of international conventions.
3. A comprehensive, mutually accessible database should be built up by accessing data from diverse sources. This should be extended to, small and difficult to access, publications in the various countries. This database should be used for making benefit-sharing claims. Data on available germplasm for breeding and cultivation should be made available to researchers and farmers of the region.

Appropriate steps must be taken to prepare computerised and easily accessible inventories based on reliable data in the following areas:

a) Traditional conservation techniques
b) Indigenous drugs and medical prescriptions
c) Traditional agricultural practices

Technical jargon should be converted to easily comprehensible information in regional languages. This information should be made widely available to enable participatory decision-making in this important area.

4. As per the laws of the countries of the region, oral documentation should be treated at par with coded documentation, for all purposes. This should be admissible as evidence of knowledge existing in the public domain and as grounds for challenging unauthorised use. Austrian law has already established a working precedence in this regard.
5. Capacity building, with easy transfer of technology, exchange of materials, experience and skills, especially in fields like conservation, should be a priority exercise. There should be exchange of expertise in taxonomy and the different ways in which countries manage and utilise their biological resources. The region should identify priority technologies for itself and either acquire or develop them jointly. A multination consortium in the Asian region should set up a state-of-the art institute in the field of biological taxonomy.

   Certain facilities, such as DNA fingerprinting facilities for animals and plants in Hyderabad and Delhi, should be open to all countries of the Asian region at the same cost and with the same conditions, as would be valid for the host country itself.
6. The countries of the region should have an exchange programme of resource people with special skills in various areas, especially Intellectual Property Rights, Farmers and Community Rights and Benefit sharing mechanisms. These people should be available for educational and campaign activities.
7. The Region should set up a common forum to challenge infringements and violations pertaining to the use of biological resources and indigenous knowledge.
8. The Region should have a mechanism, like an Inter-regional Standing Committee, for early warning, containment and emergency response to accidents, like unintended release of Genetically Modified Organisms (GMOs).
9. The Region should have a common position in international negotiations, and intervene strongly in follow–up negotiations, like the Conference of Parties (COP) of the Biodiversity Convention and similar meetings of the WTO. The heads of national delegations should interact frequently and develop a joint strategy, so that one mutually agreed Asia level position is represented by all the nations, as is already happening in ASEAN.
10. The indigenous technology of the area should be collected and ownership should be established over this and over the resource itself. Appropriate laws (should be passed) that ensure and honour the rights of a community over its traditional practices, for example, in medicine and agriculture. Any patent or extensive use of these practices (including the materials that they may involve) by anyone else must be prohibited, unless there is appropriate and adequate compensation to the community for such use.

11. Countries should exercise control over the export of their biological resources, on the same level as export controls over "dual–use" technology for national security and foreign policy purposes.
12. The voluntary sector must be strengthened. Governments should be persuaded to find NGOs to build partnerships, attend meetings, and develop joint strategies and forward planning for the region. Regional activist groups should exchange information and hold campaign workshops in each other's countries.
13. Constitutionally guaranteed rights over should be given to all communities that have been living for, say a hundred years, in forests and other reserved areas, sanctuaries and national parks. The management of these areas must involve these communities in a way that these residents have a major voice in the management of the area.

There must be concerted social, political and legal action, against all encroachments in the above-mentioned areas, by non-resident outsiders, unless such an 'encroachment' is done openly after an appropriate debate, in the larger interests of the people resident in the area and the people of the country, and after obtaining the concurrence of the people resident in the area and providing them with appropriate compensation; thus, prior approval of the people, as well as of the government must be obligatory for any such encroachment in the larger public interest. The responsibility for implementing the above mentioned sanctions would lie equally with the government and with the people represented by NGOs.

14. Every organization, governmental or non-governmental, funded, wholly or partially by public money, must be transparent and accountable for its statements and actions, and must be taken to task if any information that it presents is shown to be substantially wrong.
15. The patent laws should be revised, where necessary, to prohibit the patenting of any living form (microorganism, plant or animal) or of any product made directly by or from the living form. (This provision would thus prohibit the patenting of any genetically engineered life form, or a product, such as azadirachtin derived from neem.)
16. Asian countries, including India, should devise and design their own sui generis systems for the protection of plant varieties, farmers' rights and breeders' rights. They should not accept the provisions of UPOV in totality, but adopt IPR regimes that reflect the strength and compulsions of gene rich nations.
17. Appropriate steps should be taken, both by the government and NGOs to make the people of the Asian region aware of their bioresources and biotechnology; of the legal, social, moral, ethical, political and economic implications of modern biotechnology; and of the role that biotechnology can play today, in both the conservation of bioresources and their utilisation for development.

    An appropriate policy with regard to the above-mentioned uses of biotechnology must be evolved and reviewed continuously in real time. It should be recognised that many new areas of biotechnology have the potential of, on the one hand, improving agricultural productivity per individual and per unit time and money spent, and on the other, providing additional employment to those who, today, spend all their time in agriculture-related activities because of low productivity per unit time, money and labour spent.

18. As a rule, no foreign aid should be accepted for work, in sensitive areas, that relates to biodiversity and/or its conservation, unless it is ensured with full transparency that the aid being given has no conditions attached that would not be in national interest. This would obviously not apply to genuine bilateral or multilateral collaboration that permits free use and publication of the results of the research.
19. Asian countries must come out with viable and sensible scientific, technological and agricultural policies. A mechanism should be set up to periodically examine these policies of the Asian countries, taken together, to ensure maximal benefit to all the countries from each others experience.
20. Organisations such as parliamentary scientific committees can and should play an important role in the evolution of the above-mentioned policies, by providing an interface between professional scientists and parliamentarians.

Chapter 3

Sustainable Agriculture for Enduring Food Security in the North-Eastern States of India

U.C. Sharma
ICAR Research complex for NEH Region
Umiam - 793 103, Meghalaya

The North-Eastern Region, comprising the States of Arunachal Pradesh, Assam, Manipur, Meghalaya, Mizoram, Nagaland and Tripura, has an area of 255,090 sq km, out of which about 72% is hilly. Though the region is endowed with rich natural resources of soil, water and vegetation, it has remained economically backward due to certain reasons. Despite having low population density (122 persons per sq km, as against 291 in the country), the region is deficient in food grains. Interestingly, the shortage gap in food grains is increasing progressively as the Annual Compound Growth Rate of population is higher (2.43%) in the region than that of the productivity of food grains (2.18%). The productivity of food grains has to be improved as it forms the essential component of economic development. There is need and justification for applied and adaptive research, which is beneficial to the farming community, particularly society in general, in order to improve agricultural production for realizing of the goal of self-sufficiency in food grains and economic prosperity of the region. For sustainability of agricultural production systems, common features like extrapolation of yield trends, assessing the depletion or pollution of natural resources and holistic and heuristic concept of soil health, need to be looked into.

INTRODUCTION

During the last twenty years or so, the traditional role of agriculture, i.e., food, fibre and energy, has combined with the new requirement of conservation of the environment. In addition, there is the requirement of sustaining the economic viability of agricultural activities and promoting the quality of life of the people and society, as a whole. Changes in attitude, at the level of government policies, institutions, farmers and the society are necessary (Schaffert, 1997). This would involve the following:

(i) Investment in science and technology that would reduce the degree of uncertainty in agriculture production activities in the environment and promote economic growth.
(ii) Investment in monitoring the environment to generate better data and sound information on the effects of production; for use in prediction models, and
(iii) Create an effective mechanism to provide incentives to institutions to respond to the challenge of sustainable production.

The breakthrough in wheat and rice production in India in the late sixties, which came to be known as the 'Green Revolution', started the process of using agricultural science to develop modern techniques for higher production. The green revolution remained confined to North-Western India, where basic infrastructural facilities existed. This was necessary at that time since the need was to produce enough food-grains for the ever-increasing population of the country and not to depend on imports. The country succeeded in this endeavour and became self-sufficient in food grains. However, of late, the productivity of food grains has reached its plateau in the areas which witnessed the green revolution. About 30% of the cultivated area in the country produces 55% of the total food grain production. The rest, 70% of the cultivated area has not reaped the fruits of advancements in modern technologies. The North-Eastern region is one such area. It has, now, become inevitable to enhance agricultural production, not only to make the region self-reliant in food grains, but also to produce surplus for the central pool. Though it is very much achievable, it would require a change in the attitude of the people and in Government policies. Four issues are of particular concern, such as productivity, sustainability, investment focus and financial discipline, and resource management. In addition, technical innovations and structural reforms are required for the intensification of farming and obtaining high levels of production. The emergence and acceptance of the concept of sustainable development has brought in the general realisation that societal perceptions must shift towards ecological determinism, so as to achieve quantitative as well as qualitative growth within the carrying capacity of the ecosystem. The only way for agriculture to keep pace with the population and alleviate hunger is to increase the intensity of production. The increase in yield must be achieved through the application of available technologies.

It is worthwhile to understand the present position of the states of the region with regard to food grains production and the strategies to improve productivity, so as to achieve the goal of self-sufficiency.

PRESENT SITUATION

Rice is the most important crop of the North-Eastern (NE) region, covering about 93% of the total area under cereals and 94% of the total production. The people of the region are predominantly rice eaters. Increase in the productivity of rice can ensure self-reliance in food grains. Maximum annual Compound Growth Rate (CGR) of rice production between 1980-81 and 1993-94 was found in Mizoram (3.60%), followed by Nagaland (3.14%) and Manipur (3.10%), (Table 3.1). Meghalaya showed a decline in productivity (-1.53% annual CGR) during this period. The area under rice decreased in Manipur and Tripura, however, it increased in all the other states. Expect Meghalaya, production of rice increased in all the other states. The area, production and productivity increased at an annual CGR of 0.69%, 2.20% and 1.50% against the country's annual CGR of 0.35%, 3.02% and 2.66%, respectively. The increase in annual CGR of area under food grains was the highest in Mizoram (6.61%), followed by Arunachal (3.30%), and Nagaland (3.05%), however, though annual CGR of yield of food grains was maximum (3.32%) however, annual CGR of yield of food grains was maximum (3.32%) in Mizoram, followed by Manipur and Nagaland (Table 3.1). The yield of food grains showed an annual CGR of 1.49 % in the region, as against 2.92% in the country, between 1980-81 and 1993-94. The annual CGR of population during the corresponding period was 2.43% in the NE region and 2.16% in the country. Lower annual CGR of productivity than that of population indicated that increase in the yield of food grains has not kept pace with the population in the region. Even the annual CGR of food grains production is less than the population growth, showing a widening of the gap in food shortage in the future.

Table 3.1: Annual Compound Growth Rate (%) of rice, food grains and population (1980-91 to 1993-94)

	Rice			*Food grains*		
State	*Area*	*Production*	*Productivity*	*Production*	*Productivity*	*Population*
Arunachal Pradesh	2.97	3.84	0.80	4.28	0.94	3.19
Assam	0.80	2.23	1.41	2.07	1.42	2.19
Manipur	-1.16	1.85	3.10	1.55	3.14	2.60
Meghalaya	0.38	-1.15	-1.53	-0.03	-0.47	2.88
Mizoram	6.08	9.91	3.60	10.15	3.32	3.40
Nagaland	2.15	5.37	3.14	6.15	3.01	4.55
Tripura	-1.33	0.89	2.26	0.95	2.00	3.00
NE Region	0.69	2.20	1.50	2.18	1.49	2.43
All India	0.35	3.02	2.66	2.65	2.92	2.16

The area of production and productivity of pulses in the region increased at an annual CGR of 0.79%, 3.94% and 3.13%, that of oilseeds 3.20%, 3.88% and 0.88%, respectively, between 1980-81 and 1993-94 (Table 3.2). The area under oilseeds is going up at a fast rate, however, the productivity rate has been low compared to the country as whole. The increase in the annual CGR of pulses in the region was higher than the average CGR of the country. Highest annual CGR in the production of pulses was recorded in Nagaland (11.96%), followed by Arunachal Pradesh (8.52%) and Tripura (8.32%), between 1980-81 and 1993-94. Annual CGR of productivity of pulses was highest in Arunachal Pradesh (6.62%), followed by

Nagaland (5.05%). Except Manipur and Meghalaya, all other states of the region have shown encouraging results in oilseeds.

Table 3.2: Annual Compound Growth Rate (%) of Pulses and Oilseeds (1980-81 to 1993-94)

	Pulses			*Oilseeds*		
State	*Area*	*Production*	*Productivity*	*Area*	*Production*	*Productivity*
Arunachal Pradesh	1.80	8.52	6.62	7.64	10.90	3.20
Assam	-0.27	1.47	1.76	2.32	3.26	1.05
Manipur	-	-	-	- 0.86	-1.12	-0.26
Meghalaya	3.16	7.30	4.01	0.92	1.19	0.25
Mizoram	-	-	-	11.25	20.60	8.46
Nagaland	6.50	11.96	5.05	16.19	23.16	7.01
Tripura	6.04	8.32	2.14	6.13	7.26	1.07
NE Region	0.79	3.94	3.13	3.20	3.88	0.84
All India	-0.01	1.62	1.63	2.91	6.11	3.12

The annual CGR of area, production and productivity of various crops has been shown in Table 3.3. Annual CGR of area indicates that the area under food grains, maize, rice and oilseeds has increased in the region, while the area under wheat has decreased over the years. The productivity of all the above crops has shown an upward trend, however, the production of wheat has gone down due to the decrease in area under crop. As per estimates, the region was deficient in food grains by about 1.93 million tons during 1997-98 (Table 3.4).

Table 3.3: Annual Compound growth rate (%) of different crops (1980-81 to 1993-94) in NE Region

Crop	*Area*	*Production*	*Productivity*
Rice	0.69	2.20	1.50
Wheat	-1.52	-0.63	2.01
Maize	1.30	3.07	1.75
Pulses	0.79	3.94	3.13
Food grains	0.69	2.18	1.49
Oilseeds	3.20	3.88	0.84

Table 3.4: Approximate deficiency / surplus of food grains in different states of the Region (1997-98) (000' tons)

State	*Def. (-) / Surplus (+)*
Arunachal Pradesh	+ 49.7
Assam	- 1427.7
Manipur	- 58.2
Meghalaya	- 277.7
Mizoram	- 6.8
Nagaland	- 51.5
Tripura	- 165.1
NE Region	- 1937.3

LEISA AND HEISA

Kieft (1993) has explained the Low External Input Sustainable Agriculture (LEISA) and High External Input Sustainable Agriculture (HEISA). While LEISA is a small scale, self-supporting and labour intensive farming system; HEISA is a large scale, market oriented and capital-intensive farming system. Other attributes of these two land use systems have been explained in Table 3.5. These two land use systems could be followed as per availability of land, financial conditions of the farmers and market accessibility.

Table 3.5: Differences between LEISA and HEISA

		LEISA	*HEISA*
1.	Location	Heterogeneous, poor Infrastructure	Homogenous, good infrastructure
2.	Farming	Small scale, Self–supporting, Labour-intensive	Large scale, Market-oriented Capital intensive
3.	Diversity	High	Low
4.	External Inputs	Low	High
5.	Strategy	Minimising risk	Maximising yields / income
6.	Output	Low-medium	High-medium

Kieft (1993)

AGRO-ECOSYSTEMS

The concept of agro-ecosystems is useful in defining research needs and development opportunities, particularly in less favoured situations with high spatial and/or temporal heterogeneity. Conway (1987) has defined an agro-ecosystem as an ecological system modified by humans to produce agricultural products. The use of an agro-ecosystem approach in agriculture requires the recognition of the existence of interactions among biological and natural resources variables in all, but the very lowest level of social and economic considerations, as well. Highly productive agro-ecosystems must be able to meet greater demands for food and fibre production in the future. At the same time, they should meet the standards for environmental impact.

CONSTRAINTS

The major constraints in higher agricultural productivity are the prevalence of shifting cultivation in its present form, land tenure system, small holding size, lack of finance, marketing facilities and proper communication system, acceptable and appropriate technologies, high humidity favouring diseases, pests, weeds, etc. and above all, the necessary will for the implementation of the desired programme. Shifting cultivation in its present form, involving deforestation, resulting in soil and nutrient erosion, environmental degradation and floods, is not suitable. The shifting cultivators have either to be settled for permanent cultivation with appropriate farming systems or the practice has to be improved for sustainable production, without affecting the environment and to maintain ecological balance at all costs. It is presumed that circumstances will change the attitude of the shifting cultivators and time will see the elimination of this practice in the near future.

The land tenure system prevailing in the region, should not in any way, to a stretchable extent, affect the improvement of agriculture in the region. Even though it is of great importance in determining what agricultural systems are practiced, changes in it will test the ability of agricultural systems to adapt to changes. Many, if not most, agricultural systems are the results of a particular social structure. Some times, the social structure may accept the new technologies, if risk factors are covered and the gains are prominent.

Proper marketing facilities are necessary so that farmers get encouraged by selling their produce at suitable prices. Long-term easy loans must be provided so that cultivators can buy and use inputs adequately. Pests, diseases and weeds have to be managed with appropriate methods.

More linkages and transfer of technology programmes are needed to make the farmers aware of new technologies for the improvement of agricultural productivity. Suitable and feasible agricultural practices and farming systems, etc. have to be dealt with suitably, along with the available technical know-how and integrated, but optimal, mix of traditional wisdom and new techniques.

STRATEGIES

Nutrient management on acid soils

More than 92% of the soils of the North Eastern region are acidic in reaction. Due to the high content of exchangeable aluminum, these soils are low in available phosphorus. Phosphorus is the most limiting factor in these soils for good crop production. Secondly, due to heavy rainfall during the main crop season (March – September), there are chances of leaching of plants nutrients, particularly nitrogen and potassium (Sharma, 1989, 1990). For a balanced supply of nutrients, it is necessary to make an integrated use of nutrient sources, such as inorganic, organic and biological sources, their method and time of application to reduce leaching losses.

The best strategy to improve the efficient used nutrients is through the adopting Integrated Plant Nutrient Management System (IPNMS). The IPNMS is defined as a package of practices for the manipulation of plant growth environment to supply essential nutrients to a crop in an adequate amount and proportion, for optimum production, without degrading the natural resources, as shown in Fig 3.1 (Roy, 1995; Baligar and Fageria, 1997). The system highlights the need to understand the numerous interactions and feedbacks that vary as the climate and management change. All the factors involved should be evaluated and their thresholds, in relation to crop productivity, should be determined.

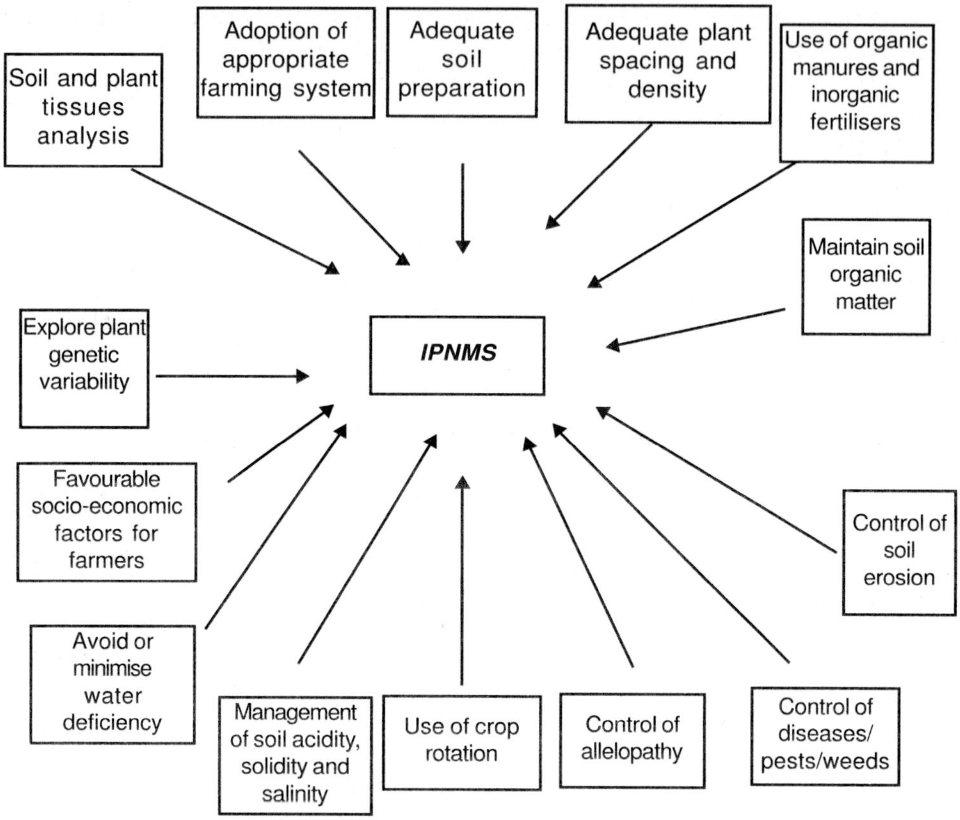

Fig. 3.1: Integrated Plant Nutrient Management System

Known yield potentials are two to four times greater than the average yield potentials obtained for various crops and are at the upper limit of the possible yield increase. The nutrient uptake by roots, growing in soil, depends on both the nutrient supply of the soil and the nutrient influx characteristics of the root (Barber, 1995). Soil chemicals (pH, Al, Mn, Ca Mg, P) and physical (aeration, moisture, bulk density) factors influence the availability of nutrients and the roots physiological and morphological parameters.

The most important acid soil constraints can be grouped into four sets:

(i) high H^+ and Al^{3+} activities in soil solution;
(ii) low soil cation exchange capacity;
(iii) high capacity of soil solid phase to adsorb anions, especially phosphate ions, and
(iv) low activities of the organic and biological soil fraction. The efficient use of nutrients in these soils will always be strongly dependent on the partial or complete solution of these constraints, involving environment, soil and plant management.

It is evident from the data given in Table 3.6, that no state in the region can fully meet the crop nutrients from organic sources. The total crop removal of nutrients for optimum yield and their replenishment through organic sources might be possible only up to 10.3% in the region, varying between 5.5% in Nagaland to 22.3% in Manipur (Table 3.6). It is, therefore, imperative to apply the nutrients, through inorganic sources, for higher productivity. The time has come when the farmers have to use the land resources with care and be responsibile not to snatch the rights of future generations by indiscriminate use of this natural resource. Principle No. 2 of the World Soil Charter states " For recognising the paramount importance of land resources for the survival and welfare of people, economic independence of the countries and also the rapidly increasing need for food production, it is imperative to give high priority to promoting optimum land use, for maintaining and improving soil productivity and for conserving soil resources". A good stewardship for land management in the region is necessary for higher productivity and self-reliance in food grains.

Table 3.6. Annual expected nutrient supply through organic sources in the North-Eastern states

State	*Source*	*Nutrients (000'tonnes)*				*Replenishment %*
		N	*P_2O_5*	*K_2O*	*Total*	
Arunachal Pradesh	Animals	2.21	0.99	1.19	4.39	–
	Compost/ FW/GM/FL	0.18	0.08	0.14	0.40	–
	Kg ha^{-1}	11.87	5.28	6.56	23.71	10.6
Assam	Animals	50.71	23.73	27.32	101.76	–
	Compost/ FW/GM/FL	0.72	0.33	0.64	1.69	–
	Kg ha^{-1}	13.82	6.50	7.56	27.88	9.4
Manipur	Animals	5.91	2.75	3.20	11.86	–
	Compost/ FW/GM/FL	0.13	0.06	0.13	0.32	–
	Kg ha^{-1}	33.51	16.62	18.04	68.17	23.3
Meghalaya	Animals	4.08	1.88	2.22	8.18	–
	Compost/ FW/GM/FL	0.18	0.07	0.14	0.39	–
	Kg ha^{-1}	20.22	9.31	11.23	40.76	16.6
Mizoram	Animals	0.39	0.19	0.22	0.80	–
	Compost/ FW/GM/FL	0.14	0.05	0.12	0.31	–
	Kg ha^{-1}	6.62	3.01	4.25	13.88	5.9
Nagaland	Animals	1.55	0.68	0.87	3.10	–
	Compost/ FW/GM/FL	0.14	0.17	0.13	0.44	–
	Kg ha^{-1}	6.26	3.14	3.70	13.10	5.5
Tripura	Animals	5.53	2.57	3.00	11.10	–
	Compost/ FW/GM/FL	0.19	0.09	0.15	0.43	–
	Kg ha^{-1}	15.45	7.18	8.51	31.14	10.5
NE Region	Animals	70.38	32.89	38.02	141.29	–
	Compost/ FW/GM/FL	1.68	0.75	1.45	3.88	–
	Kg ha^{-1}	14.34	6.69	7.85	28.88	10.83

** FW- Farm wastes; GM- Green manure; FL- Forest litter

Water management

Water is a limiting factor for crop production and its judicious management is the key to the development of sustainable agriculture, both for irrigated and rainfed areas. The green revolution in the country was possible due to the integrated effect of irrigation, fertilisers, high yielding varieties and pest management. In the states of the North- Eastern region, the scope for increasing the area under irrigation is limited, due to the hilly terrain and high costs involved for executing irrigation projects, with relatively low areas available for irrigation in the vicinity. Only minor irrigation projects could be taken up, such as ground water extraction, lift irrigation, etc. The region being seismologically explosive, the construction of big dams may be dangerous for the adjoining areas. Harvesting of rain water and its proper use appears to be a feasible solution for providing life-saving irrigation to crops, or irrigation at critical stages of crop growth. The rainwater could be utilised through:

(i) Retaining rain water *in situ* and minimising the run-off. This would also reduce soil erosion and nutrient loss in run-off water and also floods in the plains,
(ii) Recycling the run-off water after water harvesting and its judicious use for raising crops during off-season,
(iii) Recharging the ground water,
(iv) Reducing evaporation in relation to transpiration,
(v) Adopting the micro-watershed approach, and
(vi) Using crop varieties that fit well with the rainfall pattern.

Elaborate multidisciplinary studies undertaken through the micro-watershed approach revealed, that on steep sloppy lands (65% slope) under bare fallow conditions, 200 mm of water per ha was lost as run-off, along with a sediment loss of 84 t per annum (Awasthi *et al.*, 1990). By adopting mechanical measures, such as contour bunds, bench terraces, half-moon terraces and grassed waterways, up to 90% water retention *in situ* could be achieved and soil loss considerably checked (Singh, 1987). Grass cover was found to reduce run-off and soil loss by 82% and 88%, as compared to uncultivated bare fallow plots.

Crop management

The farming system approach on micro-watersheds will prove useful in the North-Eastern region, which is predominantly a hilly terrain. In plain and valley areas, suitable cropping systems for different agroclimatic situations have to be identified for higher production. Mono cropping is predominant in the North-Eastern states, and the introduction of a second crop during rabi season is necessary to enhance the production of food grains. For this, there is a need to stop free range grazing during the winter season, so that one or more crops could be introduced during this period. The following measures are recommended for judicious crop management for higher yields:

(i) Procurement of food quality seeds of a suitable crop variety
(ii) Planting the crop with intra- and inter-row spacing
(iii) Recommended dose of nutrients should be applied
(iv) Interculturing at proper stages to remove weeds
(v) Following adequate plant protection measures, and
(vi) Maintaining optimum soil moisture for proper crop growth

Soil conservation measures

Due to undulating topographical features and faulty agricultural practices prevalent in the region, the soils are highly subjected to erosion hazards. Sharma and Prasad (1995) reported that about 88.3 million tons of soil, 10.4, 0.37 and 6.05 thousand tons of N, P_2O_5 and K_2O, respectively are lost every year due to the prevalence of shifting cultivation. To avoid and reduce all these losses, it is essential to follow soil conservation measures according to the land physiographical features and crop requirement. Bench terracing, contour bunding, half-moon terraces (for horticultural plantation), grassed-water ways, etc., are required for reducing water run-off, moisture conservation and for arresting soil and nutrient erosion. Crop scientists can play a major role in developing suitable techniques to conserve soil. No practical crop management systems are more effective in reducing soil and water run-off than those that provide vegetative cover to the soil. In economically sound management systems, the farmer selects such cultivars that can provide vegetative cover. Such grasses and shrubs should be selected which can provide live borders to terraces. The selection and management of crops can have significant effects on water use efficiency.

The problems of soil degradation are the most significant threats to sustainable productivity. A combination of soil erosion, deterioration of soil structure, loss of nutrients and acidification has led to the destruction of the productive capacity of the soil. It has been estimated that about 2 billion hectares, out of a total of 8.7 billion hectares of agricultural land in the world (including forest land), have been degraded (Oldeman *et al.*, 1990). In the North-Eastern states, about 7623 thousand hectares (29.8%) of land has been subjected to degradation, in one form or the other. There is an urgent need to stop further land degradation and improve the wastelands to make them fit for agricultural crops, agroforestry and afforestation.

Half-moon terraces are circular beds with a 1.0 to 1.5 m diameter, cut into a half-moon shape on sloppy lands. These beds are suitable for fruit plants and other plantation crops like areca nut, etc. The grassed waterways are channels laid out, preferrably on natural drainage lines in the watershed. These channels have a vegetative cover to reduce soil loss, along with run-off water.

By retention of the maximum proportion of rainwater *in situ*, the soil and nutrient loss can be reduced up to a large extent. For this, suitable land use systems such as agriculture, horticulture, agri-horti, agri-horti-silvi-pastoral should be followed as per agroclimatic situations.

Above all it is, probably, the Government policies and concerted efforts of the research and extension workers, which would bring a change in the farmers' attitude towards newer technologies, and would be suitable and helpful in increasing productivity. In the backdrop of the socio-economic milieu in the region, we will have to keep the population under limits for self-sufficiency, as the land cannot produce more than its carrying capacity even with the available technologies.

REFERENCES

Awasthi R.P., Singh R.D. and Singh A.K. (1990). Proc workshop on research needs of land and water management for enhancing agricultural production in the NEH Region, WTC for Eastern region, Bhubaneswar p 55-69.

Baligar V.C. and Fageria N.K. (1997). Nutrient use efficiency in acid soils: Nutrient management and plant use efficiency. Plant-Soil Interactions at Low pH. Brazilian Soil Science Society, Brazil p 75-95.

Barber S.A. (1995). Soil nutrient bioavailability: A mechanistic approach, 2nd edn. John Wiley & Sons, N.C. New york, NY p 414.

Conway G.R. (1987). The properties of agro-ecosystems. Agricultural systems 24: 95-117.

Kieft H. (1993). The potentials of low external input agriculture in Sub-Saharan Africa. In: The Role of Plant Nutrients for sustainable Food Crop Production in Sub-Saharan Africa (Van Relaer H and Prins WH eds), Dutch Assoc of Fertiliser Produces, the Netherlands p 127-140.

Oldeman L.R., Hakkeling T.T.A. and SomBroek W.G. (1990). World map of the status of human induced soil degradation: An Explanatory Note (revised ed), U N Environment Programme, Nairobi, Kenya.

Roy R.N. (1995). FAO-Soil fertility and integrated plant nutrition management programmes. In: Nuclear Techniques in Soil-Plant Studies for Sustainable Agriculture and Environment Preservation, Int p 107-129.

Schaffert R.E. (1997). Sustainable agriculture on acid soils—the challenge. Plant-Soil Interaction at Low pH. Brazilian Soil Science Society, Brazil.

Sharma U.C. (1989). Influence of soil texture and rainfall on leaching of potassium and its recovery by potato (*Solanum tubersum* L). Indian J Agric Sci 59: 713-717.

Sharma U.C. (1990). Studies on nitrogen leaching due to precipitation in Alfibols of Meghalaya. J Indian Soc Sci 38: 218-223.

Sharma U.C. and Prasad R.N. (1995). Socio-economic aspects of acid soils management and alternate land use systems for the North Eastern states of India. Kluwer Academic Press, Netherlands.

Singh A. (1987). Application of research findings for the management of land and water resources in the Eastern Himalayan Region, Bull-ICAR Res Comp, For NEH Region, Shillong.

Chapter **4**

Agrobiodiversity in North-East India

B.K. Sharma and H.S. Gupta*
Division of Plant Breeding,
ICAR Research Complex for NEH region, Umiam-793 103, Meghalaya
** Director, Vivekanand Parvatiya Krishi Anusandhan Sansthan,*
Almora - 263 601, Uttranchal

The North-Eastern region of India is a part of the centre of megadiversity, which is considered to be the home of some crops, and the secondary centre of origin for others. Rice is supposed to have originated in South-East Asia, including this region. Consequently, we have got rich diversity in the rice germplasm of this region. Since genetic diversity is utilised mainly by plant breeders, until recently the issues related to collection and preservation were primarily of interest to plant breeders and conservationists; however, after World War II, the threat of genetic erosion, due to destruction of habitats where genetic variability exists, led to the growing concern, and subsequently, concerted efforts for *in situ* (biosphere) and *ex situ* (gene banks) conservation were made.

INTRODUCTION

The availability of a wide gene pool in the form of genetic diversity is a prerequisite for crop improvement. Vavilov (1951) was the first to show the importance of genetic variability in crop improvement. He developed a systematic plan of action for collecting genetic materials from primary centres of origin of crop plants and also explained the relationship between genetic diversity and centres of origin. Realising the importance of genetic conservation, the Consultative Group on International Agricultural Research (CGIAR) established an International Board of Plant Genetic Resources (IBPGR) in 1974, for promoting coordinated efforts in the collection, conservation and utilisation of plant genetic resources (PGR). The resources will not only help in avoiding genetic vulnerability due to pests and diseases, but also help in the advances made in the field of molecular biology and genetic engineering in transferring useful genes across sexual barriers (Swaminathan, 1988). The food requirement of developing countries, which constitute the most important centres of origin of many crop plants, can be met only by increasing crop productivity and cropping intensity — the twin pathways for realising higher yields per day. In addition, incorporating genes of higher biomass from wild species can break the yield plateau in field crops. This, in turn, calls for efficient utilisation of genetic variability for tailoring new varieties of crop plants, possessing higher yield potential coupled with resistance /tolerance to biotic stresses.

It is in this context that the collection, maintenance and evaluation of genetic diversity present in North-Eastern (NE) India assumes great importance. Subsequently, these collections may be utilised for evolving suitable varieties for different farming systems of the North-East India. Wide variations in topography and physiography, coupled with favourable agroclimatic conditions (wet tropics with rain forest), provide a conducive atmosphere for the growth of a large number of crops; consequently, there is enormous genetic diversity in the region. Poor accessibility has helped preserve this diversity. The contiguous border with China, Myanmar, Bhutan and Bangladesh has helped in microspeciation. The region is considered to be the home of many crops and the secondary source of origin of some other crops (Borthakur, 1988). The Eastern Himalayas being a floristically rich home to many economic plants, are recognised as one of the 18 'hotspots' of biodiversity in the world (Khoshoo, 1996). The present paper attempts to collate the information available on the genetic diversity of various crops in North-East India.

CEREALS

Rice

The general trend in the evolution of crops is: wild perennial – wild annual – cultivated domesticated annual (Chang, 1995). Annual ancestral forms of Asian cultigen of rice began to appear on the Southern border of the Himalayas in the Neothermal age (8000 to 13,000 years before the present). The early cultigens were the *indica* ecogeographic race, which matured earlier than their wild progenitors and, therefore thrived under drier and fluctuating climates. From the primary tropical *indica* race, a temperate zone race *sinica (keng)* evolved inside China as *japonica.* From the eastern coast of South-East Asia, the tall and large forms were brought to the Indonesian islands and they got differentiated into *javanica* (Chang, 1985).

As a result of domestication, many morphological and physiological characters, like tillering, leaf size, culm length and thickness, panicle size, number of secondary panicles, grain weight and harvest index, increased. Simultaneously, there was a decrease in pigmentation, rhizome formation, leaf interval, awning, shattering, grain dormancy, photoperiod response, sensitivity to low temperature and frequency of cross pollination. The plant underwent further changes as it was exposed to stress when disseminated from humid tropics to sub-tropics and subsequently, to the temperate zone. With the passage of time, it was exposed to natural and human selection, diverse climate, varied soil and cultural practices, which led to the great ecological diversity presently, found in *Oryza sativa.*

Of the total 35,000 collections of rice germplasm from India, around 8,000 come from this region. The Indian Council of Agricultural Research and PL 480 Fund of the United States Agency jointly supported the first systematic attempt made to collect rice germplasm from the North-East region for International development (USAID), in 1967-69. This collection was made from the erstwhile state of Assam, which included seven states of NE India. The collection was exhaustive from Arunachal Pradesh, Nagaland, Manipur and Meghalaya. However, due to inaccessibility of the region, present Assam and Tripura were only partly covered, whereas Mizoram was left out. The total races were 6630 and were assigned with the prefix 'ARC' – Assam Rice Collection (Sharma, 1982). Earlier collections from Assam and Manipur were included in the Catalogue of National Germplasm Collection of Rice (Padmanabhan *et al.*, 1971).

Rice collections from Sikkim were described by Kihara (1968) and Arora (1982), from Nagaland by Kulkarni (1962), Arunachal by Kocchar (1987), Tripura by Sasikumar and Sardana (1987), Meghalaya by Asthana and Majumdar (1981) and Mizoram (Anon., 1986). The rice genetic diversity of the North-East region, their status, conservation strategy and utilisation were well documented by Gupta *et al.* (1985). Vairavan *et al.* (1973) classified 190 land races, collected from high altitude areas of the North-East region, into nine clusters. The study showed that some clusters resembled the *japonica* plant and grain type and others occupied intermediate position between *japonica* and *indica*. Srivastava (1977, 1978) classified rice collections from North-East India into five groups. Seetharaman *et al.*, (1974) reported the presence of not only few, but a constellation of *japonica* characters in the ARC collection. These studies strengthen the observation of Kihara (1968) that racial differentiation in *O. sativa* has taken place in this region. The detailed study of the ARC shows that they possess many desirable characters viz. dwarfness, tolerance to lodging, low temperature, low phosphorus (Prasad Rao and Srinivasan, 1978), resistance to various insect pests and diseases (Shastry *et al.*, 1971; Krishnamurthy *et al.*, 1975; Chatterjee *et. al.*, 1977; Devanath and Rao, 1978), high protein content (Srivastava Nanda, 1977) and the fine grained types (Sharma, 1982). Asthana and Majumdar (1981) studied and catalogued them into the variability present in rice collection from the NE hills in 56 characters. The variability existing in this is collated in Table 4.1.

Table 4.1: Variability in rice germplasm collections from the NE hills region

Character	Range
Plant height (cm)	56 - 220
Flag leaf length (cm)	16.8 - 56.0
Flag leaf width (cm)	1.1 - 2.8
Leaf area/plant (cm^2)	180 - 6433
Days to flower	94 - 163
Number of ear bearing tillers	2.0 - 20.2
Panicle length (cm)	14.3 - 13.1
Branches/panicle	3 - 16
Percentage of sound grains/ panicle	3 - 99
Grains/panicle	31 - 329
1000 grain weight (g)	11.6 - 34.0
Grain density/cm	1.7 - 15.7
Seed yield/plant (g)	0.4 - 24.0
Leaf blast	3 - 9
Neck blast	3 - 9

Source: Asthana and Majumdar (1981)

The studies based on morphological, genetic and some chemical characters indicated that a large number of germplasm were of intermediate types between *japonica* and *indica*. Ghosh *et al.* (1981) identified Ngoba as a new alternative source of the dwarfing gene. Mehra and Arora (1982) reported that collections made from North-East India provide valuable genes for disease resistance, adaptability and grain types. Glabrous husk and glutinous forms occur in the Garo hills and areas adjacent to the Myanmar border. Collections from Mizoram possess aromatic types in both glutinous and non-glutinous forms. Pandey and Gupta (1995) observed tolerance to low temperature in the vegetative and reproductive phases in the high altitude germplasm. Table 4.2 shows the break up of rice germplasm collected from the North-East region.

The foregoing discussions show that an enormous amount of genetic diversity is present in the rice germplasm of the region, which has been studied at both national and international levels (Japan and IRRI, Manila). The germplasm possessing desirable genes can be converged while breeding suitable varieties for particular agroclimatic conditions.

Table 4.2: Number of rice germplasm collections from NE India

State	*ARC series IARI / CRRI*	*ICAR Complex, Umiam*	*NBPGR, Shillong*	*Total*
Arunachal Pradesh	1527	150	189	1866
Assam	593	-	279	872
Manipur	898	50	51	999
Meghalaya	1399	932	259	1890
Mizoram	-	38	204	242
Nagaland	849	50	286	1185
Sikkim	-	40	124	164
Tripura	246	155	306	707
Total	5512	715	1698	7925

Maize

In contrast to the Indo-Gangetic Plains, extensive variability in different characters of maize (*Zea mays* L.) becomes apparent in the North-Eastern Himalayan Region (Singh, 1977). The wide variability of maize genetic resources of the North-Eastern region has been reported by a number of workers (Anderson, 1945; Stonor and Anderson, 1949; Ono and Suzuki, 1956; Dhawan, 1964). Because of the impressive variability in the North-Eastern material, Anderson (1945) and Stonor and Anderson (1949) proposed an Asiatic origin in maize. Sachan and Sarkar (1986) have presented evidence, which showed that the prototype of the prehistoric wild corn is very much present in the form of the Sikkim primitive maize in the North-Eastern Himalayas. Based on extensive studies on the maize of the North-Eastern region, it has been concluded that there have been at least two major movements of maize in Asia – one early post-Colombian (Sarkar *et al.*, 1974) and the other pre-Colombian. The pre-Colombian accounts for the maize varieties in isolation in the NE hills. Thapa (1966) reported a marked resemblance of NE materials, in general, plant morphology to that of the hypothetical progenitors of maize described by Mangelsdorf (1964). Dhawan (1964) reported strikingly primitive features of maize collections from Sikkim, known as Sikkim primitive 1 (SP 1) and SP 2. Sachan and Sarkar (1986) identified 13 different strains from the Himalayan germplasm of maize. All the collections appeared to be derivatives of a single widespread popcorn variety, grown and selected by different ethnic groups of the North-Eastern hills. Observations on plant type (Mukherjee *et al.*, 1971; Singh, 1977), pachytene analysis (Gupta and Jain, 1971), chromosome bonding (Mohan and Raut, 1980; Sachan and Sarkar, 1982), biochemical assay (Pareira *et al.*, 1982) and maternal effects (Bhat and Dhawan, 1969) have been reported in SP maize. So far, 546 maize germplasm were collected from Meghalaya and Sikkim (379), Assam (48), Nagaland (20), Mizoram (20), Arunachal Pradesh (36), Manipur (21) and Tripura (12).

The North-Eastern materials exhibit rich variability in plant, tassel, ear and kernel characters. High tillering habit and prolific cob production characterise the popcorn types of Sikkim primitive materials, resembling the Argentinean pop. Waxy types in Mizoram, sweet corn in Manipur and cold adaptable types occur in Sikkim. Primitive and obsolete types are still found in some pockets of Nagaland in juhm cultivation (Hore *et al.*, 1989). Singh (1977) identified 15 distinct races in the North-Eastern materials, each with characteristic complex traits associated with it. According to Dhawan (1964) and Thapa (1966), the primitive group comprises several races of popcorn. In view of the wide variability, the Sikkim materials were grouped on the basis of seed colour viz. White (12), Yellow (12), Red (10), Black (2) and SP type. Mahajan (1988) reported that at least three races were prevalent in Mizoram, which were quite different from the Meghalaya Local White and Yellow, with respect to various agronomic attributes.

Minor Millets

Minor millets like finger millet (*Eleusine coracana*), Kodo millet (*Paspalum scorficuatum*), barnyard millet (*Echinochloa swarn*) and foxtail millet (*Setaria italica*) are the third most important cereal crops of the North-East region (Abraham *et al.*, 1989). These late maturing types display much variability and differ from the plain types. More localised variability also prevails in foxtail millet, Coix-lacryma jobi and *Digitaria cruciata* var. *esculenta*, a millet endemic to the Shillong plateau (Mehra and Arora, 1982). The aboriginal inhabitants grow the thin-shelled forms for food. It is widely distributed in the Garo Hills of Meghalaya and Nagaland. In the interior parts of Nagaland, it is used as a substitute for rice and is grown quite abundantly in those areas which are generally not accessible, where coix constitutes the main cereal used both as food for humans and feed for poultry (Handique and Mallick, 1986). Ability to grow under water stress conditions such as hill slopes at higher elevation, is a feature that makes it a useful crop for the North-Eastern region.

PULSES

North-East India is also a seat of diversity of several grain legumes, viz. urdbean (*Vigna mungo*), cowpea (*Vigna unguiculata*), rice bean (*Vigna umbellata*), pigeonpea (*Cajanas cajan*), etc. (Mehra and Arora, 1982). Diverse genotypes of rice bean and different species of *Vigna* have been collected from different indigenous farming systems of the region (Borthakur, 1989). Rice bean, wild forms of moth bean (*V. acontifolia*), *V. trilibata* and *V.capensis* also occur in the North-Eastern Hills and the sub-Himalayan tract. The pulse crop germplasm collected from North-Eastern region are shown in Table 4.3.

Mung Bean (*Vigna radiata*)

Local collections of mung bean are semi-bushy to bushy types, exhibiting variations in leaf size, inflorescence nature and pod type. Considerable variation also prevails in grain colour (green, yellow, brown and mottled), the common colour being green (Sarma *et al.*, 1981).

Urd Bean (*Vigna mungo*)

Urd Bean is one of the most extensively grown pulses in North-Eastern India, especially in Assam. Local collections are semi-spreading types. Much variability in days to mature occurs in usually medium to late maturity groups. The indigenous types are observed to be photoperiod sensitive.

Cowpea (*Vigna unguiculata*)

Cowpea is rarely grown as a pulse crop in the North-Eastern region. It is mostly grown for vegetable purposes. This native genetic resource possesses much wider variability in growth habits, photoperiodic and photothermic sensivity, seed size, pod shape and colour. Varying selection pressures resulted in enriching the variability of grain, vegetable and fodder types of cowpea in the North-Eastern region (Mehra and Arora, 1982).

Pigeonpea (*Cajanus cajan*)

A wide array of diversity in *Atylosia* sp., a wild relative of pigeonpea, occurs in India, with maximum concentration in the Western Ghat and in the North-Eastern region. Out of 22 species of *Atylosia*, 4 species viz. *A. barbata, A. villosa, A.volubilis* and *A. scarabaraides are* often found in the North-Eastern region (Hazarika, 1987). In the North-Eastern region, perennial types are commonly grown, especially in the Garo hills of Meghalaya and the Karbi Anglong district of Assam. There are variations in seed size and colour in the local perennial germplasm. Materials from districts of the north of the Brahmaputra produced more yields on an average, than those from the south of the river (Medhi *et al.*, 1980).

French bean (*Phaseolus vulgaris*)

The *Phaseolus* group contains a number of important pulses. It is now generally agreed that the *Phaseolus* originated in the New World and that the Old World *Phaseolus* is designated as *Vigna. P. vulgaris*, commonly known as common bean, haricot bean, navy bean, French bean, kidney bean, etc., all generally grown as vegetables in the North-Eastern region. A lot of variation occurs in pod length, grain size and colour of this crop, with both dwarf and pole types suited as vegetables or dual purposes (grain as rajmah). In higher elevations (2100-2700 m) *P. coccineous* (runner or scarlet bean) is grown (Hazarika, 1987).

Rice Bean (*Vigna umbellata*)

Vavilov (1951) places rice bean in the Indian Centre of Origin, including Assam and Myanmar. Genetic resources of rice bean have largely been collected from the North-Eastern region. Enormous variability occurs in this less known pulse crop. Much information on the North-Eastern material on rice bean has now become available (Arora *et al.*, 1980; Chandel *et al.*,1988; Sarma *et al.*,1995). Vine types of rice bean occur but some semi-bushy types have also been collected from Assam, Sikkim, Manipur and Meghalaya. Wide variability in pod length, growth habit, seed size and colour and maturity occur in the North-Eastern materials (Sarma *et al.*, 1991). The materials are also found resistant to the pod boring, weevil (*Apion clavipes*), a major insect pest and rust (*Uromyces appendiculatus*), and the major disease of rice bean (Sarma *et al.*, 1991, 1995). In view of the enormous variability, high yield, tolerance to diseases and pests and wider adaptability, this underexploited legume, grown predominately in mixed farming systems, has a great potential to become an important pulse crop of the region.

Minor Pulses (*Vigna angularis*)

Adzuki bean, more akin to rice bean, is collected from the Bomdilla ranges of Arunachal Pradesh. Vine forms possessing thinner pods like mung bean with grain colour being red, green and creamish, also prevail. Much variation in pod and grain size of the broad bean (*Vicia faba*) from the North-Eastern region is observed. Local collections of moth bean (*Vigna aconitifolia*) possess good variations in the growth habit (spreading types predominant), leaf location, grain size, pod and seed colour. Largely, local *Khesari* (*Lathyrus sativus*) germplasm has a spreading growth habit and is well adapted to drought. *Vigna vaxillata* is another less known potential pulse-cum-tuber crop of the North-Eastern region, resembling cowpea and producing both edible seeds and tubers. At an earlier stage, the pods are used as a vegetable and at maturity, the seeds are used as *dal* by tribals. The tubers are also used as a substitute for tapioca in animal feed

(Arora, 1981; Sasikumar and Sardana, 1988). Both the seeds and tubers are very rich in protein content (Bhattacharya *et al.*, 1984). Out of the 9 wild species of *Mucuna* available in India, 7 species viz., *M. atropurpurea, M. bracteeta, M. capitata, M. nigicans, M. proriena, M. macrocarpa* and *M. monosparma* are found in the North-Eastern region (Hazarika, 1987).

Table 4.3: Pulses germplasm collected from NE India

Crop	*Assam*	*Meghalaya*	*Manipur*	*Mizoram*	*Nagaland*	*Arunachal*	*Sikkim*	*Tripura*	*Total*
Mung bean	10	29						3	42
Urad bean	100	1					10	10	121
Cowpea	3	35		8	4			17	67
Pigeon pea	59			1				4	64
Rice bean	6	66	20	7	42	6	40	3	230
Sem bean	7	7		6		2	14		29
French bean	104	10	2	3				2	111
Pea	4	37					15		56
Chick pea	2	3	2	1					8

OILSEEDS

The important edible oilseeds of the North-Eastern region are rapeseed-mustard, sesame and soybean. The diversity observed in oilseed genetic resources are summarised below:

Rapeseed-mustard (*Brassica* sp.)

Toria is, by far the dominant ecotype cultivated in North-Eastern India. Mustard and yellow *sarson* are the two other types cultivated on a limited scale (Doloi, 1987). In *Brassica,* wide diversity occurs in *B. juncea* and *B. campestris* from Assam and Manipur. Variations in plant height, seed size, maturity in toria and siliqua and seed size in yellow sarson have been observed. Highly leafy types occur throughout the region.

Sesame (*Sesamum indicum*)

Sesame is the second most important oilseed crop in the region, grown in both *kharif* (rainy season) and *rabi* (winter). The local collections of sesame are mostly tall types, branched abundantly, late and shy in bearing with, variations in days to flower, fruiting nodes, capsules/plant, maturity, plant type, seed size, pigmentation, seed colour and yield.

Niger (*Guizotia abyssinica*)

Cold adaptable germplasm in Niger, sporadically occurs in Sikkim (Mehra and Arora, 1982).

Soybean (*Glycine max*)

Soybean is widely cultivated in Sikkim, Nagaland, Meghalaya and Arunachal Pradesh. Considerable genetic variation exists in the North-Eastern materials. Very small sized soyabean occurs in Nagaland and Mizoram. The collections from the North-Eastern region, along with materials from other parts of India and exotic collections, were evaluated and catalogued for 18 morphoagronomic traits (Bhatia *et al.*, 1983). Large variations in the North-Eastern materials were also reported in morphoagronomic attributes (Sarma and Abraham, 1988; Asthana *et al.*, 1984), disease tolerance/resistance to rust (*Phakospora pachyrhizi*), frog-eye-leaf-spot (*Cercospora sojini*) and purple seed stain (*C. kikuchi*) by Kumar *et al.*,1987; Chandra *et al.*, 1987; Sharma *et al.*, 1989. Azad Thakur (1988) reported that a number of North Eastern materials were free from the attack of leaf roller (*Nacoleia vulgaris* and *N. diemenalis*) and stem fly (*Melanogromyza* sp.) insects.

Perilla (Perilla frutescense)

Perilla is another promising non-traditional oilseed crop of the North-Eastern region, which may be exploited for both edible and industrial purposes. It is cultivated for its small seeds, in Nagaland, Meghalaya, Mizoram and Manipur, mostly in kitchen gardens. The seeds yield aromatic oil. The plant grows well from an elevation of 300 m to 3500 m above MSL. The tribals with sohphlang *(Moghania vestita)* tubers use the seeds. Considerable variability in morphological and biochemical traits has been reported by Abraham *et al.* (1989). The oilseed germplasm collected from North-East India are presented in Table 4.4.

Table 4.4: Oilseeds germplasm collected from the NE India

State	*Soybean*	*Other Oilseeds*	*Total*
Arunachal pradesh	14	12	26
Assam	5	33	38
Manipur	43	4	47
Meghalaya	26	95	121
Mizoram	24	190	214
Nagaland	4	26	30
Sikkim	47	5	52
Tripura	-	22	22
Total	163	387	550

FIBRE CROPS

The major fibre crops of the region are jute, ramie and cotton. Variability present in the crops is described below:

Jute (*Corchorus capsularis* and *C. olitorius*)

Jute plays a considerable role in the economy of the region, more particularly in Assam. The jute and mesta (*Hibiscus sabdariffa*) have their origin in Africa. However, the primary centre of origin of *C. olitorius* is African and that of *C. capsularis* is Indo-Burma (Kundu, 1951). The fibre crops germplasm status from North-East India is given in Table 4.5. Some wild types of jute also prevail in the Eastern Himalaya. Both the types are distributed in Assam, Tripura and the Garo Hills of Meghalaya. Wide variability has been reported, in morphological (plant type, branching habit, leaf size and shape, root type and length, calyx, corolla, pod shape and size, seed colour), physiological (photoperiod sensivity, leaf area index, net assimilation rate, fibre content, harvest index, relative growth rate), anatomical (schlerenchyma size and shape of fibre bundles, ultimate fibre cell, cambial activity) characters, yield, quality of fibres and reaction to diseases and pests, by various workers (Singh and Basak, 1977; Singh, 1987; Sasikumar and Sardana, 1989). Singh and Singh (1982) found variability in wild forms, cultivated types, semi-cultivated types and improved types from material collected from North-East India.

Table 4.5: Collections of Fibre crops germplasm from NE India

Crop	*Species*	*Number*	*Institution*
Jute	*C. capsularis*	344	CRIJAF, AAU, ICAR Complex
	C. olitorius	210	CRIJAF, AAU, ICAR Complex
Mesta	*H. cannabinus*	33	CRIJAF, AAU
	H. sabdariffa	28	CRIJAF, AAU
Sunhemp	*C. juncea*	12	CRIJAF, AAU
Ramie	*B. nivea*	54	CRIJAF
Cotton	*G. arboretum*	157	CICR
	G. barbadense	5	CICR
	G. hirsutum	1	CICR

Ramie (*Boehmeria nivea*)

It is an important fibre yielding the plant of Urticaceae, which yields the strongest known vegetable fibre (Dempsey, 1975; Sarma and Ganguly, 1982). In many parts of North-East India it is found in wild state, as a naturally occurring plant. Ramie was first collected from Goalpara in Assam. The genus *Boehmeria* has more than 100 species. About 45 species are reported to occur in India, of which 19 species have been reported from North-East India (Kanjilal *et al.*, 1940). The widely distributed species in the North-Eastern region are: *B. nivea, B. malabarica, B. sidifolia, B. macrophylla, B. platyphylla, B. hamiltoni, B. polystachya, B. rugulosa,*

B. kurzi, B. rotundifolia, B. didymogyne and *B. halfori*. Despite the unique characteristic of ramie, it is still a minor fibre crop in world trade mainly because of problems emanating from production and processing (Sarma, 1989). So far 54 collections have been made and wide variability in leaf size, petiole length, bark weight, wood weight, tillers/clump, plant height, base diameter, fibre content and fibre quality has been observed (Sarma, 1984). The indigenous types generally have large leaves with profuse leaves/clumps which may be used for cattle feed, especially for pigs in tribal areas, as used in Vietnam and Colombia (Blasco and Bohorgnez, 1967; Gomez, 1968), since ramie leaves are rich in proteins, vitamin A and calcium.

Cotton (*Gossypium* sp.)

The *Gossypium* to which cotton belongs contains several species – both wild and cultivated. Hutchinson and Ghosh (1937) identified 4 species of cotton in India: *Gossypium barbadens, G. hirsutum, G. herbaceum* and *G. arboratum;* of which *G. arboratum* is the most widely cultivated. The *typicum* and *neglectum* varieties of *G. arboratum* occur mainly in Assam, whereas the *cernuum* variety is mainly confined to East Bengal and Assam. The variety possesses very long bolls, but the staple is very short and coarse. However, its ginning percentage is very high (43-50%). The variety goes under the trade name of *Comilla* (Pugh and Dutta, 1940). The *neglectum* variety spreads from the hilly tracts of Assam and East Bengal, almost throughout India. Singh *et al.*, collected 101 cotton germplasm from the Garo hills of Meghalaya grown under jhum cultivation, and reported genetic variability in boll weight, seeds/three loculi, ginning outtern, lint index and boll worm tolerance. The Garo hills cotton belongs to *cernum*, which is locally known as *Kill.* Mehra and Arora (1982) reported a hardy, drought tolerant and disease and pest resistant material from the Garo hills. The *Khaki* cotton of Mizoram is another important cotton genetic resource of the North-Eastern region.

SUGARCANE (*Saccharum* sp.)

Sugarcane, in the North-Eastern region, is mainly cultivated in Assam, although limited cultivation is done in Nagaland and Mizoram. There are 3 cultivated and 2 wild species of the genus *Saccharum*. The three cultivated species are: *Saccharum officinarum, S. barbari* and *S. sinense* and the two wild species are: *S. spontaneum* and *S. robustum*. Barbar (1988) is of the opinion that the Indian canes probably originated in Assam and Bengal from some plants closely related to *S. spontaneum.* Pugh and Dutt (1940) also stated that the Sunnabila group of *S. sinense* was very widely distributed in the North-Eastern region. Several allied genera and species which could be of valuable breeding material are found in the North-Eastern region. Out of 24 known related wild species and genera in the country, 12 have been reported to occur in the North-East. These are: *Saccharum revennee, S. spontaneum, S. longisectosum, S. procerum, S. arundinaceum, S. rufipilum, S. aikkimense, Nerange porphyrocerna, N. nepalensis, Fuscus, Sclerostachya* and *Imperata cylindrica.* Large number of races of *S. spontaneum* are found in Manipur and the Naga hills. Five species of *Erianthus,* viz., *E. elephantinus, E. fulvus, E. longosetotus, E. chrysothrix* and *E. hookeri* also occur in the North-Eastern region (Bor, 1960; Arora *et al.,* 1983; Sharma, 1987).

The enormous genetic variability present in cereals, pulses, oilseeds, fibre and other field crops of economic importance in the region, provide a vast gene pool, which can be converged in developing suitable varieties for various agroclimatic conditions. Following the exciting developments in the field of molecular biology and biotechnology, the genetic resources could be used to a much greater extent than what is presently possible. With the development of techniques in gene isolation, cloning and transfer (mediated through vectors or by direct gene transfer), it would now be possible to introduce a specific gene into a target plant without disturbing the genetic make-up of the adopted genome. The new developments also cut across sexual barriers for the transfer of useful genes. However, all this would be possible only, if we conserve the valuable genetic diversity through short, medium or long term conservation methods.

REFERENCES

Abraham M.J., Gupta A.S. and Sarma B.K. (1989). Indian Journal of Agricultural Sciences 50: 579-81.

Arora R.K. (1981). Glimpses of Indian Ethnobotany (Jain SK ed), New Delhi p 91-106.

Asthana A.N. and Majumdar N.D. (1981). Res Bull No 11, ICAR Complex, Shillong.

Bor N.L. (1960). The grasses of Burma, Ceylon, India and Pakistan Periodical Export.

Chandel K.P.S., Arora R.K. and Pant K.C. (1988). Rice Bean NBPGR Sci Mongr New Delhi.

Dempsey J.M. (1975). Fibre Crops 128 Florida Univ Press, Gainsville p 90.

Gupta H.S., Sharma B.K., Pandey, D.K. and Prasad R.N. (1995). Indian J Hill Farmg 8: 111-125.

Khoshoo T.N. (1986). Diversity 12: 24-25.

Mengelsdorf P.C. (1974). Corn, its Origin, Evolution and Improvement. Harvard Univ. Press, Cambridge p 262.

Ono T. and Suzuki H. (1956). Chromosomes of maize In: Land and Crops of Nepal Himalaya (Kihara H ed), Vol 2, Fauna and Flora Res Soc Kyto, Japan p 365-372.

Padmanabhan S.Y. (1971). National Germplasm Collection of Rice CRRI, Cuttack p 235.

Pugh B.M. and Dutt C.P. (1986). Principles and Practice of Crop Production in India. Allahabad Agrl Inst, Allahabad.

Sarkar K.R., Mukherjee B.K., Gupta D. and Jain H.K. (1974). Maize In: Evolutionary trend in World Crops (Hutchinson JB ed), Cambridge Univ Press p 121-127.

Sarma B.K., Singh M., Gupta H.S., Singh G. and Srivastava S.L. (1995). Studies in Rice Bean Germplasm. Res Bull 34 ICAR Complex for NEH Region, Umiam p 70.

Sharma S.D. (1992). Collection and Evaluation of Rice from the NE. AGP/ PGR Rome p 62-69.

Srivastava D.P. (1977). Proc 64th Session of Indian Sci Congress Assoc Sect X, part III, p 770.

Swanminathan M.S. (1988). Indian J Pl Genetic Resources 1: 6-22.

Watt G. (1889). A dictionary of the economic products of India, Vol VI (ii) Periodical Excerpts, Delhi.

Chapter **5**

Mathematical Modelling of Carbon Fluxes Using Remote Sensing Data

Yogesh Kant, V. Krishna Prasad, T. Rajagopal[*] and K.V.S. Badarinath
National Remote Sensing Agency,
Dept. of Space, Govt. of India, Balanagar, Hyderabad - 500 037
[*]Taxonomy Lab., Osmania University, Hyderabad - 500 007

Quantification of carbon fluxes and understanding the human induced global climatic changes, forms one of the major studies with respect to the International Geosphere Biosphere Programme. In the present study, an attempt has been made to quantify carbon fluxes through mathematical modelling of land use/land cover changes (LU/LC). Digital change image from satellite data, over two time periods, has been used as an input in studying LU/LC changes and thereby, carbon dynamics. Information pertaining to LU/LC has been mathematically modelled through land use transition matrix, with leading diagonal elements representing the land use of retained category of carbon accumulation and off-diagonal elements representing the changes in land use over a period of time. Total carbon flux has been modelled by taking areal estimates of LU/LC changes of two time periods and carbon factors for soil and standing biomass. Trace gases from shifting cultivation areas have been quantified, based on the ratios of non-carbondioxide trace gases to the total carbon released by burning. Results of the study, through land use transition matrix, suggest that over a period of eight years, 16.1% of dense forest and 10.02% of scrub forests were converted into agricultural lands. In the conversion process, there has been a loss of 47.97 Mt of carbon from forest lands. Dry matter burnt through shifting cultivation has been found to be 16.81 Mt of greenhouse gas emissions, calculated from the total carbon released associated with shifting cultivation, suggesting that CO_2, CH_4, CO, NO and NO_x emissions accounted for about 8.405 Mt, 0.315 Mt, 1.77 Mt, 9.24 X 10^{-4} Mt and 3.34 X 10^{-3} Mt, respectively from the study area. Finally, for the restoration of carbon and biodiversity in the study area, economically important species have been suggested for ecorestoration programmes.

INTRODUCTION

Rapid population growth, together with faster industrial development and unrestricted utilisation of natural resources, has been causing drastic changes in vital ecological processes of the earth. The greatest crisis ever faced by mankind is global warming, caused due to increased Greenhouse Gases (GHG's), the major sources being from burning of fossil fuels, forest degradation and destruction, forest fires and accelerated decay of organic matter in the soil. Knowledge of the carbon cycle over the past few decades suggests, increased levels of atmospheric CO_2 from nearly 280 ppmv in pre-industrial times, to the present concentrations of nearly 380 ppmv (Tolbert, 1997). King *et al.* (1990) estimates that carbon content for the terrestrial ecosystem ranges from 530-830 Gt of carbon. Forest biomass has been estimated to contain more than 80% of the carbon stored in terrestrial vegetation and forest soils, more than 70% of the world's soil carbon pool (Olsen *et al.*, 1983; Post *et al.*, 1990). Recent studies indicate that CO_2 emissions from fossil fuels for India increased from 117 Tg C/Yr in 1986-87 to 153 Tg C/Yr in 1989-90, while the projections for 2000 were estimated to be 222 and 289 Tg C/Yr i.e., an approximate growth rate of 9% from 1986-87 to 1989-1990 and projected growth rate of 5-7% in the next 20 years. Difficulties in quantifying the GHG emissions arise due to complexities in carbon storage in different ecosystems and the effects of human land use and land management. The different GHG's associated with land use are CO_2, CH_4, and NO_2, with some of the trace gases like Chloroflurocarbons and Ozone, implicated in green house gas effect, having no biological sinks (Adger *et al.*, 1994). The predictions of climate change, which are highly dependent on land use, land cover changes (LU/LC) are shrouded with uncertainty, as over a period of time they change continuously, either naturally or by anthropogenic factors. To assess the implications for the causes of GHG emissions, accurate data with respect to LU/LC changes over different time scales are required. Remote sensing, with it's potentiality to capture spatial and spectral information, provides valuable information with respect to LU/LC changes over a period of time, which can be effectively used for the quantification of GHG emissions and carbon fluxes. In the present study, an attempt has been made to quantify the carbon dynamics and GHG emissions from land use changes over the Vamsadhara river basin - Eastern Ghats, from satellite remote sensing data. Studies with respect to the quantification of GHG's and carbon pools acquire importance in the context of increasing 'Global Warming' and in studies related to the International Geosphere Biosphere Programme (IGBP), which is mainly concerned to reduce uncertainty in global changes and involved in understanding and evaluating the anthropogenic caused global changes.

STUDY AREA

Vamsadhara river basin, covering parts of the hilly northern parts of Srikakulam district - Andhra Pradesh and Orissa state (Eastern Ghats), has been taken for the study. Vamsadhara river is an interstate river, originating in the Eastern Ghats in the extreme north of Bissumcuttack of Koraput district in Orissa state, and after flowing 154 Km in Orissa, the river flows along the boundary between Orissa and Andhra Pradesh state for 29 Km, and finally empties into the Bay of Bengal, near Kalingapatnam port in Andhra Pradesh State. Vamsadhara basin has a total drainage area of 21592.698 sq Km and the terrain is mostly hilly. The hills of the region are steep and rugged and devoid of plateaus. The average annual rainfall in the basin ranges from 1397 mm to 1041 mm. The predominant rock types are Khondalites and Charconites.

Vegetation of Study Area

Luxuriant forests of Sal mixed, and mixed Sal have been seen in pockets near Durapanga, Telanga padar of Bissumcuttack - Orissa areas. The vegetation is mostly composed of mixed dry deciduous species with patches of dry deciduous and scrub forests. Vegetation is represented by elements like *Dalbergia paniculata, Chloroxylon swietenia, Canthium parviflorum, Adina cordifolia, Diospyros melanoxylon, Lannea coromandelica, Anogeissus latifolia, Mangifera indica, Pterocarpus marsupium, Schleichera oleosa, Syzygium cuminii, Xylia xylocarpa, Woodfordia fruticosa, Strychnos nuxvomica, Murraya paniculata, Pavetta indica, Mimusops elengi, Phyllanthes emblica, Sterculia urens, Terminalia alata, Terminalia arjuna, Mallotus phillippensis,* etc. Scrub vegetation is represented by elements like *Dichrostachys cinerea, Alangium salvifolium, Atlantia monophylla, Barleria prionitis, Cassia occidentalis, Cipadessa baccifera, Jatropha curcas, Phoenix sylvestris, Zizyphus rotundifolia, Achyranthes aspera,* etc. Field studies suggest that the forests are highly perturbed due to a single large land use practice of shifting cultivation, which is locally known as 'Podu'.

DATA SETS AND METHODOLOGY

Satellite Data

IRS-1B LISS I data of 25th and 26th November 1994 and LANDSAT MSS data of 23rd and 24th October 1986 have been taken for analysis. The data sets used in the present study are shown in Table 5.1. The total study area is covered in four different scenes and they have been mosaiced using EASI/PACE Image Processing software, on IBM RISC 6000 system. The satellite data mosaic, thus generated, has been geometrically registered to the Survey of India (SOI) toposheets on a 1:250,000 scale, for the extraction of information comparable on a map scale. Representative areas of different land use land cover classes are used as training areas and classification has been done using supervised classification approach with maximum likelihood algorithm. Thematic images have been prepared. Area statistics of LU/LC in two different years have been computed for the entire Vamsadhara basin, watershed wise for two different years in Khariff (1986) and Khariff (1994). Different LU/LC classes include dense forest, degraded scrub, fallow lands, agricultural lands, and water bodies. Digital change images of the study area have been generated and carbon fluxes for the study area have been evaluated.

Table 5.1: Satellite data sets used

S.NO.	Sensor	Date of Pass	PATH/ROW
1.	IRS 1B LISS-I	25.11.1994	21 / 54
		25.11.1994	21 / 55
		26.11.1994	22 / 54
		26.11.1994	22 / 55
2.	IRS 1B LISS-I	06.04.1995	21 / 54
		06.04.1995	21 / 55
		07.04.1995	22 / 54
		07.04.1995	22 / 55
3.	LANDSAT MSS	23.10.1986	141 / 46
		23.10.1986	141 / 47
		24.10.1986	140 / 46
		24.10.1986	140 / 47

Modelling Land Use Dynamics

In the present study, an attempt has been made to model the land use change dynamics over two time periods derived from satellite data. The information pertaining to land use dynamics can be represented in the form of a matrix, enabling the isolation of various elements of land use change, i.e., land retained for the same use, and land changing into and out of a particular category (Adger *et al.*, 1994).

If the land use transition matrix is ALu, then

$$\text{Alu} \quad \begin{matrix} a_{11} \; a_{12} \; \cdots\cdots\cdots \; a_{ln} \\ a_{21} \; a_{22} \cdots\cdots\cdots \; a_{2n} \\ \cdots \; \cdots\cdots \; \cdots\cdots \; \cdots \\ a_{m1} \; am_2 \; a_{mn} \end{matrix}$$

where m = n; and i = 1,2,........n land use categories. The matrix can be interpreted in the form of Leading diagonal elements and Off diagonal elements.

The Leading diagonal elements are assumed to represent the land use of the retained category, i.e., they represent the land use of the same category throughout the transitional period. The fluxes associated with soil and biomass carbon for these land uses are equilibrium accumulators.

Mathematically, Leading diagonal elements can be defined as β_I

$$\beta_I = a_{ij} \qquad i = j$$

The Off diagonal elements are assumed to represent changes in land use over a period of time. In these elements, large fluxes of carbon occur when transition occurs from one land use category to another and when the equilibrium of carbon fluxes shifts from one position to another.

The conversion impact for changing land use (δ) into each land use category is given by

$$\delta_I = \Sigma\, a_{ij} \qquad a_{ij} : i = j$$

The net fluxes of land use retained in the same use over a period, and of land converting into other categories, are seen in the sum of the columns, which is the land in a particular category 'i' at the end of the period (γ).

For each category, it is given by

$$\gamma_I = \Sigma\, a_{ij}$$

Thus

$$\gamma_I = \beta_I + \delta_I$$

where γ_I = area in category 'i' at the end of the period. The carbon fluxes associated with different land use categories have been estimated as

$$TF = \Sigma\, \gamma\, (Sf_i + Bf_i)$$

or

$$TF = \Sigma\beta_I\, (SF_i^{\beta} + B\, F_i^{\beta}) + \Sigma\, \delta\, (SF_i^{\delta} + B\, F_i^{\delta})$$

where TF = total carbon flux (mt C year $^{-1}$)

i = 1,2,....n land use categories

$\Sigma\, \gamma_I$ = Land area of study area

SF_i = Carbon factor for soil under land use category I (t C ha^{-1} year $^{-1}$)

BF_i = Carbon factor for standing biomass of land use category i (t C ha^{-1} year^{-1})

In the construction of land use transition matrix, the satellite derived areal estimates for two different time periods have been used and the total carbon flux has been estimated by following the above approach.

CARBON STORAGE AND ESTIMATION OF NON-TRACE GASES

Extensive taxonomic and phytosociologic studies have been reformed to study the community characteristics. Biomass values for the particular vegetation type have been taken from the literature cited values for mixed dry deciduous, dry deciduous and scrub forests (Singh and Misra, 1979; Negi *et al.*, 1991; Singh *et al.*, 1993). Carbon content has been taken as 47% of the dry biomass (Westlake, 1963). The carbon factors used for biomass (Bf_i) and Soil (Sf_i) in calculating total carbon flux (TF) are shown in Table 5.2. Carbon storage for sand and water has been ignored.

Table 5.2: Carbon storage factors for different vegetation types

S.NO.	*Category*	*Biomass (t C ha^{-1}yr^{-1})*	*Soils ((t Cha^{-1}yr^{-1})*
1.	Dense forest	82.5	75.3
2.	Degraded forest	25	30
3.	Barren areas	2	8
4.	Fallow land	15	30
5.	Agricultural land	7.5	40

Biomass burnt areas delineated from the satellite data have been used to find the areal estimates of the biomass burnt. The amount of dry matter burnt in grams has been calculated, using the equation given by Sieler and Crutzen (1980).

$$M = A_i \times B_i \times a_i \times b_i$$

where

M = Mass of dry matter burnt in g

i = type of basic vegetation burnt

A_i = area of vegetation type burnt in Km

B_i = above ground biomass

a_i = fraction of biomass over soil in vegetation

b_i = combustion factor for vegetation i.

The fraction of biomass over soil in vegetation and combustion factor for vegetation type has been taken to be 0.3 and 0.7, respectively. Trace gases have been estimated, based on the ratio's of non carbon dioxide trace gases to the total carbon released by burning. The emissions of CH_4 and CO have been estimated by multiplying the emission ratio's of CH_4 and CO, relative to the emissions of total carbon. The emission ratio's for open burning for CH_4, CO, NO and NO_x have been taken as 0.012, 0.06, 0.007 and 0.121, respectively.

To calculate the emissions of NO and NO_x, the ratio of total carbon released is multiplied by the estimated N/C ratio of the fuel by weight (default value 0.001, following Crutzen and Andreae, 1990) to yield the total amount of nitrogen (N) released. The total nitrogen released is then multiplied by the ratios of the emissions of NO and NO_x, relative to the total N released from the fuel to yield emissions of NO and NO_x.

RESULTS AND DISCUSSION

Land use and land cover, watershed-wise analysis of the study area, through the analysis of aggregated transition land use matrix (Table 5.3) suggests that the total Vamsadhara basin area of 21592.698 sq Km has undergone a 23.38% change with respect to categories, like dense forests, degraded forests and fallow lands.

Digital change image generated over Kharif, for two different time periods (1986 and 1994). In the matrix, the right-hand column represents the distribution of land use in 1986, whilst the columns totals give the distribution for 1994. The Leading diagonal elements indicate the percentage of land, which was found to be in the same category in both of these years. The Off diagonal elements indicate the transition between the classes of land use cover (over a period of eight years) that have occurred in the study area. Thus, for example, it can be seen from row 5th of the matrix, that, the agricultural areas accounted for 39.96% in 1986, of which 35.42% was also under agriculture in 1994. Column 5 shows that the total amount of agricultural land in 1994 was 47.61%, which amounted to an increase of 7.65% of agricultural land over a period of eight years. The dense forest, which was 29.03% in 1986, has been reduced to 13.194% over a period of time. Likewise, is the case with degraded forests where a percentage change of nearly 10.02% had occurred with the decrease in forest cover. In case of fallow lands, there had been an increase of 12.53% in 1994, when compared to 1986. Analysis of the change image suggests that the forested areas have undergone considerable change into agricultural lands by the way of shifting cultivation.

Table 5.3: Land use transition matrix Vamsadhara Basin (Area in sq Km)

Dense Forest	2848.43	1353.40	0	685.23	1380.17	0	0	6267.234
Degraded Scrub	0	0	1353.619	824.455	1468.04	44.9047	0	4501.2689
Rocky Barren	0	0	0	0	0	0	0	-
Fallow	0	0	0	1627.797	0	0	0	1627.797
Agricultue	0	0	0	1196.02	7430.87	0	0	8626.8924
Sand	0	0	0	0	0	0	0	-
Water	0	0	457.683	0	0	79.015	31.0879	567.786
1994 (Km²)	2848.430	2165.358	1811.30	4333.52	10279.084	123.919	31.087	21592.698

The total carbon flux calculated for different time periods, for different land use categories, by area are shown in Table 5.4. The carbon fluctuations and storage in land retained in the same category, leading to carbon accumulation towards equilibrium position, and those areas where land use change has occurred, have shown different profiles. The effect of land use change on both the soil and biomass organic carbon showed decreased rates of organic matter accumulation over a period of time. The area weighted carbon dynamics (Table 5.4) for the study area suggest that the total biomass, which was 97.81 Mt in 1986, had been reduced to 19.85 Mt over a period of eight years. Likewise, is the case with soil carbon, which was 100 Mt, and was reduced to 70.36 Mt. The total carbon flux (TF), which was 208.33 Mt in 1984, has been reduced to 153.38 Mt. The chief land use practice of shifting cultivation, at the cost of conversion of dense forests and degraded scrubs into agricultural lands, leads to these changes in carbon fluxes. When the land changes form the category of dense forests to agricultural lands through the way of Podu cultivation, by biomass burning, there is an instantaneous release of CO_2, as vegetation is cut and burnt. Assuming that soil carbon was previously in equilibrium, the drawing down of soil carbon occurs only after secondary succession, due to the growth of plants from the dispersal of propagules from undisturbed adjacent areas and regeneration of unaffected stumps of previous vegetation. Although the secondary successor species accumulate organic matter in the first few years, part of this is offset by a reduction in soil organic matter of the previous forested land. Thus, if the same patch is revisited again for podu cultivation, the carbon accumulation can never reach the initial levels and it continues to decrease over a period of time.

Table 5.4: Area Weighted Carbon dynamics of the study area (Mt C)

S.NO	Category	1994 (Biomass)	1994 (Soil)	1986 (Biomass)	1986 (Soil)
1	Dense forest	23.499547	21.448677	51.704680	47.192272
2	Degraded forest	5.413395	6.496074	37.135468	13.503806
3	Fallow land	6.500280	1.300560	24.416955	4.883391
4	Agricultural areas	7.709313	41.116336	64.701693	34.507569
5.	Total	19.857983	70.361647	97.812012	100..08703

Moreover studies on shifting cultivation (Ramakrishnan, 1992) suggest that, when a forest is cleared, soil properties are drastically altered. Soil temperature increases, and due to high insulation, moisture levels fluctuate and soil pH increases and much of the soil nitrogen is lost in the form of volatilisation. The organic carbon content of the soil is highly dependent on the fall of huge amount of litter and it's rapid decomposition, due to favourable conditions like moisture and temperature. Due to shifting cultivation, rapid loss of organic matter in the first year of burning takes place due to oxidation of humidified material, which gets much reduced in the second year and so on (Nye and Greenland, 1964; Banerjee and Chand, 1988). Thus, during the course of time, due to great disturbances in various agricultural practices, the ecosystem loses its ability to hold nutrients in the soil. Biomass burning is also accompanied by the release of trace gases. Dry matter burnt through shifting cultivation has been estimated to be 16.81 Mt. CO_2 and trace gases, CH_4, CO, NO and NO_x, released over a period of time, calculated from carbon released over a period of eight years, have been found to be 8.405 Mt, 0.315 Mt, 1.77 Mt, 9.24×10^{-4} Mt and 3.34×10^{-3} Mt, respectively. Finally, for the restoration of carbon and biodiversity in the study area, species that are threatened in the study area, along with species that serve multiple end uses, viz., food, fuel and fibre, species that improve soil fertility and the species that are economically beneficial and ecologically suitable for the study area for ecorestoration have been suggested (Table 5.5).

Table 5.5: Multipurpose tree species suggested for ecorestoration programmes

Timber and fuel wood species	*Oil seed bearing species*	*Wild species of multi-purpose importance*	*Nitrogen fixing species*
Albizzia lebbeck	*Bassia latifolia*	*Gyrocarpus americanus*	*Acacia auriculariforimis*
Bauhinia purpurea	*Pongamia pinnata*	*Bassia latifolia*	*Acacia leucophloea*
Dalbergia sissoo	*Hevea braziliensis*	*Sterculia urens*	*Acacia nilotica*
Eucalyptus camuldulensis	*Mesua ferrea*	*Caryota urens*	*Albizzia lebbeck*
Melia azedarach	*Calophyllum inophyllum*	*Gmelina arborea*	*Casuarina equisetifolia*
Toona ciliata	*Garcinia indica*	*Dillenia pentagyna*	*Dalbergia sissoo*
Gmelina arborea	*Jatropha curcas*	*Pterocarpus marsupium*	*Dalbergia latifolia*
Lagaestroemia speciosa	*Citrullus colycinthus*	*Garuga pinnata*	*Erythrina varigata*
Sesbania grandiflora	*Butea monosperma*	*Soymida febrifuga*	*Hardwickia binata*
	Azadirachta indica	*Givotia rottleriformis*	*Leucaena leucocephala*
			Gliricidia sepium
			Pithecolobium dulce
			Prosopis julifera
			Sesbania grandiflora
			Sesbania sesban

Though it may be seen from the study, that the overall carbon released together with trace gas emissions, over the study area, are relatively minor, satellite derived LU/LC estimates in conjunction with mathematical modeling, provide a novel approach in studying carbon fluxes over the study area.

ACKNOWLEDGEMENTS

Authors are thankful to Dr. D. P. Rao, N.R.S.A, for providing all the facilities. Yogesh Kant and V. Krishna Prasad thank ISRO-GBP Committee for providing fellowship.

REFERENCES

Adger W. Neil and Katrina Brown (1994). Land Use and the Causes of Global Warming. John Wiley Sons, England.

Banerjee and Chand (1988). Physio-chemical properties and moisture characteristics of soils as influenced by forest fires. Indian Forester 107(3): 178-182.

Crutzen P.J. and Andreae M.O. (1990). Biomass burning in the tropics: Impact on atmospheric chemistry and biogeochemical cycles. Science 250: 1669-1678.

King G.A., Winjum I.K., Dixon R.K. and Arnaut L.Y. (eds) (1990). Response and feedbacks of forest systems to global climate change. US Environmental Protection Agency, Cor fallis, p 156.

Negi J.D.S., Sharma S.C. and Lepacha S.T.S. (1991). Biomass and nutrient accumulation (Inventory) in shrub land ecosystem of Garhwal. Indian Forester 117(8): 635-641.

Nye P.H. and Greenland D.J. (1964). Changes in the soil after clearing tropical forests. Plant and Soil: 21(1): 101-112.

Olsen, J.S., Watts J.A. and Allison L.J. (1983). Carbon in live vegetation of the major world ecosystem. Rep ORNL-5862, Oak Ridge, Natl Lab Oak Ridge, Tenn.

Post W.M., Peng T.H., Emanuel W., King A.W., Dale V.H. and Delnglis (1990). The global carbon cycle. American Scientist 78: 310-326.

Ramakrishnan P.S. (1992). Shifting agriculture and sustainable development. An Interdisciplinary study from North-Eastern India. Man and Biosphere Series, UNESCO Pub, Paris, Vol 10 p 415.

Singh K.P. and Misra R. (1979). Structure and Functioning of Natural, Modified and Silvicultural Ecosystems in Eastern Uttar Pradesh. Final Technical Report (1975-1978).

Singh J., George M. and Varghese G. (1993). Nutrient cycling in moist deciduous forest litter production and nutrients return. Indian Forester 119(4): 1004-1010.

Tolbert N.E. (1997). The C_2 oxidative photosynthetic carbon cycle. Annu Rev Plant Physiology and Plant Mol Biol 48: 1-25.

Westlake D.F. (1963). Comparisons of plant productivity. Biol Rev 38: 385-425.

Chapter 6

Sulphur in the Environment

Ishrat Khan, Altaf Ahmad and Muhammad Iqbal
Department of Botany, Hamdard University, New Delhi - 110 062

INTRODUCTION

The fundamental importance of sulphur as one of the three philosophic essentials goes back to alchemy. At the start of the 20th century, the production of sulphuric acid was a gauge for the industrial state of a nation. Less than 80 years later, sulphur was stigmatised as a yellow poison, after being identified as one of the major reasons for the acid rain problem. Later, after Clean Air Acts came into force, severe sulphur deficiency became a major nutrient disorder in agricultural production. An insufficient sulphur supply reduces crop productivity, diminishes crop quality, impairs nitrogen utilisation efficiency and thus, increases undesired nitrogen losses to the environment. Understanding the mechanisms and processes involved in sulphur cycling at different scales and its links to the plant environment is an important contribution for sustainable agricultural production and environmental protection. In this chapter, an overview of sulphur in the environment has been discussed.

FORMS OF SULPHUR IN THE ENVIRONMENT

Sulphur is found in valence states, ranging from +6 in sulphate to -2 in sulphides. There are numerous inorganic and organic sulphur compounds found abundantly in different spheres on the earth. Important inorganic and organic sulphur compounds of anthropogenic, geogenic and biogenic origin, which are found in the atmosphere, hydrosphere, lithosphere and biosphere are listed in table 6.1 (Haneklaus *et al.*, 2003).

Major terrestrial sulphur pools are volcanoes and rocks with 50-106 and 2.4-1010 Tg sulphur and significant marine reservoirs are oceanic sediments and sea water with 3-108 and 1.3-109 Tg sulphur, respectively. Substantially, the sulphur reservoir of soil and land biota (3-105 Tg sulphur), lakes (300 Tg sulphur), soils (120-130 Tg sulphur) and atmosphere (5 Tg sulphur) is lower. Values for sulphur found in plant biomass vary between 0.6-103 Tg sulphur (Trudinger, 1979) and 8.5-103 Tg sulphur (Galloway, 1996). Sulphur inherent to the world population, constitutes the smallest sulphur reservoir, with 0.84 Tg sulphur. It can be expected that by 2025, with an estimated world population of 8 billion people, about 1.1 Tg sulphur will be found in human bodies. Clark (2002) estimates an intake of 310 g sulphur yr^{-1} to be sufficient to satisfy the human sulphur demand. Sulphur represents about 0.25% (140 g sulphur) of the total body weight (Clark, 2002). This equals an annual release of 0.008 Tg sulphur at an actual mortality rate of 54-106 people (Anon, 2002).

SULPHUR IN THE ATMOSPHERE

This section briefly describes the atmospheric sulphur cycle, its processes, pathways and fluxes.

Emission of sulphur in atmosphere

Natural emissions exhibit high spatial and temporal variability. Approximately, one-half occur in the marine environment. Terrestrial biogenic emissions are correlated with temperature, and thus emission rates tend to be higher towards the tropics. Also, emissions from plants and soils are highly dependent on specific plant and soil types, and are thus, highly variable geographically, often on a small scale. Volcanic emissions are of two kinds: those during non-eruptive periods contribute about 90% of the total sulphur, while those during eruptions contribute significantly less, on an average.

Estimates of global anthropogenic sulphur emissions range from 60 to 110 Tg (S) a^{-1} (2 to 3 Tmol a^{-1}) for the years 1976 to 1985 (Moller, 1984; Hameed and Dignon, 1988), comparable to, but probably somewhat larger than, natural emissions (excluding seasalt spray). The combustion of fossil fuels accounts for 80 to 85% of the total, with the remainder coming from the smelting of ores and other industrial processes and burning. The approximate 1980 regional apportionment of fossil-fuel sulphur emissions is: China and Japan, 16%; USA and Canada, 28%; Europe, 34%; the remainder of the Northern Hemisphere, 15%; and the Southern Hemisphere, 7% (Hameed and Dignon, 1988). Annual emissions of sulphur in North America and Europe, east to the Ural Mountains, are known with an uncertainty of about, 10-15% as a result of detailed emission inventories developed in recent acid rain programmes (Dovland and Saltbones, 1986; NAPAP, 1990). In these regions, emissions are resolved into a grid of approximately 100 km on a side. For other regions of the world, the spatial distributions of anthropogenic sulphur emissions are not as well known.

Of great interest is the extent to which sulphur emissions in different regions of the world will increase in the coming decades. Emissions from Asia, South America, and Africa may well increase dramatically. Galloway (1989) used United Nations estimates of population growth for 1980 to 2020 to estimate future sulphur emissions. In one scenario, the per capita emissions were held constant at the 1980 levels, while population alone increased. In a second scenario, per capita emissions were allowed to increase by a factor of four, relative to 1980 values, along with population growth. The conclusion arrived at was that dramatic increases in sulphur emissions are likely from these continents in the next few decades, and, as a result, globally. As Galloway points out, the consequences are likely to be severely increased acidification in tropical regions, and significant disruption of the biogeochemical sulphur cycle, particularly in the tropics and sub-tropics, with reference to biogeochemical cycling.

Transport and transformation in atmosphere

From the perspective of the study of chemical cycling and surface-atmosphere exchange in terrestrial and wetland ecosystems, the atmosphere serves as a delivery system bringing material, to and taking material away, from the immediate vicinity of an ecosystem. The transport of chemical constituents through the atmosphere is affected by several different scales of motion in the atmosphere, each dominated by different physical processes. On the global scale, transport is controlled by the general circulation of the atmosphere, the westerlies of the mid-latitudes and the easterlies of the subtropics. The time and distance scales for such transport are several days and several thousand kilometres, respectively. On a next smaller scale, transport is governed, primarily by synoptic disturbances, such as the travelling low-and high-pressure systems that migrate across the mid-latitudes of the Northern Hemisphere. Characteristic time and distance scales are of the order of a few days and a few thousand kilometres, respectively. Extensive reference material is available on meteorological regimes and factors affecting the transport of chemical constituents through the atmosphere (Hasse, 1983; Merrill, 1986; Whelpdale and Moody, 1989).

The reduced sulphur compounds undergo oxidation to sulphur dioxide, in the atmosphere primarily through reaction with the hydroxyl radical. The expected atmospheric residence time in most cases is a day or even less. Sulphur dioxide undergoes oxidation in the gas phase, primarily with the hydroxyl radical, on the surface of particles, or in precipitation elements, through oxidation with the peroxyl radical or ozone, or in the presence of metal catalysts. The product in all cases is sulphate, either in the form of particles or as cloud and precipitation droplets.

Deposition

Chemical constituents in the atmosphere can be brought to the surface by a variety of processes. Those for which precipitation is the delivery mechanism are termed as wet deposition processes. Gases, such as sulphur dioxide, can dissolve in cloud and rain drops or adsorb on to frozen precipitation elements. The efficiency of wet removal of sulphur species depends on the form and characteristics of the precipitation in a given location: type, intensity, duration, frequency, etc. In principle, wet deposition is easily measured. With proper attention to the sampling location, cleanliness, sample handling and chemical analysis procedures, reliable measurements of wet sulphur deposition can be routinely obtained.

Those deposition processes which do not involve precipitation are collectively termed as dry deposition. Particles larger than about 10 μm in diameter may be removed by gravitational sedimentation. Smaller particles and gases, however, are more efficiently brought to the near-surface region by turbulent atmospheric motions, where they may subsequently be brought into contact with surface elements by molecular-scale processes. Actual uptake is accomplished by chemical reaction, dissolution, adsorption, etc. The efficiency of transfer to the near-surface region depends, primarily on the turbulence intensity several meters above the surface. Transfer to the surface depends on the characteristics of the surface (e.g., roughness, wetness, chemical properties) and on the properties of the constituent (e.g., state, reactivity, solubility). Sulphur dioxide is dry deposited, more efficient than particulate sulphate because it is more readily taken up at the surface. In contrast to wet, dry deposition is not easily measured directly. It can be measured in a research mode by the flux-gradient or eddy-correlation approach; but these techniques are not used for routine network measurements. The more usual approach is to parameterise the dry deposition process, using a 'deposition velocity', and calculate the deposition as a product with a measured or modelled ambient concentration.

A third category of the deposition process includes mass transfer to the surface by impaction of fog or cloud droplets, and by riming. These are not usually included as either wet or dry deposition. However, in high-elevation forested ecosystems or in areas with frequent fog, they can be very efficient. Techniques to quantify such depositions are limited, although some progress is being made by examining throughfall in forests.

SULPHUR IN THE HYDROSPHERE

Volatile sulphur gases are produced biologically (CH_3SCH_3, CS_2, CH_3SH, CH_3SSCH_3) and/or phytochemically (H_2S, COS) on the surface water of the ocean (Andreae, 1986). Dimethyl sulphide (DMS) concentrations vary between 0.5 to 5 nmol L^{-1} and are strongly dependent on region and season (Bates *et al.*, 1987). In comparison, the DMS concentration in freshwater is only about 0.1-1.0 nmol L^{-1} (Nriagu and Holdway, 1989). Carbon disulphide (CS_2) concentrations are about 16 pmol L^{-1} and carbonyl sulphide (COS) concentrations range from 10 to 100 pmol L^{-1} (Bates *et al.*, 1987).

SULPHUR IN THE PEDOSPHERE

The concentration of sulphur in parent materials ranges from 0.026% to 1% sulphur with igneous rocks < metamorphic rocks < magmatic rocks of upper continental crust << limestones < sedimentary rocks (sulphides) < shales < sedimentary rocks (sulphates) << coal (Bowen, 1966; Wedepohl, 1968; Friend, 1973). The typical range of sulphur in agricultural soils of humid and semihumid regions is 100 to 500 μg g^{-1}, or 0.01 to 0.05% sulphur. This equals 224 to 1,120 kg sulphur ha^{-1} in the Ah-horizon (Stevenson, 1986). The total sulphur content of soils may be as low as 20 μg g^{-1} (0.002%) in highly leached and weathered soils of humid regions or as high as 35000 μg g^{-1} (3.5%) in marine marsh soils, and up to 50,000 μg g^{-1} (5%) in calcareous and saline soils of arid and semi-arid regions (Stevenson, 1986). Tropical soils generally contain low amounts of sulphur due to their low organic matter content. Under humid conditions, Haehndel and Isermann (1993) determined sulphate concentrations of 592 to 7,006 kg SO_4^{2-}-S ha^{-1} in sub-soils and values of 67 to 233 mg SO_4^{2-}-S L^{-1} in the soil solution at the seam of the mean groundwater table.

The C:N:S ratio varies, depending on the parent material, leaching and fertilisation, and is on an average, 130:10:1.3. It may be as narrow as 90:8:1 on mollisols or as wide as 200:12:1 on more leached spodosols (Agripedia, 2002). The C:S ratio in soils from different regions in the world ranged from 85 on cultivated land up to 250 on grassland (Stevenson and Cole, 1999). The N:S ratio on these sites varied between 6.7 and 11.1 (Stevenson and Cole, 1999). The critical C:S ratios for mineralisation and immobilisation are 200:1 and 400:1, respectively (Agripedia, 2002).

The edaphon constitutes about 5% of the total organic matter in soils (Topp, 1981). The Ah horizon of soils typically comprises soil biota in the following amounts (dry matter m^{-2} in a no-till farm soil): bacteria (50 g), fungi (100 g), amoeba (5 g), nematodes (0.2 g), arthropods (0.5 g) and worms (1-20 g) (Seafriends, 2002). Farm stock amounts to 50 g m^{-2} and is, thus, comparable to bacteria and earthworms, while fungi outweigh this number considerably (Seafriends, 2002). Sulphur bacteria have sulphur globules in their cells and are capable of oxidising/reducing sulphide/sulphate. These comprise the following groups: thiosulphate bacteria, denitrifying thiosulphate bacteria, sulphate reducing bacteria and saprophytic bacteria (Ellis, 1932). Microbial biomass is in the range of 146 - 968 μg g^{-1} soil (Roembke *et al.*, 2002) with sulphur content of 928-1355 μg S g^{-1} (Saggar *et al.*, 1981). Sulphur found in microbial biomass amounts to about 1-3% of the total organic sulphur in agricultural soils (Saggar *et al.*, 1981; Chapman, 1987; Wu *et al.*, 1994). Hu *et al.*, (2002) found higher values with 4.4 to 4.9% in non-vegetated soils and 5.2 and 8.8% in vegetated soils. The turnover of the soil microbial biomass is fundamental for the incorporation of SO_4^{2-}-S into soil organic matter (O'Donnell *et al.*, 1994).

SULPHUR IN PLANTS

The smallest scale of sulphur cycling is present in plants. Plants are able to assimilate atmospheric SO_2 and H_2S, but this fraction contributes only a minor way to their sulphur nutrition. Plants satisfy their sulphur requirement, predominately by sulphate uptake by roots (Hawkesford *et al.*, 2003). The next metabolic step is the reduction of sulphate. Metabolic intermediates of sulphate assimilation are activated adenosine phosphosulphate, sulphite, sulphide and sulphur containing amino acids, such as methionine and cysteine (Kopriva and

Koprivova, 2003). Cysteine is the first stable organic compound, in which sulphur, taken up by roots or leaves, appears in the plant metabolism. Up to 90% of the total sulphur in plants is bound in the amino acids, cysteine and methionine (Giovanelli, 1990). All reduced sulphur containing compounds in primary (e.g., methionine and glutathione) and secondary (e.g., glucosinolates) metabolism are derived from cysteine. Plants coevolve with their environment and develop strategies to cope with excessive sulphur (Ernst, 1993). During the reduction of sulphate, taken up by plant roots, significant amounts of sulphur are released to the atmosphere, of which H_2S is the predominant form (Schroeder, 1993). It has been suggested that the release of H_2S regulates, homeostatically, the size of the cysteine pool and, thus, maintains it at a low level because of its cytotoxicity (Rennenberg, 1984).

SULPHUR AS AN ESSENTIAL NUTRIENT

The ecological interest in nitrogen and phosphorus stems primarily from their role as plant nutrients. Sulphur is also a plant nutrient, although it is often less limiting than either nitrogen or phosphorus. Sulphur is an essential element for all living plants, animals, and microorganisms. Three amino acids found in almost all proteins (cysteine, cystine, and methionine) contain sulphur. Sulphur is also found in sulpholipids, some vitamins, sulphate esters, and a variety of other compounds. Biogeochemists often distinguish between sulphate-ester sulphur, where the sulphur atom is bonded to oxygen, and carbon-bonded sulphur (which includes the sulphur in amino acids). Usually, most of the sulphur in vascular plants is in amino acids, resulting in a tight and predictable coupling between sulphur, carbon, nitrogen, and phosphorus in plant tissues (Johnson, 1984; Maynard *et al.*, 1984)

SULPHUR AS AN ELECTRON ACCEPTOR AND DONOR

Sulphate can serve as a terminal electron acceptor for bacterial respiration in the absence of oxygen, and so dissimilatory sulphate reduction is coupled with carbon oxidation and nutrient mineralisation in anoxic environments. Sulphate-reducing bacteria, themselves metabolise only a limited range of relatively low-molecular-weight organic substrates (Sorensen *et al.*, 1981; Ivanov *et al.*, 1989). Other bacteria ferment the complex organic matter in sediments, and their fermentation products become the substrates for the sulphate-reducing bacteria. Thus, it is a bacterial assemblage of fermenters and sulphate reducers which decompose organic carbon. Overall, two moles of carbon are oxidised to carbon dioxide, for every mole of sulphate reduced (Berner, 1984). The mineralisation of other elements in organic matter is stoichiometrically related to the mineralisation of carbon.

Energy is conserved in the reduced sulphur compounds formed during sulphate reduction (Fenchel and Jorgensen, 1977; Howarth, 1984). Relative to aerobic respiration, approximately 75% of the potential energy consumed in organic matter is transferred and stored as reduced sulphur during sulphate reduction (Howarth and Teal, 1980). When these reduced sulphur compounds are re-oxidised to sulphate, the energy is available to support bacterial growth. A variety of chemosynthetic bacteria can oxidise reduced sulphur compounds, using oxygen or nitrate as electron acceptors, and fix carbon dioxide (Kelly, 1982; Jorgensen, 1982). Chemosynthetic sulphur oxidisers are commonly found at the oxic/anoxic interface in sediments, where both reduced sulphur and oxygen are abundant.

FOOD PRODUCTION AND PLANT NUTRIENT SULPHUR

Food from agriculture supports around six billion people inhabiting this planet today. The population is expected to increase to about 7.2 and 8.3 billion, respectively, by 2015 and 2030. The major increase is expected to be in the developing countries. This will be despite the continuous decline in growth rates. Each year there will be about 77 million added to the already burgeoning population (FAO, 2002). This scenario requires increased food production. This will certainly have to come through intensification of cultivation as there are limited land resources. This implies more pressure on soil and water resources, and is of particular concern because in many regions, already there are "exhaustion symptoms", because of unsustained use of natural resources (Abrol *et al.*, 2002). For their sustained use, management of soil fertility with efficient utilisation of available water resources will be a pre-requisite. Thus, additional production will be highly input sensitive, and mineral fertilisers will play a key role in realising the potential of future crop cultivars, evolved by wide breeding and transgenesis, besides conventional breeding. In a holistic scenario, mineral fertilisers accounted for around 30% of total increase in crop production in the last four decades, during which world cereal production doubled (Roy, 2001). Mineral fertilisers will play a vital role and they have to be used in a more efficient manner. However, there is a possibility of some misuse of fertilisers, when trying to compensate poor growth caused by excessive losses on unsuitable soils or lack of certain nutrients (Luc Maene, 2002). Efficient plant nutrient use requires balanced fertilisation, optimisation of nutrient replenishment and minimum nutrient losses to the environment.

Importance of sulphur as a plant nutrient is becoming more imminent due to its effect on crop productivity and quality (Pasricha and Abrol, 2003). In the past, inputs of sulphur from atmospheric deposition and fertilisers were more than sufficient to meet the requirements in the industrially developed countries of Western Europe and North America. In Asia and other developing world countries, the amount of sulphur made available from mineralisation of soil organic sulphur was generally sufficient because of the low level of crop harvests. Even low level of SO_2 deposition from the atmosphere, because of industrial backwardness, was no obstacle for crop production. However, rapid population growth necessitated increased food production by adopting high yielding, fertiliser responsive, short duration crop cultivars grown intensively under irrigated conditions. Due to increased harvest levels, demand for all nutrients, including sulphur increased and the soil supply could no longer keep pace with the increased demands. This has resulted in widespread sulphur deficiency and responses to its application. Meanwhile, traditional sulphur containing fertilisers, such as ammonium sulphate and single super phosphate are no longer widely used, although, still more than 75% of the total sulphur for crop nutrition comes from these two sources. Their use is decreasing further with increased levels of nutrients being applied from S-free high analysis mineral fertilisers. Thus, the gap between total amount of sulphur removed in crop harvests and that applied through fertilisers is increasing with time. A major portion of the worldwide 60 million tons of total sulphur produced annually, worldwide goes to the fertiliser industry for the manufacture of phosphate fertilisers. Although, the demand and supply scenario shows a surplus in the world market, this does not mean that adequate amount of sulphur is being used as a plant nutrient. In order to check stock piling of sulphur produced, it is necessary to make its use in fertiliser formulations to meet the increasing needs of this element in crop production.

Table 6.1: Relevant inorganic and organic S compounds in the atmosphere, hydropshere, lithosphere and biosphere

Inorganic compounds		*Organic compounds*	
Name	*Formula*	*Name*	*Functional group*
Hydrogen Sulphide	H_2S	*Reduced*	
Sulphate	SO_4^{2}	Carbonyl sulphide	COS
	plant: lipid sulphates	Carbon disulphide	CS_2
	minerals:$CaSO_4$*$2H_2O$,	Thiol/Mercaptanes	R-SH
	Na_2SO_4*10 H_2O,	sulphide (dimethyl	R-S-R
	MgSO4*H_2O, $BaSO_4$,	sulphide, DMS)	
	$SrSO_4$		
Thiosulphate	$S_2O_3^{2-}$	Disulphide (dimethyldisulphide, DMDS)	R-S-S-R
Sulphite	SO_3^{2-}	Polysulphide	$R\text{-}S_x^-$
Polythionate	$S_xO_6^{2-}$	*Oxidised*	
Polysulphide	S_x^{2-}	Sulphoxide	R_2SO
		Sulphone	R_2SO_2
Sulphur dioxide	SO_2	Sulphonic acid	$R\text{-}SO_3H$
Sulphide	minerals: PbS, ZnS,	Thiosulphate	$R\text{-}S_2O_3^-$
	FeS, FeS_2, Fe_3S_4,	Sulphonium ion	R_3S^+
	HgS, $CuFeS_2$, FeAsS,	Sulphate ester	$R\text{-}O\text{-}SO_3^-$
	Cu_2S, MoS_2, Ag_2S,		
	Sb_2S_3, As_4S_4		
Elemental S	S_8		

Source: Haneklaus *et al.*, 2003

REFERENCES

Abrol Y.P., Sangwan S. and Tiwari M.K. (eds) (2002). Land Use-Historical Perspective–Focus on Indo-Gangetic Plains. Allied Publishers, New Delhi.

Agripedia (2002). Introductory Soil Microbiology. http://www.ca.uky.edu/Agripedia/classes/pls566/class22.htm.

Anon (2002a). http://www.demographie.de/info/epub/pdfdateien/SWITodesursachen.

Bates T.S., Cline J.D., Gammon R.H. and Kelly-Hansen S.R. (1987). Regional and seasonal variations in the flux of oceanic dimethylsulphide to the atmosphere. J Geophys Res 92: 2930-2938.

Bowen H.J.M. (ed) (1966). Trace Elements in Biochemistry. Academic Press, London.

Chapman S.J. (1987). Microbial sulphur in some Scottish soils. Soil Biol Biochem 19: 301-305.

Clark T.J. (2002). Sulfur. http://www.vitamins-minerals.tv/minerals/sulfur.htm.

Dovland H. and Saltbones J. (1986). Emissions of sulphur dioxide in Europe in 1980 and 1983. EMEP/CCC-Report: 1/86, Norwegian Institute for Air Research, Lillestrøm p 34.

Ellis D. (ed) (1932). Sulphur bacteria - a monograph. Longmans, Green and Co, London.

Ernst WHO (1993). Ecological aspects of sulphur in higher plants: the impact of SO_2 and the evolution of the biosynthesis of organic sulphur compounds on populations and ecosystems. In: De Kok LJ, Stulen I, Rennenberg H, Brunold C and Rauser WE (eds) Sulphur Nutrition and Assimilation in Higher Plants, SPB Academic Publishers, The Hague p 295-313.

Fenchel T. and Jorgensen B.B. (1977). Detritus food chains of aquatic ecosystems: The role of bacteria. Adv Microb Ecol 1: 1-57.

Friend J.P. (1973). The global sulphur cycle. In: Rasool SI (ed) Chemistry of the Lower Atmosphere, 177-201. Plenum Press, New York.

Galloway J.N. (1989). Atmospheric acidification: projections for the future. Ambio 18: 161-166.

Galloway J.N. (1996). Anthropogenic mobilisation of sulphur and nitrogen: Immediate and delayed consequences. Annu Rev Energy Environ 21: 261-292.

Giovanelli J. (1990). Regulatory aspects of cysteine and methionine biosynthesis. In: Rennenberg H, Brunold C, De Kok LJ and Stulen I (eds) Sulphur Nutrition and Sulphur Assimilation in Higher Plants; Fundamental, Environmental and Agricultural Aspects, SPB Academic Publishers, The Hague p 33-48.

Haehndel R. and Isermann K (1993). Soluble nitrogen and carbon present underground sports fields. Proceedings of 7th International Turfgrass Research Conference, Palm Beach, Florida p 1-6.

Hameed S. and Dignon J. (1988). Changes in the geographical distributions of global emissions of NO_x and SO_x from fossil-fuel combustion between 1966 and 1980. Atmospheric Environment 22: 441-449.

Haneklaus S., Bloem E. and Schnug E. (2003). The global sulphur cycle and its links to plant environment. In: Sulphur in Plant (Abrol YP and Ahmad A eds), Kluwer Academic Publishers, p 1-28.

Hasse L. (1983). Introductory meteorology and fluid mechanics. Chapter 1. In: Air-Sea Exchange of Gases and Particles (Liss PS and Slinn WGN eds), Reidel, Boston p 1-51.

Hawkesford M.J., Buchner P., Hopkins L. and Howarth J.R. (2003). Sulphate uptake and transport. In: Sulphur in Plant (Abrol Y.P. and Ahmad A. eds), Kluwer Academic Publishers, p 71-86.

Howarth R.W. (1984). The ecological significance of sulphur in the energy dynamics of salt marsh and marine sediments. Biogeochemistry, 1: 5-27.

Howarth R.W. and Teal J.M. (1980). Energy flow in a salt marsh ecosystem: the role of reduced inorganic sulphur compounds. Am Nat 116: 862-72.

Ivanov M.V., Lein A. Yu Reeburgh M.S. and Skyring G.W. (1989). Interaction of sulphur and carbon cycles in marine sediments. In: Evolution of the Global Biogeochemical Sulphur Cycle (Brimblecombe P. and Lein A. Yu eds), Wiley, Chichester p 125-179.

Johnson D.W. (1984). Sulphur cycling in forests. Biogeochemistry 1: 29-43.

Kopriva S. and Koprivova A. (2003). Sulphate assimilation: A pathway which likes to surprise. In: Sulphur in Plant (Abrol Y.P. and Ahmad A. eds), Kluwer Academic Publishers, p 87-112.

Maynard D.G., Stewart J.W.B. and Bettany J.R. (1984). Sulphur cycling in grassland and parkland soils. Biogeochemistry 1: 97-111.

Merrill J.T. (1986). Atmospheric pathways to the oceans. In: The Role of Air-Sea Exchange in Geochemical Cycling (Buat- Ménard ed), Reidel, Boston p 35-63.

Moller D. (1984). Estimation of the global man-made sulphur emissions. Atmos Environ 18: 19-27.

NAPAP (1990). Emissions involved in acidic deposition processes. State-of-Science/Technology Report 1, January 1990 draft, National Acid Precipitation Assessment Program, Washington DC.

Nriagu J.O. and Holdway D.A. (1989). Production and release of dimethyl sulphide from the Great Lakes. Tellus 41B: 161-169.

O'Donnell A.G., Wu J. and Syers J.K. (1994). Sulphate-S amendments in soil and their effects on the transformation of soil sulphur. Soil Biol Biochem 26: 507-514.

Pasriha N.S. and Abrol Y.P. (2003). Food production and plant nutrient sulphur. In: Sulphur in Plants (Abrol YP and Ahmad A eds), Kluwer Academic Publishers, p 29-44.

Rennenberg H. (1984). The fate of excess sulphur in higher plants. Annu Rev Plant Physiol 35: 121-153.

Roembke J., Beck L., Foerster B., Fruend H.C., Horak F., Ruf A., Rosciczweski C., Scheurig M. and Woas S. (2002). Boden als Lebensraum fuer Mikroorganismen. Eine Literaturstudie. http://www.uvm.baden-wuerttemberg.de/bofaweb/berichte/tbb04/tbb04.htm.

Saggar S., Bettany J.R. and Stewart J.W.B. (1981). Measurement of microbial sulphur in soil. Soil Biol Biochem 13: 493-498.

Schroeder P. (1993). Plants as sources of atmospheric sulphur. In: Sulphur Nutrition and Assimilation in Higher Plants (De Kok LJ, Stulen I, Rennenberg H, Brunold C and Rauser WE eds), SPB Academic Publishers, The Hague p 253-270.

Seafriends (2002). Soil ecology. http://seafriends.org.nz/enviro/soil/ecology.htm.

Sorensen J., Christensen D. and Jørgensen B.B. (1981). Volatile fatty acids and hydrogen as substrates for sulphate-reducing bacteria in anaerobic marine sediment. Appl Environ Microbiol 42: 5-11.

Stevenson F.J. (ed) (1986). Cycle of Soil Carbon, Nitrogen, Phosphorus, Sulphur, Micronutrients. John Wiley and Sons, New York.

Stevenson F.J. and Cole M.A. (eds) (1999). Cycles of soil Carbon, Nitrogen, Phosphorus, Sulphur and Micronutrients. 2nd ed, John Wiley and Sons, New York.

Topp W. (1981). Biologie der Bodenorganismen. UTB, Heildelberg.

Trudinger P.A. (1979). The biological sulphur cycle. In: Biogeochemical Cycling of Mineral-Forming Elements (Trudinger PA and Swaine DJ eds), Elsevier Science Publisher, Amsterdam p 293-313.

Wedepohl K.H. (1968). Chemical fractionation in the sedimentary environment. In: Origin and Distribution of the Elements (Ahrens LH ed), Vol 30 of Earth Sciences, Pergamon Press, London.

Whelpdale D.M. and Moody J.L. (1989). Large-scale meteorological regimes and transport processes. Chapter 2. In: The Long Range Atmospheric Transport of Substances from Continents to Continents and Continents to Oceans (Knap AH ed), Reidel, Boston p 3-36.

Wu J., O'Donnell A.G., He Z.L. and Syers J.K. (1994). Fumigation-extraction method for the measurement of soil microbial biomass-S. Soil Biol Biochem 26: 117-125.

Chapter 7

Climate, Climate Variability and Climate Change for North-East India

D. R. Sikka
40 Mausam Vihar, Delhi - 110 051

The climate of North-East India which has sustained the rich biodiversity of this humid tropical region of India is its resource. The summer and winter monsoons control the climate of the region. It is also modulated by the peculiar geography of the region. Climatic aspects of the region are briefly described for different seasons of the year. During the summer season, NE India is pivtol region. Being a source of heat, it maintains the monsoons over India by releasing of heavy rainfall over the region. There is considerable inter-annual variability over NE India in each season, with respect to the climatic elements. This variability is presented as the pre-monsoon and monsoon seasons, with respect to the variability of rainfall. The incidence of floods and droughts over NE India has been studied for the last 125 years or so as to provide a background under which the climate change, if any, is to be examined in the next 100 years or so under anthropogenically caused global warning scenarios. Uncertainty in the climate change scenarios over the region resulting from general circulation modelling experiments is also discussed.

INTRODUCTION

Different parts of India exhibit vastly different climatic conditions over the year. These range from the extremely cold snow-clad Himalayan region in winter to the extremely hot desert of Rajasthan in summer, and from the very high moisture (rainfall) region over the North-Eastern states and the West coast of India to the drought-prone dry tracts with meagre rainfall, over West Rajasthan and Saurashtra and Kutch during the summer monsoon rainy season. Climate is a rich natural resource, which controls the biodiversity of a region. Man has exploited this resource to his advantage for centuries. Climatic conditions over a region show variability on different temporal scales, such as sub-seasonal (weeks), inter-seasonal, inter-annual, decadal centuries and larger scales. There are epochs of cold and dry conditions interwoven with other epochs of rather warm and humid conditions due to natural causes. These epochs result in variability in the socio-economic conditions of the people living in the region. Moreover, there is a constant threat that due to anthropogenically caused global warming, climate may be differentially affected over different parts of the world.

Certain aspects of climate, climate variability and climate change are discussed in this paper with respect to the North-East Indian states. This is the region known for its rich biodiversity. Section 2 provides a discussion on the climatic elements of the region and highlights its importance as a pivotal region for the maintenance of monsoon circulation over India as a whole. Section 3 discusses the climate variability over the region, with respect to rains during the pre-monsoon (March-April-May) and monsoon (June to September) seasons. It also emphasises the incidence of excess (flood) and deficit (drought) rainfall on a seasonal scale, using the instrumental period rainfall data on the North-East region and its two major sub-regions, North Assam and South Assam, which comprise 6 states in the region (Assam, Manipur, Meghalaya, Mizoram, Nagaland and Tripura). The seventh state, Arunachal Pradesh, that is a predominantly hilly region is not included. Section 4 briefly discusses the subject of the anthropogenically caused threat of global warming and climate change, in general. Inferences to short-term climatic fluctuations, and changes, based on the available records of the few last decades for the region are provided. It also examines the results of large-scale general circulation modelling experiments with all their implied uncertainties, with respect to North-East India during the winter and monsoon seasons. The possible impact of climate changes for agriculture over the region are also touched upon in section 4.

CLIMATE OF NORTH-EAST INDIA AND ITS IMPORTANCE FOR THE CLIMATE OF INDIA

1. Physiography of the Region

India is located in South Asia within the latitudes, 8° to 37° N and longitudes, 68° to 97° E. It extends about 3200 Km from north to south and about 3000 Km from east to west. It has a long coastline of about 5700 Km. Such a vast area, covering sub-continental dimensions and extending from the deep tropical belt to temperate latitudes, enjoys different climatic regimes during the course of the year. North-East India, consisting of 7 states, Arunachal Pradesh, Assam, Manipur, Meghalaya, Mizoram, Nagaland and Tripura approximately lies between latitudes, 22° to 29° N and longitudes, 88° to 97° E. It is a land-locked region, surrounded to the north by high Himalayan ranges, and to the south and west by the alluvial plains of Bangladesh. Further south lies the Bay of Bengal, which is a warm water region. Several rivers, the major ones being the Brahamaputra and Barak, traverse North-East India. The region covers the massive hilly tracts of Arunachal Pradesh and Meghalaya, the Garo and Khasi-Jaintia hills, fertile valleys of the Brahmaputra, Barak and other rivers and the plain parts of Assam and Tripura. Thus, the topographic features of this region are quite complex, with height above mean sea level (anisl) ranging from just about

10 m in Tripura plains to over 2000m in the hilly tracts. The hills of Garoland Khasi-Jaintia lie almost straight in the path of the south-west monsoon winds, and the steep rise in the hills results in the lifting of the moisture rich air mass, producing copious rains. The world's highest annual rainfall stations, like Cherapunjee, lie in this belt. The varying topography of the region (Khasi-Jaintia, Garo, Anglong and Cachar Hills) provides a variety of annual rainfall patterns, with rainfall ranging from about 1200 mm to over 10000 mm over different stations.

Environmentally too, the region offers a rich variety in flora and fauna. Whereas alpine and sub-alpine forest types exist in the hilly tracts of the region, the valleys are rich in evergreen tropical rain forest. The region is very rich in biodiversity, which is the greatest natural asset, not only for the region but also for the world. Soil characteristics of the region also show rich diversity, such as old mountain type, silty clay-to-clay types, red alluvial and riverine types, etc. Agriculture and horticulture is practiced in the river valley and floral plains. Large parts of the region are also covered with tea plantations along hilly terraces. The major crops of the region are rice, wheat, maize and pulses which are cultivated depending upon the types of soil and the climatic conditions. Fruits such as pineapple, orange, bananas are also grown. Thus, agriculture, tea gardening and horticulture are important professions of the people. The irrigation scenario of the region is not very complex, and rainfall is plentiful. Environmentally destructive practices like shifting cultivation (Jhum) and deforestation are followed by the tribal population in several parts of the hilly belts in the region. Except for petroleum refineries, there are no large industries present in the area, thus the threat of air pollution is not yet present. However, the land surface conditions are getting affected locally, due to the pressure of habitations, small-scale industrialization and communication, etc.

2. Climate of North-East India and its Controls

The climate of North-East India is controlled by the winter monsoons (NE monsoon) in winter (January-February) and summer monsoons (SE monsoon) in summer (June to September) like the rest of India. The intervening seasons are known as the pre-monsoon season (March-April-May) and the post-monsoon season (October-November-December). The NE monsoon wind system sets up in mid-October after the withdrawal of the SW monsoon wind system. It is part of the great NE monsoon wind system of Asia, through locally modulated by the hilly tracts of NE India. The period, November to February is generally dry. Almost 22-25% of the annual rainfall occurs during the pre-monsoon season over different parts of the region. This is primarily due to the favourable location of the region with respect to the southerly winds from the adjoining Bay of Bengal in the near surface levels, and the prevalence of connective instability which triggers spells of widespread thunderstorm activity over the region under favourable positioning of the sub-tropical westerly jet stream and troughs in the mid-tropospheric westerlies. The thunderstorms become locally destructive at times (severe local thunderstorms or tornadoes). The pre-monsoon thunderstorm rains merge with the commencement of the monsoon rains in early June. Sometimes it is difficult to clearly identify the burst of the summer monsoon over the region, unlike the West Coast of India which witnesses the dramatic burst (onset) of summer monsoons every year. During the summer monsoon (henceforth called monsoon) season, NE India receives heavy rains and nearly 66-68% of the annual rainfall of the area is received in the 4 months of the monsoon season. Obviously, these heavy rains produce floods in all the rivers of the region and hence, the region is known for its vast floral variety.

During the winter season, the average daily temperature over different parts of the region varies considerably due to topography. The plains and valleys enjoy a maximum temperature of about 15 -20 °C and minimum of about 5 –10 °C. For the hilly regions maximum and minimum temperatures are lower and they depend upon the altitude of the hills of the region - for Shillong region, minimum temperature during cold spells can go below freezing point. Summer heating begins to be felt from early March and the maximum day temperature in the plains and valleys reaches up to 35 – 40 °C in April and May. The region is also affected by three to four heat waves, each of 3-5 days duration during March-April-May. However, over the hilly regions, the pre-monsoon season temperatures are lower and pleasant during the day (25 – 30 °C). During the monsoon season, the day temperature is moderate and diversual variability in the temperature is low due to extensive clouding. The daily mean temperature also does not vary much on the sub-seasonal scale, remaining comfortable within 20 - 25 °C over different parts of the region. There is a short spell of increase in temperature in October, before seasonal cooling sets up in November. Mean daily evaporation during winter is 1-2 mm per day and it reaches to about 4-5 mm/day in April-May. It reduces considerably to about 1-3 mm/day during the monsoon season and increases slightly to 2-4 mm/day during the post monsoon season. There is considerable day-to-day variability in evaporation, depending upon the temperature, wind and moisture conditions, in all the seasons. The annual evaporation is about 1000-1250 mm over different parts of NE India and is rather low, about 300-400 mm for the monsoon season. The net annual water balance is positive for the region, as annual rainfall exceeds annual evapotranspiration. Average annual rainfall is over 2000 mm for the region with considerable special variability depending upon the location of the stations with respect to aspect of the hills and topography of the place. Annually, about 100-150 days are designated as rainy days over different stations in the region. Considerable rainfall occurs in the pre-monsoon season too. Thus, the rainy season is rather long, from late March or early April to the end of October. Annually, nearly 50 days witness thunderstorms; most of the thunderstorms occur during May to September. Duration of sunshine also varies significantly on an annual basis ranging between 5-8 hours in winter and pre-monsoon and post-monsoon seasons to about 4-5 hours in the monsoon season. Solar radiation also has large inter-seasonal variability over the region.

3. Climatological Behaviour of Pentad Rainfall for Some Stations in NE India

Ananthakrishnan and Pathan (1971) have published plots of 5-days (pentad) mean rainfall of a number of meteorological observations in India based on 1901-

1950 data. This publication includes data for 10 stations in North-East India (Cherrapunji, Dhubri, Dibrugarh, Gawhati, Kalimpong, Lumbding, Shillong, Silchar, Sibasagar and Tezpur). Most of the stations are located in Assam and Meghalaya. We have reproduced in figures, plots for 8 of the stations, except for two stations, Sibasagar and Silchar. These plots also include, as inset, monthly progress of the mean rainfall for each station. These plots show that the pentad rainfall begins to increase in most of the stations from the 16th pentad (17-21 March). For Lumbding, the increase is noticed three pentads later, in the 19th pentad (1 to 5 April). There are several stations where the increase in pentad rainfall is rapid in May (Cherrapunji, Dubri, Dibrugarh, Kalimpong, Shillong, Sibasagar, Silchar) and the increase in rainfall is somewhat spread over 3-4 adjoining pentads and between pentad 32 (June 5-9) and pentad 40 (July 15-19). The rainfall remains quite high up to pentad 55-56 (28 September-17 October) and it gradually tapers off by pentad 66 (28 October-1 November). Thus, the rainy season lasts from pentad 16th to pentad 66th–a period of about 250 days from mid of March to the end of October, which makes the region rich in biodiversity and very humid. For most of the stations, June-July-August are the months of maximum rainfall, which occurs in the monsoon season within a prolonged rainy period between March to October. Lumbding and Guwahati are the two stations where the annual rainfall does not exceed 1700 mm and the maximum rainfall in any pentad for these two stations is less than half of the values received in other stations, with annual rainfall exceeding 2000 mm. The mean picture of the pentad rainfall shows considerable inter-annual variations for all pentads - in respect of episodes, like the beginning of the pre-monsoon rains, onset of monsoon rains, increase in the rainfall, sharp decrease in the rainfall. On a year to year basis, there could be several pentads during June to September, when the maximum rainfall is achieved to be followed by a decrease and increase in the later pentads as the monsoon passes through different intra-seasonal episodes.

The above summary about the climate of the states of NE India is based on the data compiled by the IMD in their climatological table of observations in India (1960 and 1999). The IMD maintains about 27 meteorological observations (Appendix I) in 7 states of the region, with data varying from 20 to nearly 100 years.

4. Importance of NE India to the Weather and Climate of India

India's climate is dominated by the winter and summer monsoons. Winter monsoons, during January-February, are generally dry except for occasional rains over NW India, under the influence of the eastward moving western disturbances from the Mediterranean Sea and Caspian Sea regions. On rare occasions, these western disturbances produce secondary disturbances, which move across the Gangetic valley and affect the weather of NE India. During the winter months, NE India and the adjoining Tibetan region which maintains the NE flow over the Bay of Bengal are cold source regions. By March-April, peninsular India begins to get heated and the south to south west flow prevails over Central and North Bay of Bengal, and penetrates into NE India. The connective instability in the lower troposphere, existing over NE India in the pre-monsoon season, is released in the form of widespread to local thunderstorms over the region under favourable middle and upper troposphere flow patterns, associated with the mid-latitude troughs and the sub-tropical jet stream, respectively. These thunderstorms are responsible for the gradual building up of the heat source over NE India in April-May, which becomes very intense during the SW monsoon season as a result of abundant moisture and widespread rainfall (Flohn, 1968). Thus, by May end, NE India and the adjoining regions becomes the seat of a mid and upper tropospheric anticyclone (the Tibetan High). This is a region of diverging easterly flows in the upper troposphere, which attains the strengths of the upper tropospheric jet stream, known as the Tropical Easterly Jet (TEJ) stream (Koteswarm, 1958). TEJ is an important component of the summer monsoons over India (Krishnamuri and Bhalme, 1976). The Tibetan anticyclone is centred just to the north of NE India, where the heaviest monsoon rains occur. Das (1962) has emphasised that an east-west circulation prevails over northern India, with a rising motion over NE India and a sinking motion over the Rajasthan desert. This lateral circulation is an important dynamic component of the monsoon. Another important component of the summer monsoon is the Hadley cell circulation with an upward motion over northern India and a sinking motion over south India, which is harmful as the monsoon cell of India (Koteswaram, 1960) with the easterlies and westerlies at a surface opposite to Hadley circulation in the rest of tropics, but similar in a thermal sense. During the summer monsoon, persistent winds blow northward from the south Indian Ocean (Joseph and Raman, 1966; Findlalter, 1969). This cross-equatorial flow in the lower troposphere is very strong off the Somalia Coast and is known as the Low Land Jet (LLJ) stream, which is, recognised as another important component of the monsoons (Krishnamrurti and Bhalme, 1976). It swings across the Arabian Sea into Central Bay of Bengal and is the major artery supplying moisture to the monsoon rains over India. The precipitable water over North Bay of Bengal is about 8 gm/wt, which is perhaps the highest in the world.

All these components of the monsoon, along with the monsoon trough over India, fluctuate on a sub-seasonal scales (synoptic 3-5 days, super synoptic 10-20 days and low frequency 30-50 days modes of the monsoon), which are responsible for the intra-seasonal variability of the monsoon rains over different parts of the country. Such episodes are known as active (more than normal rain) and 'break' monsoon epochs. During the 'break' monsoon period, the monsoon trough moves toward the foot of the Himalayas and causes a heavy rainfall spell over NE India, as well as on the foothills of Himalayas. However, the central parts of the country get much less rain or go dry during the 'break' monsoon period. During active monsoon, heavy rains occur over central India in association with the western movement of monsoon disturbances (monsoon lows and depressions) forming over the North Bay of Bengal and the adjoining land areas and traversing the Indian area parallel to, but usually south of, the mean portion of the monsoon trough. During the 'break' monsoon, the Tibetan anticyclone which is usually less active, shifts eastward and TEJ is also weakened. Lower tropospheric anticyclonic activity develops during the break monsoon over Central India, whereas in the active monsoon period, Central India is the seat of well-marked cyclonic activity (Sikka and Gadgil, 1978). On a 30–50 day scale, a monsoon cloud band moves from the near-equatorial

Bay of Bengal and SE Arabian Sea, towards the foot of Himalayas. Such low frequency meridional oscillations in the monsoon are dynamically very important for major rainfall fluctuations (Sikka and Gadgil, 1980; Yasunari, 1980). Bhide *et al.*, (1997) have demonstrated that sub-seasonal fluctuations and westward migration of heat sources and sinks from NE India to Gujarat and Rajasthan are important dynamical fluctuations of the regional monsoon system. All the above mentioned aspects demonstrate the pivotal role played by NE India during the pre-monsoon systems in building up of the Indian monsoons, their seasonal maintenance and sub-seasonal fluctuations, though such a pivotal role is yet to be fully understood. Thus, there is a need for organising phenomenological important climate-related process studies over NE India during these two seasons.

CLIMATE VARIABILITY OVER NORTH-EAST INDIA IN PRE-MONSOON AND MONSOON SEASONS

Rainfall is the most important climatic element in pre-monsoon and monsoon seasons over NE India. Therefore, we have analysed the variability of rainfall over this region on inter-annual and decadal scales in terms of the occurrence of extreme events, like seasonally excess (flood) and deficit (drought) rainfall. Parthasarathy *et al.* (1995) have provided a standard rainfall series from 1871 to 1994 for India as a whole, 5 homogeneous rainfall regions of India and 29 meteorological sub-divisions. For our analysis, we use their data for the NE India Homogeneous Region and North Assam and South Assam sub-regions. NE India comprises the meteorological sub-divisions of Gangetic West Bengal, Sub-Himalayan region, West Bengal and the sub-region of North Assam and South Assam in this rainfall series. North Assam and South Assam consist of six states in NE India (Assam, Manipur, Meghalaya, Mizoram, Nagaland and Tripura). Rainfall data for the 7th state, Arunachal Pradesh has not been included in the Parthasarathy *et al.* (1995) series, as it is mainly a hilly sub-division.

Table 7.1 gives the rainfall statistics for NE India Homogenous Region and its meteorological sub-divisions.

Table 7.1: Rainfall statistics for NE India Homogenous Region and its four meteorological sub-divisions (Parthsarathy et al., 1995)

Mean Rainfall and Stand deviation (in brackets)

Region/Meteorological Sub-division	*Annual mm*	*Winter Jan-Feb mm*	*Pre-monsoon M-A-M mm*	*Monsoon JJAS mm*	*Post-Monsoon O-N-D mm*
NE India Homogeneous Region	2066.0 (177.7)	63.0 (22.5)	425.8 (83.8)	1419.2 (121.3)	178.1 (73.1)
Gangetic West Bengal sub-division	1519.0 (220.9)	38.2 (34.6)	183.4 (79.9)	1145.2 (167.8)	157.2 (83.0)
Sub-Himalayan West Bengal Sub-division	2505.5 (328.7)	26.5 (21.5)	336.0 (95.5)	2001.2 (300.1)	141.8 (104.4)
North Assam (Sub-region)	2227.2 (224.5)	53.8 (22.3)	569.8 (130.0)	1453.5 (164.3)	149.9 (68.5)
South Assam (Sub-region)	2210.6 (210.3)	43.7 (28.6)	506.3 (119.2)	1449.1 (193.5)	211.5 (97.5)

According to the IMD data, the annual and monsoon season rainfall and percentage of annual rainfall in the monsoon season, for the 3 meteorological sub-divisions in the 7 states comprising NE India, are given in Table 7.2.

Table 7.2: Rainfall Statistics for annual and monsoon season for meteorological sub-divisions in NE India (as per IMD data)

Meteorological Sub-division	*Annual mm*	*Rainfall Statistics Monsoon Season mm*	*Percentage of Monsoon Rainfall to Annual Rainfall*
Arunachal Pradesh	4142	2743	66
Assam and Meghalaya	2735	1838	67
Manipur, Mizoram, Nagaland and Tripura	2435	1610	66

For NE India Homogenous Region, the pre-monsoon season contributes about 21% and the monsoon season contributes about 69 % to the total annual rainfall. Similarly, for North Assam and South Assam, the corresponding figures are: (i) 26% and 65%, and (ii) 23% and 66%, respectively. The coefficients of variability of rainfall for these areas vary between 9% and 13 % for monsoon rainfall and between 20 % to 23% for the pre-monsoon season rainfall. The coefficient of variability for the pre-monsoon season is about twice the value of the monsoon season rainfall. Thus, even though the pre-monsoon season contributes about 23-26% to the annual rainfall as against about 65% for the monsoon season rainfall, the pre-monsoon season rainfall over North Assam and South Assam is much more variable and hence inter-annual variability is large.

1. Inter-annual Variability of Excess and Deficit Rainfall in the Pre-Monsoon Season Incidence of Excess and Deficit Rainfall in the Pre-monsoon Season

Even though NE India Region and its two meteorological sub-regions (North Assam and South Assam) are humid regions, it is important to study the incidence of excess and deficit rainfall seasons as extreme events in rainfall on a seasonal scale impact on the regions economic and social systems, as well as on its ecological and natural vegetation. In some deficit rainfall years, the deficiency in pre-monsoon season's rains would be 25% or lower. Standard deviation (σ) provides a good measure to study the incidence of excess/deficit monsoon year. If the rainfall in a season is more than (less than) one standard deviation, we assign the season to an excess (deficit) rainfall category, or in other words floods (drought). Based on the Parthasarathy *et al.* (1995) data, Table 7.3 provides the statistics for the incidence of excess (flood) and deficit (drought) rainfall during the pre-monsoon season.

Table 7.3: Years of incidence of excess (flood) and deficit (drought) seasons (Period 1871-1994) during the pre-monsoon

Region	Years of rainfall equal to or excess within different S.D. limits			Years of rainfall equal to or less within different S.D. limits		
	≥1.0σ	≥1.50σ	≥2.0σ	≥1.0σ	≥1.50σ	≥2.0σ
NE India	1881,1892	1892,1893	1915	1873,1879	1897,1901	1897
	1893,1904	1915,1929	1948	1897,1901	1902,1922	1979
	1915,1921	1936,1948	1949	1902,1908	1935,1960	
	1925,1929	1949,1956	1993	1917,1919	1979	
	1936,1941	1977,1993		1922,1935		
	1948,1949			1939,1950		
	1952,1956			1957,1960		
	1977,1990			1962,1965		
	1993			1966,1975		
				1979,1992		
	17 years	10 years	04 years	20 years	7 years	2 years
North Assam	1874	1874,1992	1874	1884,1890	1890,1917	1979
	1888,1892	1904,1913	1892	1901,1903	1939,1978	
	1904,1913	1915,1921	1929	1917,1918	1979	
	1915,1921	1929,1948		1922,1935		
	1925,1929	1943,1949	1948	1937,1939		
	1941,1942	1953,1956	1956	1952,1960		
				1962,1967		
	1943,1948	1977,1994	1994	1971,1978		
	1949,1953			1979,1986		
	1956,1977			1989,1992		
	1988,1994					
	19 years	14 years	6 years	20 years	5 years	1 year
South Assam	1875,1881	1893,1915	1915	1873,1879	1897,1898	1898
	1883,1888	1929,1936	1936	1885,1898	1903,1908	
	1892,1893	1941,1946	1948	1901,1903	1919,1935	
	1894,1904	1948,1977	1993	1908,1909	1960,1979	
	1915,1929	1991,1993		1919,1922		
	1936,1941			1935,1950		
	1946,1948			1954,1957		
	1949, 1952			1960,1965		
	1956,1961			1966,1979		
	1977,1991			1987		
	1993					
	21 years	10 years	4 years	19 years	8 years	1 year

For NE India Homogeneous Region, the highest rainfall in the pre-monsoon season has been recorded in the year 1948 (656.8 mm) and the lowest rainfall in the year 1979 (226.7). The contrast in the quantum of extreme pre-monsoon season rainfall is very striking. For North Assam, the maximum rainfall occurred in the pre-monsoon season during 1948 (926.8 mm) and the minimum in 1979 (300.3 mm), again sharing large contents in the two extreme events. Similarly, for South Assam, the maximum rainfall in the pre-monsoon season occurred in 1993 (872.3 mm) and minimum in 1898 (288.7 mm). During the pre-monsoon season, rice is extensively cultivated in the states of NE India and the impact of moderate and severe moderate drought (rainfall deficit between 1-50 to 1-990) and severe drought (rainfall deficit equal to or less than 2.00) becomes very serious to agriculture, as with such drought in the season, the deficiency in the seasonal rainfall is below

25%. These are also accompanied with moderate to severe heat waves, with the weekly evapotranspiration approaching the values of 0.50 mm or even more, making the impact of drought quite serious for the pre-monsoon crop of rice.

2. Interannual Variability of Excess and Deficit Rainfall in Monsoon Season-Incidence Excess and Deficit Rainfall Years

Table 7.4 provides the data on the years of incidence of excess (flood) and deficit (drought) during the monsoon season.

Table 7.4: Years of incidence of excess (flood) and deficit (drought) rainfall during the monsoon season (Period 1871-1994)

Region	Years of rainfall equal to or excess within different S.D. limits			Years of rainfall equal to or less within different S.D. limits		
	≥1.0σ	≥1.50σ	≥2.0σ	≥1.0σ	≥1.50σ	≥2.0σ
NE India	1875,1878	1878,1879	1878	1873,1884	1873,1884	1873
	1879,1880	1886,1899	1886	1891,1892	1891,1896	1884
	1886,1893	1918,1947	1918	1896,1900	1959,1972	1896
	1899,1905	1956,1974	1974	1925,1957	1980,1981	1992
	1918,1905	1988,1993	1993	1958,1959	1982,1992	
	1918,1947			1961,1962	1994	
	1956,1966			1967,1972		
	1968,1974			1975,1980		
	1987,1988			1981,1982		
	1989,1993			1986,1992		
				1994		
	18 years	10 years	05 years	21 years	11 years	04 years
North Assam	1871,1972	1878,1874	1878	1873,1877	1884,1888	1884
	1878,1879	1889,1902	1879	1882,1883	1891,1896	1891
	1889,1899	1917,1918	1918	1884,1888	1900	1896
	1902,1915	1947,1974	1974	1891,1893		1900
	1917,1918	1987,1993	1987	1896,1900		
	1935,1947		1993	1913,1956		
	1974,1987			1958,1959		
	1993			1961,1962		
				1967,1971		
				1975,1978		
				1979,1986		
				1992,1994		
	15 years	10 years	06 years	24 years	05 years	04 years
South Assam	1875,1878	1875,1878	1886	1871,1873	1896,1925	1980
	1879,1880	1879,1886	1918	1884,1886	1961,1971	
	1886,1893	1893,1918	1947	1900,1914	1972,1980	
	1899,1905	1947,1956	1956	1918,1919	1981,1986	
	1906,1910	1966,1974		1925,1928	1990,1992	
	1918,1934	1988		1957,1958		
	1947,1951			1959,1961		
	1953,1956			1967,1971		
	1960,1966			1972,1975		
	1969,1974			1980,1981		
	1988,1993			1982,1985		
				1986,1990		
				1992,1994		
	22 years	11 years	04 years	26 years	10 years	01 year

Maximum monsoon rainfall was witnessed over the NE Region of India during 1918 and minimum rainfall during 1992 (1137 mm). Similar figures have been observed for North Assam in 1993 (2055 mm) and for South Assam in 1918 (1892 mm) and in 1980 (1032 mm). During the monsoon season, due to the prevalence of undesired (extensively) cloudiness over the states of NE India, heat wave conditions cannot prevail, even under deficit rain conditions. The mean weekly potential evaporation privation being between 0.20 to 0.30 mm, the adverse impact of drought conditions is minimal as the deficiency in monsoon season rainfall is never below 95% of the normal rainfall. However, it is the excess rain (flood) conditions on a daily or weekly scale, which destroys the standing crops. Individual flood events are the biggest means disrupting social life in the region and causing immense economic losses. The states of NE India are most susceptible to individual flood events, causing the rivers and rivulets to flow above the danger level (Dhar and Nandargi, 1998). There are multiple flood events from April to September. The pre-monsoon season floods are due to severe thunderstorms, which may occur in spells of 3-5 days. The monsoon floods occur due to heavy rains in the catchment areas under the influence of low-pressure systems, as well as the northward movement of the monsoon trough. Several studies have been made on floods in the states of NE India and the details have been reported in the Indian Journal of Meteorology and Geophysics (Dhar and Changrani, 1966) and others. The Brahamaputra River Commission is responsible for the hydrometerological studies of the region, and the Regional Flood Forecasting Office of IMD and the Central Water Commission deal with the seasonal floods. The run-off of the heavy rainfall in NE India is carried by two major river systems: the Brahmaputra and the Barak with their several tributaries and riverlets. As per the Irrigation Commission report, the average annual flow of the Brahmaputra is 499.91 km^3 and of the Barak system, 90.81 km^3.

Based on the data presented in Tables 7.3 and 7.4, the following broad inferences can be drawn:

i. Over North Assam, two successive years of excess (flood) monsoon seasons have occurred (1878 and 1879, 1917 and 1918). The events of 1878 and 1879 were of a phenomenal category ($>/2.0\sigma$).

ii. For South Assam, 3 successive years of excess (flood) monsoon season have occurred (1878 and 1879, 1879 and 1880,1905 and 1906), of which one event was for 3 successive seasons (1878-1879-1880). However, none of these successive monsoon seasons reached the phenomenal category.

iii. Two successive years of drought events have occurred over North Assam on 5 occasions (1891 and 1892, 1957 and 1958, 1958 and 1959, 1961 and 1962, 1978 and 1979), of which drought occured for 3 successive monsoon seasons during 1957 to 1959. None of the successive drought sequences reached the phenomenal category ($</2.0\sigma$).

iv. Simultaneous occurrence of flood and drought monsoon seasons for North and South Assam is rather intra quest as it occurred for only 5 years (1878,1879,1899,1918,1974) out of the total number of years when flood / drought occurred over one of the two regions.

v. Significantly, North Assam has not witnessed any moderate or severe drought during the monsoon season since 1900 (or in the last 100 years), but South Assam has faced 9 such droughts since 1900.
vi. Through North and South Assam are high rainfall sub-regions and their rainfall patterns are homogeneous, yet simultaneous occurrence of floods or drought over both regions is not frequent. Hardly 20-40 % of the extreme event seasons have occurred in the same year over both areas.

3. Incidence of Excess (Flood) and Deficit (Drought) Seasonal Rainfall on Decadal Scale and Climatic Fluctuations

Pre-monsoon season, normally 30 years duration, is considered adequate to represent a climatic fluctuation. In the records under study (1871-1994), we pass through 4 such climatic durations of 30 years each, such as 1871-1900, 1901-1930, 1931-1960 and 1961-1990. We analyse the occurrence of excess (flood) and deficit (drought) events in these 4 periods for the pre-monsoon and monsoon seasons to determine whether any change is noticeable. For this purpose we would consider only 1.0σ as the limit for defining floods or drought:

Table 7.5 shows that considerable multi-decadal epochal fluctuations in the incidence of excess (flood) and deficit (drought) events have occurred on a 30 year scale. The most noticeable fluctuation occurred in the case of excess rainfall seasons in the epoch 1931-1960, when 10 such events occurred as against only 2 such events in the recent epoch 1961-1990. For North Assam, only 2 deficit rainfall counts occurred in the epoch 1871-1900, while the same figure reached the number of 7 in the recent epoch 1961-1990. Is it a definite change being caused by sub-regional scale changes taking place in the land surface processes? It is very difficult to reliably answer this question. For South Assam, the excess rainfall events were only 2 in the first epoch 1871-1900, while they decreased only 2 in the recent epoch 1961-1990.

Table 7.5: Pre-monsoon season extreme seasonal rainfall events in 30 years epoch

Region/ Sub-Division	*Types of Event*	*No. of Events* 1871-1900	1901-1930	1931-1960	1961-1990
NE India Homogenous Region	(i) Excess (Flood)	3	5	6	2
	(ii) Deficit (Drought)	3	5	6	5
North Assam	(i) Excess (Flood)	3	3	10	2
	(ii) Deficit (Drought)	2	5	5	7
South Assam	(i) Excess (Flood)	7	3	7	2
	(ii) Deficit (Drought)	4	6	5	4

MONSOON SEASON

Extreme Rainfall events in 30 years epoch. Table 7.6 presents this data.

Table 7.6: Incidence of monsoon season extreme rainfall in 30 years epochs

Region/ Sub-Division	*Rainfall Event*	*No. of Events in Epochs*			
		1871-1900	*1901-1930*	*1931-1960*	*1961-1990*
NE India Homogenous Region	(i) Excess (Flood) (ii) Deficit (Drought)	7 6	2 1	2 3	6 4
North Assam	(i) Excess (Flood) (ii) Deficit (Drought)	6	4	2	2
South Assam	(i) Excess (Flood) (ii) Deficit (Drought)	7	4	6	4

Again we notice considerable inter-epochal variations in the incidence of excess (flood)-deficit (drought) which occurs in the monsoon seasons too. Significant differences in the deficit rainfall events found for have been the NE India Homogenous Region between the 1901-1930 and 1961-1990 epochs. For North Assam, the number of excess rainfall events appear to have decreased from the 1871-1900 to the 1901-1990 epoch. For South Assam, no significant difference is noticeable on an inter-epochal scale.

Any long-term trend in the incidence of extreme events in a region of rich biodiversity like the NE India a states has to be marked with great concern. Even though it is too early to say whether changes in land surface characteristics (like shifting cultivation, deforestation, regional agricultural patterns, human habitation, etc.) may cause long-term fluctuations in extreme rainfall regions, the best solution is to empower the NE India states with information so that remedial measures, if any, could be adopted. For this purpose, it is necessary that events be carefully monitored on regional and sub-regional spatial scales. Since there are considerable sharp variations in the rainfall, it is also imperative that water resource planning for the region should be done for each region on a hydrological basis, as whole or as a sub-basis. Inter-state water disputes for the states of NE India could take place in the future, particularly in the drought season. These water disputes do not exist at present in any serious manner among the NE Indian states. However, they could occur in the future, particularly if frequent droughts begin to occur.

CLIMATE CHANGE OVER NORTH-EAST INDIA - OBSERVATIONS AND MODELLING RESULTS

1. Climate Change — Natural and Anthropogenic

The climate of a region cannot be considered permanent for ever, as it fluctuates due to natural causes on different scales, temporal, inter annual, decadal, centennial, millennia and interglacial. The interactions of different components of the climate system, atmosphere, ocean cryosphere, biosphere and geosphere are involved in natural climatic fluctuations on different temporal and spatial scales. Since the instrumental records for India extend only to about the past 150 years, investigations have used different proxy data to determine possible climatic fluctuations over India for several millennia. Such investigations are based on pollen studies, lake sediment studies, tree-ring studies, etc., all of which have been used for the study of natural climatic fluctuations over NE India.

Another kind of climate change, which is being caused by human activities due to excessive use of florid fuel for energy and fertilisers for crop production, is threatening humanity in the nest country. Such a climate change is, therefore, of anthropogenic origin due to the present trend of excessive energy consumption and fertiliser use and if it continues, there is concern that the level of CO_2 and oxides of nitrogen may considerably increase in the atmosphere to make an impact on global warming due to enhanced greenhouse gas warming. The global concentration of CO_2, which was only about 280 ppm in the pre-industrialised era, is expected to reach 600 ppm by the end of the 21st century. There would also be a considerable increase in other greenhouse gases, like nitrous oxide and methane, due to the increase in agricultural productivity. Although it is certain that increase in greenhouse gases would disturb the chemical state of the natural atmosphere, producing warming at the surface layer, however, the quantum of increase in surface warming is a matter of great controversy among the global climate modelling scientific community. As per the IPCC report (IPCC, 1995), equilibrium atmospheric climate models, coupled with ocean-atmosphere nodes and double CO_2 concentration and transient increase in CO_2 concentration, etc. have been used for climate anthropogenic change studies. The estimates for global warming, with double the present concentration of CO_2 from such modelling studies, show that surface temperature increase, varying from 1-5 °C to 3-5 °C. The increase is maximum in near polar latitudes and minimum in the tropics. The increase is also estimated to be maximum in winter and minimum in summer. It also has regional variations. Although global estimates of warming are more reliable, regional estimates remain very uncertain.

2. Observation Studies for India and North-East India about Climate Fluctuations in the last 100 years

Temperature records for meteorological observations over India are available for about 100 years. Several workers (Hingane *et al.*, 1985; Thapaliyal and Kulshrestha, 1991) have examined the temperature records for India and have shown that surface temperatures over several parts of India have increased by about 0.4 °C in the past 100 years. For observation over NE India, the estimates are similar. For rainfall records it is observed that the rainfall of India does not show any long-term trend, either for annual or for the southwest monsoon season. However, there are multi-decadal (30 years) fluctuations in the monsoon season rainfall for India (Joseph, 1976; Sikka, 1980). There is no such study for the multi-decadal fluctuation of rainfall for NE India. For the data provided by Parthasarathy *et al.*, (1995) for NE India Homogeneous Region. Assam and Assam subregions, the rainfall values for four 30 years epochs from 1871 to 1990. A 30 years epoch is comedienne adequate for forming climatic means have been given in Table 7.7.

Table 7.7: The rainfall values for four 30 years epochs from 1871 to 1990

30 year Epoch	Average Rainfall (mm)					
	NE India		North Assam		South Assam	
	Pre-Monsoon	Monsoon	Pre-Monsoon	Monsoon	Pre-monsoon	Monsoon
1871-1900	438.0	1427.5	585.9	1428.7	518.0	1483.9
1901-1930	421.3	1429.9	557.2	1488.8	497.3	1439.5
1931-1960	439.6	1449.3	602.5	1455.2	525.2	1495.8
1961-1990	400.8	1383.3	530.7	1439.0	482.5	1447.0

As would be expected, decadal scale variability is lower than inter-annual, in both the seasons. Decadal maximum and minimum rainfall in all the above areas have occurred in different decades, both for the pre-monsoon and monsoon rainfall. The data state multi-decadal scale fluctuations in rainfall. However, no definite long-term trend is describable. For the NE India Homogenous Region epoch, 1931-1960 had the maximum average rainfall and the epoch 1961-1990 had the minimum. For North Assam and for the pre-monsoon season, the epochal maximum rainfall also occurred in the epoch, 1931-1960 and the minimum in the epoch, 1961-90. Similarly, for south Assam the pre-monsoon epochal maximum rainfall occurred in the epoch, 1931-1960 and minimum in the epoch, 1961-1990. However, the monsoon maximum season for north Assam was in the epoch, 1901-1930 and minimum in the epoch, 1961-90. For south Assam, the monsoon maximum epoch is again 1961-1990, but the minimum epoch is 1901-1930. There is no definite trend. Thus, minor epochal fluctuations are observed in the data series. However, there is minor decrease in rainfall from the 1931-1960 to the 1961-1990 epoch. Hussain (1993) has examined ecological data, temperature and rainfall of Shillong, Meghalaya for the decades 1971-1980 and 1981-1990, respectively. He is of the opinion that there have been mild, but gradual changes in these data for Shillong, as follows:

i. Small increases in average annual temperature - summers have become more warm, while winter nights are colder, but days are warmer.
ii. Average rainfall for Shillong in recent decades is lower than the average for the epoch 1931-1960.
iii. Due to temperature and moisture changes, there have been several ecological changes, such as growth in microorganisms, insects and pests (termites).

3. Inferences about Climate Change over NE India from Modelling Studies

As already mentioned, global climate models are being used to draw scenarios for global and regional climate changes. The estimates for regional changes from such modelling studies are not highly reliable. For India also, several studies have been made (Sikka and Pant, 1991; Lal and Subash Chandra, 1993; Lal *et al.*, 1995) to draw inferences about climate change, (temperature, evaporation, rainfall) using different types of climate models and variations of different models. As per the studies, the following aspects are average for India as a whole and for NE India.

i. During winter, temperature over India would be warmer under double CO_2 by about 1-2 °C, but for summer and monsoon seasons, there is no appreciable warming. The warming over NE India for all the seasons is within about 1 °C and so not appreciable. When the models are run with increase in sulphate aerosols, the warming estimates are reduced for India, and for NE India warming is negligible.

ii. Under double CO_2 conditions, the monsoon season rainfall for North India would increase by about 5-10 °C, but this increase is within an uncertain range for NE India. When coupled, the ocean-atmosphere models which are the estimate for increased rainfall over India are reduced with sulphate aerosols included, thus suggesting a slight decrease (about 5%) in the total seasonal rainfall for India. For NE India, the increase or decrease in monsoon season rainfall remains uncertain in taking module-based guidelines for anthropogenic climate change, and there is hardly any indication that the rainfall over NE India would change significantly under the global warming scenario. The inter-annual and decadal scale variability of the rainfall of NE India is small, as discussed earlier. As such, there is no cause of immediate anthropogenic concern, for the rainfall of NE India would show substantial changes under increased greenhouse gases on the global scale (anthropogenic climate change).

4. Impact of Climate Change on Agriculture over North East India

CO_2 is the food for vegetation through photosynthesis. Increase in CO_2 level would lead to enhanced photosynthetic activity up to at least 600 ppm of CO_2. This would benefit food production. However, if there were corresponding increases in temperature and in evapotranspiration, this would harm food production. Since there is hardly any change being projected, either in temperature or in the rainfall of NE India under global warming due to the increase in greenhouse gases, the agriculture in NE India is not likely to be impacted adversely. The inter-annual and decadal scale fluctuations in temperature, rainfall, etc., are likely to remain dominant and hence, the agricultural production would remain to be influenced predominantly by natural climate variability rather than anthropogenically caused climate changes. As NE India is a rich biodiversity region, enhanced monitoring of the climate and climatic elements are needed for this ecologically sensitive region of India.

ACKNOWLEDGEMENT

The author is thankful to Prof. Pramod Tandon of the North-Eastern Hill University, Shillong for inviting him to the workshop and arranging typing of the MS without which this paper could not have been written.

Appendix I

List of Meteorological Observations in the States of NE India (IMD 1999)

S. No.	Name	State	Lat N°	Long E°	Ht-a m Sl (m)	Annual rainfall mm
1.	Agartala	Tripura	23° 53°	91° 15°	16	2178.6
2.	Aijal	Mizoram	23° 44°	92° 45°	1097	2244.7
3.	Chaparmukh	Assam	26° 12°	92° 31°	66	2766.6
4.	Cherapunji	Meghalaya	25° 15°	91° 44°	1313	11465.7
5.	Dhubri	Assam	26° 01°	89° 58°	35	2513.6
6.	Dibrugarh	Assam	27° 29°	95° 01°	111	2588.7
7.	Digboi	Assam	27° 23°	95° 37°	152	2601.1
8.	Goalpara	Assam	26° 11°	90° 38°	38	2514.7
9.	Gohpur	Assam	26° 50°	93° 27°	83	2477.9
10.	Golaghat	Assam	26° 31°	93° 59°	95	1808.8
11.	Gauhati	Assam	26° 06°	91° 35°	54	1717.7
12.	Haflong	Assam	25° 10°	93° 01°	682	2285.3
13.	Imphal	Manipur	24° 46°	93° 54°	981	1353.1
14.	Kaliashar	Tripura	24° 19°	97° 00°	29	2540.0
15.	Kohima	Nagaland	25° 38°	94° 10°	1406	1856.0
16.	Lumbding	Assam	25° 45°	93° 11°	149	1233.2
17.	Majbat	Assam	26° 45°	92° 21°	120	1935.3
18.	North Lakhimpur	Assam	27° 14°	94° 07°	102	3258.5
19.	Pasighat Pradesh	Arunachal	28° 06°	95° 23°	157	4376.8
20.	Rangia	Assam	26° 26°	91° 37°	60	1929.6
21.	Shillong	Meghalaya	25° 39°	91° 53°	1598	2050.5
22.	Sibsagar	Assam	26° 59°	94° 38°	97	2323.5
23.	Silchar	Assam	24° 49°	92° 48°	29	3213.7
24.	Tangla	Assam	26° 39°	91° 55°	78	1916.5
25.	Tezpur	Assam	26° 37°	92° 47°	79	1768.3
26.	Tura	Meghalaya	25° 31°	90° 14°	370	3225.5
27.	Ziro	Arunachal Pradesh	27° 55°	94° 48°	1476	1739.9

REFERENCES

Ananthakrishnan R. and Pathan J.M. (1971). Rainfall patterns over India and the adjacent seas. IMD Sci Rept 144, Institute of Tropical Meteorology, Poona.

Bhide U.V., Mujumdar V.R., Ghanekar S.P., Paul D.K., Chen T.C. and Rao G.V. (1997). Diagnostic study of heat sources and moisture sinks in the monsoon trough area during active phases of the Indian summer monsoons. Tellus 49A: 455-473.

Das P.K. (1962). Mean vertical motion and non-adiablatic heat sources over India during the monsoons. Tellus: 213-220.

Dhar O.N. and Changrani T.G. (1966). A study of meteorological situations associated with major floods in Assam during the monsoon months. Ind J Metur Geophys 17: 111-118.

Dhar O.N. and Nandargi S.S. (1998). Floods in Indian rivers and their meteorological aspects. Memoir Geological Soc India, 41: 1-25.

Flohn H. (1968). Contributions to the meterology of Tibetan Highlands, Paper No 130 Dptt Atmos Sci Colorado State University.

Hingane L., Roopa Kumar K. and Raman Murthy B.H.V. (1985). Long-term trends of surface air temperature in India. J Climateol 5: 521-528.

Hussain Z. (1993). Variations in temperature and rainfall of Shillong during 1981-1990. In: Dynamics of a Tribal Society-A Micro-Study (Dutta Ray B and Agnihotri SK eds), Uppal Pub House, New Delhi, India p 28-40.

IMD (1970). Climatological Tables of Observations in India (1931-69).

IMD (1999). Climatological Tables of Observatories in India (1950-1990).

IPCC (1995). Climate Change 1995.

Joseph P.V. (1976). Climate Cycle in Monsoons and Cyclones 1871-1974. Proc Symp Tropical Monsoon (IITM Pune): 378-389.

Joseph P.V. and Raman P.L. (1966). Existence of low-level westerly jet stream over peninsular India during July. Ind J Meteor Geophys 17: 407-410.

Koteswarm P. (1958). Easterly jet stream in the tropics. Tellus 10: 43-57.

Koteswaram P. (1960). The Asian summer monsoon and general circulation over the tropics. In: Proc Symp Monsoons of World, IMD New Delhi, India p 105-110.

Krishnamurti T.N. and Bhalme H.N. (1976). Oscillations of a monsoon system: PT I observational aspects. J Atmos Sci 33: 1937-1954.

Lal M. and Subhash Chander (1993). Potential impact of greenhouse warming on the water resources of the Indian continent. J Environ Hydrology 1(3): 3-13.

Lal M., Cubasch U., Voss R. and Waszreitz J. (1995). Effect of transient increase in greenhouse and sulphate aerosols on monsoon climate. Curr Sci 60: 752-761.

Parthasarathy P., Munot A.A. and Kothawale D.R. (1995). Monthly and seasonal rainfall series for all India Homogenous Regions and meteorological sub-divisions. IITM Rs Rept No Rr 065.

Sikka D.R. (1980). Some aspects of the large-scale fluctuations of summer monsoon rainfall over India in relation to fluctuations in the planetary and regional scale circulation parameters. Proc Ind Aca Sci (E&P Sci) 89(2): 179-195.

Sikka D.R. and Gadgil S. (1978). Large-scale rainfall over India during the summer monsoon and its relation to lower and upper tropospheric vorticity. Ind J Meteor Geophys 29: 219-231.

Sikka D.R. and Gadgil S. (1980). On the maximum cloud zone and the ITCZ over Indian longitudes during the Southwest monsoon. Mon Wea Rev 108: 1840-1853.

Sikka D.R. and Pant G.B. (1991). Global climate change regional scenarios over India. In: Proc Indo-US Workshop (Abrol YP *et al.,* eds), Oxford and IBH Pub, New Delhi, India p 551-572.

Thapaliyal V. and Kulshrestha S.M. (1991). Climate changes and trends over India. Mausam 42: 333-338.

Yasunari T. (1980). A quasi-stationary appearance of a 30-40 day period in the cloudiness fluctuations during the summer monsoons over India. J Meteor Soc Japan 58: 225-229.

Chapter **8**

Collection and Conservation of Rice Germplasm in India: Status and Future Strategies for North-East India

A. Pattanayak, Janak Kumar Singh and B.K. Sarma
Division of Plant Breeding, ICAR Research Complex for NEH Region, Umaim - 793 103, Meghalaya

North-East India is considered to be the secondary centre of the origin of rice. An estimated 9650 land races of rice are present in the region. The diversity present in the rice germplasm from the region is enormous and has contributed immensely to the development of modern varieties. The study of the rice germplasm from the region indicates the prevalence of many desirable genes for yield improvement and stress tolerance. Several agencies have collected the germplasm from the region at various times and from various locations. A review of the collection and conservation efforts by various agencies and subsequent evaluation of the collected germplasm is presented here. With the advent of new technologies and IPR, a concerted effort is required for characterisation, registration and long-term conservation of the germplasm. Possible ways of utilising the genetic diversity have been suggested and areas that need further attention have been identified.

INTRODUCTION

Over the past few decades, the high-yielding varieties (HYVs including hybrids) of crops developed by the national and international agricultural research centres have quickly displaced the vast mosaic of traditional crop varieties. The new cultivars emerging from various agricultural research centres and private companies were uniform and very few in number, compared to the great diversity and number of landraces of folk varieties. Today, farmers grow landraces only in small pockets and many have disappeared forever in the wake of modern intensive agriculture, although some have been collected and stored in gene banks. The erosion of crop genetic diversity poses a serious threat to food supplies. The Food and Agriculture Organization (FAO) estimates that, since the beginning of this century, about 75 percent of the genetic diversity of agricultural crops has been lost.

There is, at present, no systematic information on the loss of genetic diversity in Indian agriculture, but the available data indicate an advanced state of genetic uniformity in crop plants. It is estimated that just a few decades ago, Indian farmers grew more than 30,000 different varieties of rice (Mooney, 1983), but in another 15 years, this enormous diversity will be reduced to no more than 50 varieties, with the top ten accounting for over three-quarters of the rice acreage. Unless the farmers replant their seeds, these outstanding varieties will be lost forever. Many of these "heirloom" varieties (seeds passed down from generation to generation) are still maintained by gardeners, farmers and tribes in isolated rural and tribal areas, and are often available in kitchen gardens, courtyards or home gardens. Today, due to rural-urban migration, tribal acculturation and extinction and constantly shrinking rural population, elderly farmers and seed savers often cannot find anyone who will continue growing their living "heirlooms". The rapid disappearance of many traditional varieties of grain and vegetable crops has been described as a botanical holocaust (Maheshwari, 1997).

The tribal-inhabited belt is more often the centre of domestic and genetic diversity of food crops, being maintained by peasants and subsistence farmers. These areas hold unique and important genetic materials, which should be identified, collected, characterised and properly documented. The genetic diversity is held by the tribes in their door-yard gardens, *baris* (land attached to their houses and huts), kitchen gardens and fields. The primitive cultivars, including the wild species grown by farmers, are valuable sources of genetic material for modern plant breeding. IR-72, a modern variety of rice (*Oryza sativa*), was developed by crossbreeding 22 landraces from 7 nations *viz.* India, Indonesia, China, Philippines, Vietnam, Thailand and Malaysia. In 1969, *O. nivara*, a wild rice from India, was discovered to resist grassy stunt virus (GSV). It was crossed and backcrossed with IR 24 three times to produce a variety resistant to GSV and with IR 24 grain qualities.

PLANT GENETIC RESOURCES AND INDIA

India is the seventh largest (32, 90,000 km^2) and the second most populous country in the world. Its wide agro-climatic, geophysical and ethnic diversity and its diverse ecological locations add to the confluence towards making India one of the megadiversity zones of the world. It has also got two hotspots of biodiversity, one being in the North-East and the other in Western Ghats. The country has been divided into 21 agro-ecological zones, which offer nearly every ecological situation in the world, and are separated by natural features and crop growing periods (Abrol, 1999). Each of these agro-ecological zones, is in turn comprised of myriad microhabitats. It is within this diversity of habitats that an amazing variety of crops and livestock have been developed, over the millennia in Indian farming.

At least 167 crop species and 320 wild varieties of crops have their primary, secondary or regional centres of diversity in the country. To give some examples, one species of rice (*Oryza sativa*) has diversified into at least, 50,000 distinct varieties. Ethnic diversity has played a major role in the diversification of crop genetic resources in the country. The ancestors of many of our crop plants and landraces or folk varieties are interwoven with tribal cultures (Maheshwari, 1991; Khoshoo and Sharma, 1991; Paroda and Arora, 1991).

RICE GERMPLASM IN INDIA

Rice has originated from South and South-East Asia, of which India contributes a major part. In India, the North-Eastern Hills are considered to be the secondary centres of the origin of rice. It is believed that the woman folk in Arunachal Pradesh first domesticated rice. The Asian cultivated rice has evolved into three distinct eco-geographic races namely, Indica, Japonica and Javanica and three distinct cultural types, *viz.* upland, lowland and deep water, in areas where the crop is monsoon dependent. The crop then evolved over thousand of years, as the peasants and farmers selected different types to suit local cultivation practices and needs. As a result of domestication, the crop underwent many changes, morphologically and physiologically. Characters of agronomic importance like tillering, leaf size, culm length and thickness, panicle size, number of secondary panicles, grain weight and harvest index increased. Consequently, characters like pigmentation, rhizome formation, leaf interval, awaning, shattering, seed dormancy, photoperiod sensitivity, sensitivity to low temperature and frequency of cross pollination were reduced.

Weedy rice are found in directly seeded rice fields in certain regions. These might have originated from hybrids between wild and cultivated rice. The weedy population contains great gene diversity and seems to serve as a gene reservoir for cultivars. India also has abundant resources of wild rice, particularly *O. nivera, O. rufipogon, O. officinalis* and *O. granulate*. *Oryza nivera* is found in the Koraput area/Jeypore tract and Simplipal hills of Orissa, Raigarh and Raipur districts of M.P., Southern part of Bihar and Bengal, North-Eastern region and Eastern U.P. *Oryza rufipogon* is prevalent in the marshy areas of Southern West Bengal and the adjoining parts of Orissa, the North-Eastern region and the Western Ghats. *Oryza officinalis* and *O. granulate* are prevalent in the Simlipal hills of Orissa. The full spectrum of rice germplasm in the country includes:

- Germplasm developed through ages of domestication, isolation and natural selection.
- Wild *Oryza* species and related genera.
- Natural hybrids between cultigen and wild relatives and primitive cultivars of the cultigen in areas of rice diversity.
- Germplasm generated in the breeding programmes, farmer's varieties, mutants, polyploids, intergeneric and interspecific hybrids, somaclones and transgenics.
- Commercial types, obsolete varieties, minor varieties and special purpose types in the major centers of cultivation.

The status of all India rice germplasm collection is collated in Table 8.1.

Table 8.1: Rice germplasm collections in India*

S.No.	Agency/Area	Year	Collection
1.	Base collection from Coimbatore (AC)	1911 – 47	2000
2.	Jeypore Botanical Survey (JBS)	1955 – 59	1745
3.	Manipur Collection (MNP)	1965 – 70	904
4.	West Bengal Collection Chinsura, Bankura and Kalimpong	1960 – 71	5550
5.	Assam Rice Collection (ARC)	1968 – 71	6630
6.	Dr. Richharia's Collection from M.P.	1971 – 81	19000
7.	Collections from various states (Upland)	1976 – 77	1431
8.	National collaborative program (NCS)	1978 – 80	6439
9.	U.P. Hill collection by Pantnagar University and VPKAS, Almora	1978 – 90	2580
10.	Lowland and saline types from Orissa	1980 – 82	300
11.	Collections by NBPGR from different parts of India	1976 – 84	8353
12.	Collections by P.A.U., Kapurthala	1982 – 84	1178
13.	Collections by H.A.U., Kaul	1982 – 84	960
14.	Collections by NBPGR/CRRI from Goa, Gujarat, Karnataka, Sikkim, Mizoram, Arunachal Pradesh and Orissa.	1984 – 88	3092
15.	Collections from Rajasthan and Orissa	1980 – 90	3489
16.	Collections by NBPGR from Orissa, W.B.	1986 – 95	3122
17.	Collections by CRRI from Orissa	1990 – 95	717
18.	Wild rice collection by NBPGR/IRRI/ IGKVV/CRRI	1984 – 97	840
19.	Collections of ICAR Complex	1999 – 2002	1000
	TOTAL		67,760

* Basic source: Report of the Rice Germplasm Committee, Jan. 27, 1999, NBPGR, New Delhi

STATUS OF RICE GERMPLASM IN NORTH EAST INDIA

The North-Eastern region of India is considered to be one of the richest pockets of rice germplasm in the world. Of about the 40,000 rice germplasm collected from India, approximately 9650 come from this region. Wide variations in topography and physiography, coupled with favourable agroclimatic conditions (wet tropical), provide a conducive atmosphere for rice. Poor accessibility has helped in preserving this diversity.

In addition, the region includes areas (parts of Arunachal Pradesh) which are considered to be the centres of the origin of rice cultivation.

1. Collection

The first systematic effort to collect rice germplasm from the North-Eastern region was supported jointly by the Indian Council of Agricultural Research and PL – 480 of USAID. In 1967, the Indian Agricultural Institute under the leadership of Dr. M. S. Swaminathan, initiated the project and collections were made till 1972. The collection was exhaustive from Arunachal Pradesh, Nagaland, Manipur and Meghalaya. Tripura and Assam were partially covered and Mizoram was omitted. Collections were both direct (from the field) and indirect (through extension workers and village markets). During the period of 5 years, a total of 6630 land races were collected and accessions were given the prefix ARC - Assam Rice Collection (Sharma *et al.*, 1971). Occasional reports of rice collections from Sikkim, Nagaland, Arunachal Pradesh, Tripura and Mizoram were made by various workers (Asthana and Mazumdar, 1981; Sasikumar and Sardana, 1987). The Shillong station of NBPGR had under taken explorations of Assam, Arunachal Pradesh, Meghalaya, Nagaland, Mizoram and Tripura from 1986-1999 and collected about 1350 germplasm of rice.

The biggest effort, however, to collect plant biodiversity in general, and rice biodiversity in particular, from the North-Eastern region started only very recently. The Indian Council of Agricultural Research launched a new project "Sustainable Management of Plant Biodiversity" under the World Bank aided, "National Agricultural Technology Project". Under this scheme, an ambitious project on the collection and evaluation of crop diversity from NE India was started in 1999, in nine centres of NE India. The major responsibility of collecting rice biodiversity was given to the ICAR Research Complex for the NEH Region. The complex carried out 14 trips to various parts of Arunachal Pradesh, Assam (including Majuli), Manipur, Meghalaya, Mizoram, Nagaland, Sikkim and Darjeeling district of West Bengal and collected more than 1000 germplasms of rice. Areas covered and statewise collections are given in Table 8.2.

Table 8.2: Collection of rice germplasm by ICAR Research Complex from 1999–2002

	1999-2000	*2000-01*	*2001-02*	*2002-03*	*Total*
Arunchal Pradesh					
West Kameng			05		
East siang			05		
Changlang			54		
Tirap			21		
Assam					
Cachar	09				
Jorhat				40	
Barpeta			22		
Bongaigoan			38		
Majuli		95			
Manipur					
Imphal East		02	20		
Thoubal			22		
Churchandpur			03		
Bishnupur			06		
Meghalaya					
Ribhoi	04	16	04		
West Garo Hill	21	23			
East Khasi hill					
Jaintia hill			17		
Mizoram					
Kolasib	67				
Chhimtipui	10				
Champhai		21			
Serchhip		11			
Lunglei		17			
Nagaland					
Kohima		06			
West Bengal					
Darjeeling			15		
Sikkim					
East Sikkim			15		
West Sikkim			30		
South Sikkim			03		
North Sikkim			02		
Collections by ICAR Complex outside NATP					377
	111	191	282	40	1001

2. Characterisation

Varietal diversity present in earlier collections (mainly ARC series) may be summarised as follows:

The rice genotypes grown in Tawang area of Arunachal Pradesh (3000m msl) exhibit Japonica characters of thin calm, globose grain, narrow and dark green leaves, drooping flag leaf and high photosensitivity. The frequency of Japonica types gradually decreased towards the eastern side. The pubescent husk types occur frequently in the western parts of Arunachal Pradesh, while the glabrous type was more frequent in the eastern side. About 40% of the *jhum* cultivars from Nagaland were intermediate between Indica and Japonica types. Nagaland is probably the centre of distribution of glutinous and glabrous varieties because of the occurrence of the highest percentage of such cultivars. The hill collections from Manipur were similar to the Nagaland type, while the valley varieties were similar to the Assam plains. The cultivars collected from Tripura were also intermediate between Indica and Japonica type, but most of the genotypes were glutinous types and many were fine grain aromatic types. The majority of the genotypes collected from the Garo Hills, showed Japonica characters. Glabrous husk character was more prominent in the Khasi and Jaintia Hills varieties. A large number of collections [like Taichung (Native) 1] from Karbi Anglong and the North Kacachar Hills of Assam, possessed dwarf characters. The presence of waxy characters in many collections of Assam hills is a noteworthy features.

Only a part (25%) of the genotypes collected after 1999 have been characterised. Variability in collections was mainly of the Indica types – tall, medium-coarse grained, maturing in about 4 months (Sarma *et al.*, 2000). The change in genotype constitution may be attributed to the migration of people from the plains of Bangladesh with their own varieties. Variations were observed in panicle types, awn character, glume colour, grain character and kernel colour. Significant characters found in these genotypes include earliness suited to low/ medium elevation, fine grain, scented/aromatic grain, red kernel, disease and pest resistance, and water logging resistance.

3. Evaluation

Evaluation of germplasm from NE India was done mainly with the ARC series collection, and the collections were made by the ICAR Research Complex. One hundred and ninety land races from the ARC series were studied for genetic divergence and were classified into 9 clusters (Vairavan *et al.*, 1973). The study showed that some clusters were close to Japonica plant and grain type whereas others were intermediate between Indica and Japonica. Srivastava (1978) grouped the rice collections from NE India into five groups. Some land races with glaberima-like characters were also reported (Seetharaman *et al.*, 1976). These studies strengthen the observation that racial differentiation in *O. sativa* took place in this region (Kihara, 1968).

Various workers have studied the Assam Rice Collection in great detail. The collection possess many desirable characters *viz.*, dwarfness, tolerance to lodging, low temperature, iron toxicity, low phosphorus and resistance to gall midge, stem borer, green leafhopper, blast, bacterial blight, tungro virus and bacterial leaf streak (Banerjee, 1971; Chattarjee *et al.*, 1977; Devadath and Rao, 1978; Balakrishna Rao 1978). Studies indicated that the ARC collections possess many genes for resistance to various insect-pests and diseases, and the resistant types were in greater frequency in the ARC collection than in other stocks (Table 8.3). Some of the land races also possessed high yield potential, particularly under low phosphorus and low temperature (Prasad Rao and Srinivasan, 1980). Apart from this, some of the Japonica grain type and semi-dwarf types of Indica showed higher protein content (Srivastava and Nanda, 1977).

Table 8.3: Number of accessions resistant to diseases and pests in AC, JBS, ARC and ICAR Research Complex collections

Disease/pest	*AC*	*JBS*	*ARC*	*ICAR Research Complex*
Blast	12	19	250	48
Helminthosporium	10	-	231	92
Bacterial leaf blight	4	2	25	-
Stem borer	6	-	4	15
Gallmidge	9	12	23	7
Brown plant hopper	1	2	167	-
Green leaf hopper	-	-	52	-
Tungro virus	1	2	9	-
Total collections	5619	1568	6630	162

The collections of rice germplasm made by the ICAR Research Complex from the North-Eastern Hills Region, from 1976 – 1980, have been studied systematically and catalogued (Asthana and Mazumder, 1981). Variability present in this collection in 56 characters was studied and promising selections were made for desirable traits. Twenty two groups were identified, based on distinguishing characters, with the help of a metroglyph. The cultures were divided into 33 major groups, based on their morphological characters. These cultures were further classified into 9 groups and 49 sub-groups on the basis of pigmentation. The studies based on morphological, genetic and some chemical characters indicated that a large number of the germplasm were intermediate type between Japonica and Indica. Cultures with desirable characters are collated in Table 8.4.

From the collections of the NE Hills, Ngoba was identified as a new alternate source of dwarfing gene (Ghosh *et al.*, 1981). Wide range of variability was reported for many characters in 137 cultivars evaluated at the Tripura Centre (Asthana and Mazumdar, 1981). Sporadic cultivation of glabrous husk and glutinous forms occurs in the Garo Hills and areas adjacent to the Myanmar border. Collections from Mizoram possess aromatic types of glutinous and non-glutinous forms. A number of high altitude germplasm were evaluated for tolerance to low temperature, at the vegetative and reproductive phases, and it has been found that some of them possess tolerance to low temperature at both the stages, whereas others exhibit tolerance at the reproductive stage, only

(Table 8.4). These collections show variability for other characters which were associated with cold tolerance (Gupta *et al.,* 1995).

Table 8.4: Rice collections with desirable traits*

Characters	*Cultivars*
Tolerance to low temperature at vegetative stage(3)	Thebere, Ryllo Red-4, Ngoba, Namyi, Khonorullo.
Tolerance to low temperature at reproductive stage(3)	Khonorullo, Kbathangmaw, Nonglwai, Thebere, Naymi.
Good panicle exertion (1)	Abor(B), Khonorullo, Nonglwai, Kbathangmaw, Dullo-10
High-spike fertility at low temperature	Kbathangmaw, Nonglwai, Khonorullo, Ryllo Red-4, Abor(B), Dullo-10.
Tolerance to lodging	Manpa, IC 25724, Moping, Imoenkhe, Rangagellang, Bethi.
Glutinous (Waxy)	IC25682, Tura 364, Nemo (8)(Leiletbuhbam), Lynsi, Pawnbuh, Asiankel, Betlong.
Higher yield in lowland	Ngoba, Khawnowjoma, Chao, Tura 113
Dwarfness	Chanmouri,Tura 299, Chinairi, Rangellong, AR73, Kangpui.
Drought tolerance	Amiong, Addy, Changpalman, Pyare
Tolerance to iron toxicity	Nemo, Amo, Leiletbuhbam, Panghkla
Aroma	Pyapon, Horipuri, Moyatsuk.

* Source : Gupta *et al.* (1995)

Out of the 1000 germplasm collected under NATP by the ICAR Research Complex for NEH Region, only 250 have been characterised so far. These germplasm were selected to serve as representatives of the entire collection spread across eight states and to study the changes in the composition of the cultivars. For the first time in the region, the collection also included some wild rice from Assam. Initial morphological and biochemical characterisation showed that the collection included many sticky varieties of Japonica rice (Table 8.5). The representative set had a good proportion of aromatic rice (15%) and deep water rice (15%). Black kernel rice was also collected. The black kernel colour was alcohol extractable.

Table 8.5: Variability in rice germplasm collections from the NEH region

UPLAND RICE

Character	Range
Flag leaf length	20.9-70.1 cm.
Flag leaf breadth	1.12-3.24 cm.
Days to 50% flowering	77-124 days
Leaf length	23.2-75.1cm
Leaf width	1.04-2.08cm
Ear bearing tillers/plant	5-12
Plant height	28-134cm.
Panicle length	16-29cm
Branches of panicle	5.2-16.2
Yield/plant	0.7-28.6gm.
100 Seed weight	1.2-3.3gm.
Leaf blast (0-9 scale)	3-9

LOWLAND RICE

Character	*Range*
Flag leaf length	19.5 - 40.6cm.
Flag leaf breadth	0.48 - 1.98cm.
Leaf length	23.3 - 55.8cm
Leaf width	0.7 - 1.8cm
No. of leafes	3.2 - 5.8
Plant height	73.0 - 145.7cm.
Panicle length	15.6 - 26.5cm
Ear bearing tiller	4.0 - 16.0
Branches of panicle	7.6 - 13.2
Panicle weight	0.9 - 5.7gm.
Yield/plant	1.1 - 20.5gm.
Leaf blast (0-9 scale)	3 - 9

CONSERVATION OF RICE GERMPLASM FROM NORTH-EAST INDIA

Germplasm maintenance of rice must take care of viability, as well as maintenance of purity. Cold and dry storage conditions are necessary to prolong the viability of seeds. Temperatures around 15–18°C and RH around 30–40% are generally prescribed for better storage. For medium term storage (5–10 years), the storage conditions followed at IRRI is: 8–9% seed moisture, 3-4°C temperature and 60–75% RH. For long-term storage, the use of vacuum cans and below freezing temperature are closer to ideal conditions (Chang, 1976). Recently, long-term storage facilities have been created for the preservation of genetic resources, at the National level at NBPGR, New Delhi.

In the absence of cold storage facilities in the North-East region, entire rice germplasm collections require to be grown afresh every year with adequate precautions to maintain viability and purity of the accessions. Special attention is to be paid to minimise natural cross-pollination, contamination and erroneous labelling. Maintenance of seed files of original seed samples, duplicate records, nursery maps and field maps are also equally important. Recently, two medium-term storage modules, one at the ICAR Research Complex and one at the NBPGR regional station, have been installed to facilitate storage of the germplasm.

A big gap is often felt regarding conservation planning and processes in NE India. A concerted conservation programme of rice of the region should be launched to bridge the gap. Collections of the ARC series were initially maintained *ex situ* at CRRI and a set was sent to the International Rice Research Institute (IRRI) for long-term conservation. But, because of lack of long-term storage facility, a large numbers of the germplasm were lost. However, recently the IRRI has sent back seeds of the entire set of the ARC series to India. Till recently, there was no concerted effort to catalogue and conserve rice germplasm in the North-Eastern region. Germplasm collected by different centres are still maintained by the concerned centres and used for breeding purposes. Very recently, under the NATP, some centres have been identified for *ex situ* conservation. However, more sub-centres of localised importance could be started, each taking care of a particular type. The conservation process could be initiated at each sub-centre, diverting the reserve collections to the main centre for long-term preservation. Eventually, the necessary facilities for long-term preservation need to be available only at the main centre. The NBPGR is now conserving all accessions collected under the NATP for a long term. It has also taken steps to collect all pre NATP rice germplasm at one place, catalogue all the germplasm (pre-existing and collected under NATP) and store them under long-term conservation. This type of decentralised short-term and centralised long-term conservation strategy would result in the short-term and long-term utility of the germplasm.

UTILISATION OF GERMPLASM FOR INCREASING PRODUCTIVITY

As discussed already, the germplasm of rice collected from the NEH Region possess numerous valuable gene(s). Some of the cultures have been further screened at the ICAR Research Complex for various biotic and abiotic stresses prevailing in the region. Blast (leaf and neck) is a major problem in low and mid-altitude areas, whereas sheath rot causes significant damage in the higher hills. Screening of local entries indicate that Kerpu, Poppy, IC 25705, Chomju, Khamang are highly resistant to blast. Lyngsi, Ryllo red, Tangale, Namyi and Thebek were found resistant to sheath rot.

Screening against abiotic stresses, like cold (at higher altitude), iron toxicity and phosphorus deficiency showed that Khonorullo, Ryllo Red 6, Nyami, Abor(B), Meghalaya 1 are tolerant to low temperature at the reproductive phase. Mirikrak, Kbathangmaw, Pawnbuh, BG 34-8 were found tolerant to P-deficiency.

On the basis of screening for various stresses, promising cultures like Mirikrak and Khonorullo were hybridised with elite varieties IR 26, IR 29, Pusa–33, Jaya and Rasi (IET 1444) for evolving suitable varieties for mid and high altitude areas. From the above crosses RCPL 1-87-8 and RCPL 1-87-4 have been identified as promising cultures for the mid-altitudes and RCPL 1-6C, RCPL 1-1C, RCPL1-11C and RCPL1-12C for the higher altitudes (Pattanayak and Gupta 2000). Four varieties, NEH Megha Rice-1 (RCPL 1-6C), NEH Megha Rice – 2 (RCPL 1-1C), Shah Sarang 1 (RCPL1-87-8) and Lumpnah 1 (RCPL1-87-4) have already been released by the State Seed Sub-Committee of Meghalaya for commercial cultivation in Meghalaya. In addition, for boro crop in low altitude areas, a variety, TRC Bordhan 1 (TRC 246 C-10), was developed by utilising Thiumpu, a local cultivar. The Central Varietal Committee has also recommended it for release. In Manipur, utilising a local cultivar for pre-kharif and kharif, three advanced cultures were developed *viz.* RCM 5, RCM 6 and RCM 7. Apart from these, Ngoba, Nonglwai, Kbathangmaw, Meghalaya 1, Pinini, Yamuk, Bali, Nami and Namyi, possessing a diverse genetic background for evolving high yielding varieties suitable for various stress(s) prevailing in the region have, been used in conjunction with elite varieties of the IRRI.

FUTURE STRATEGIES

To achieve success in the mission of rice germplasm collection, evaluation and conservation, the creation of a sound system is required in which public sector, private sector, the NGOs and other stake holders work in a partnership mode. In addition, uniformity in generating information, documentation and creating a system of easy retrieval of information is necessary. Efforts towards collection and conservation of rice germplasm from the region have so far remained periodic and isolated. Agencies responsible for collection had no collaboration until the introduction of the NATP in 1999. The following areas need attention:

1. Collection yet to be made

Specific collections of glutinous or waxy rice and autumn or ahu rice genetic resources are yet to be made from the entire region. Glutinous rice is preferred as a staple food in all the hill states of the region. The glutinous rice germplasm collection (158 accessions), maintained at RARS, Karimganj and Titabar, do not include the materials from Manipur and other NE hills states. *Ahu* rice is one of several agricultural classes of rice grown in the Northeast. Its cultivation starts with the pre-monsoon showers in March – April and the crop is harvested in June – July. Ahu rice is grown in hilly areas as jhum, and in drought affected upland areas as a dry broadcast crop or a wet-broadcast crop, and in certain deep-water situations as a mixed crop with deep-water rice. Each of these growing situations certainly have specific types of varieties. Further diversity in *ahu* and other classes of rice has resulted from differential preference for varieties, even in similar situations, by different ethnic tribes and communities of the NEH region. The indigenous *ahu* rice varieties from the hilly regions of NE India are almost un-represented in accessions of *ahu* rice collection maintained by RARS, Karimganj and Titabar (Anonymous 2001). Besides, parts of Tripura and major parts of Sikkim and Mizoram are yet to be covered. Since rice germplasm collection from different parts of the NEH region is yet to be completed, there is scope for further collection of glutinous, *ahu* and other classes of rice from this region. A list of unexplored areas from where some useful germplasm may be obtained is given in Table 8.6. Wild rice collections from the North-Eastern region are far from complete. Isolated efforts made by the NBPGR Regional Station, Shillong, RARS – Karimganj of AAU and CAU, Imphal led to some collections of wild species (mainly *Oryza rufipogon* and *Zizania latifolia*), but these are not available with the National Gene Bank.

Table 8.6: List of areas in NE states from where useful germplasm are expected

State	*Area*	*Specific characters*
Assam	Barak valley, Dhemaji, Lakhimpur and Goalpara districts	Resistance to water logging, stem borer, BPH, BLB, Gall midge
Arunachal Pradesh	Upper Subansiri, South Siang districts, Dibang valley	Cold tolerance
Mizoram	West Aizal district, Blue Mountain	Drought tolerant upland type
Meghalaya	Ranikor, Maheskhola, Balpakhram Reserve Forest	Cold tolerance, stem borer resistance, waxy, high protein
Manipur	Tamelong and Chandel districts	Black kernel aromatic type, gall midge resistance
Nagaland	Wokha, Mon and Tuensang districts	Bulu type, aroma, black kernel, cold tolerance
Sikkim	All districts	Hill type
Tripura	West Tripura	Drought tolerance, stem borer resistance, gall midge resistance

2. Germplasm Conservation

The choice of conservation strategy should depend on the nature of the material, its reproduction/regeneration method and ecological status. Both *in situ* and *ex situ* conservation methods may be used, but priority should be given for developing sites for *in situ* conservation. The region is highly diverse in its climatic condition and, therefore, cultivars/land races are highly location specific. Such germplasm is not a good candidate for *ex situ* conservation, unique locations may be identified for developing sites for *in situ* conservation, and areas with more diversity should get priority. The glutinous round and red rice of the region holds specific preferences by the natives and is traditionally grown in abundance. With the introduction of modern varieties, such traditional cultivars are going out of cultivation to meet the ever-increasing food demands. It is imperative to systematically monitor the drift in the cultivation of particular land races over time. This also holds good for the traditional rice cultivars grown in the Apatani plateau, wet rice cultivation systems of Nagaland and Sikkim and other traditional farming systems practiced by various ethnic tribes of the region.

To sustain the rich genetic diversity of rice in India in general and its North Eastern part in particular, efforts should be made to conserve all genotypes at the gene bank and to maintain the materials under active collection at partner locations. Molecular classification of the germplasm and quality analysis for its effective use in breeding are the other necessary components for the characterisation of rice germplasm.

3. Awareness for Conservation

The North-Eastern region is rich in biodiversity and the civilisation is still biodiversity based. The people are still enviably rich in the knowledge and use of their living resources. With the spread of new technologies, the regions biodiversity and ethnic knowledge base for fruitful use of this diversity is under threat. Faced with such a situation, a campaign to prepare peoples' biodiversity register, especially on rice biodiversity, is of paramount importance. Hence, an organisation of regional awareness-cum-training is urgently needed. Emphasis has to be laid on aspects like careful planning of onsite conservation, sustainable use and restoration of eroded sites and documentation of indigenous knowledge for the judicious use of these living resources.

REFERENCES

Abrol Y.P. and Gadgil S. (1999). Rice in variable climate. APC Publications, New Delhi, India p 243.

Anonymous (2001). Annual Report 2001. ICAR Research Complex for the NEH Region, Umiam, Meghalaya, India p 123.

Asthana A.N. and Majumdar N.D. (1981). Studies in Rice Germplasm of the North Eastern Hill Region, Research Bulletin No 11, ICAR Research Complex for the NEH Region, Shillong, India.

Balakrishna Rao M.J. (1978). Recent developments in breeding approaches for varietal improvements in rice. In: Papers Natl Symp on Increasing Rice Yield in Kharif, Central Rice Research Institute, Cuttack p 5.

Banerjee S.N. (1971). Recent progress in rice insect research in India. Trop Agril Res Ser 5: 83-97.

Chang T.T. (1976). Exploitation of useful gene pools in rice through conservation and evaluation. SABRO Journal 8: 11-16.

Chatterjee S.M., Prasad K., Misra B.C. and Rajamani R. (1977). Identification of gallmidge (*Orseolia oryzae* Wood-Mason) resistant rice germplasm and their utilisation in breeding. J Entomol Res 1: 111-113.

Devadath S. and Seshagiri Rao C. (1978). Identification of sources of resistance to bacterial blight and bacterial leaf streak of rice. Oryza 15: 93-95.

Ghosh A.K., Sardana S., Bhattacharya N.R. and Asthana A.N. (1981). New non-allelic semi-dwarfs in rice (*Oryza sativa* L). Oryza 18: 214-216.

Gupta H.S., Sarma B.K., Pandey D.K. and Prasad R.N. (1995). Collection, characterisation, evaluation and conservation of rice germplasm in the North Eastern Hills of India. Indian J Hill Farmg 8: 11-25.

Khoshoo T.N. and Sharma M. (1991). Indian-Biosphere: Some aspects. Har-Anand Publications/ Vikas Publishing House Pvt Ltd, New Delhi.

Kihara H. (1968). History of biological and other sciences in Japan in retrospect. Proc Int Cong Genet 3: 49.

Maheswari J.K. (1991). Seeds of ethnobotanical importance. In : Trends in Seed Research (Mukherjee KG ed), Today and Tomorrow Printers and Publishers, New Delhi, India.

Maheswari J.K. (1997). Maintenance and conservation of "Heirloom" varieties in Indian agroecosystems. International Development Research Centre, Ottawa, Canada (Source: http//www.idrc.ca/books/focus/833/maheshwa.html).

Mooney P.R. (1983). The law of the seed: Another development and plant genetic resources. Dev Dialogue ½: 173.

Paroda R.S. and Arora R.K. (1991). Plant genetic resources conservation and management. IBPGR Regional Office, New Delhi.

Pattanayak A. and Gupta H.S. (2000). Evaluation of anther culture-derived lines under upland conditions of the North Eastern Hills of India. IRRN 25: 9-10.

Prasad Rao U. and Srinivasan T.E. (1980). Suitability of dwarf rice varieties from North East India to different sowings in one year. Indian J Agril Sci 50: 402-404.

Sarma B.K., Singh J.K., Verma D.K., Annadurai, Pattanayak A. and Laxminarayana K. (2000). Collection of crop diversity from North East India. J Assam Sci Soc 4: 326-335.

Sasikumar B. and Sradana S. (1987). Genetic variability in cold tolerant genotypes of rice. Oryza 24: 119-122.

Seetharaman R., Srivastava D.P. and Ghorai D.P. (1976). Preliminary studies in rice cultivars from North East India. Indian J Genet Plant Breed 34: 143-149.

Sharma S.D., Vellanki J.M.R., Hakim K.L. and Singh R.K. (1971). Primitive and current cultivars of rice in Assam – a rich source of valuable genes. Curr Sci 40: 126-128.

Srivastava D.P. (1978). Diversity in the late duration rice cultivars from North East India. Oryza 15: 26-33.

Srivastava D.P. and Nanda B.B. (1977). Variations in grain protein in some groups of rice varieties from the collection of North East India. Oryza 14: 45-46.

Vairavan S., Siddiq E.A., Arunachalam V. and Swaminathan M.S. (1973). A study of the nature of genetic divergence in rice from Assam and North East Himalayas. Theor Appl Genet 43: 213-221.

Chapter 9

Inland Indigenous Ornamental Fish Germplasm of North Eastern India: Status and Future Strategies

B. K. Mahapatra, K. Vinod and B. K. Mandal
Division of Fisheries, ICAR Research Complex for NEH Region, Umroi Road, Umiam - 793 103, Meghalaya

Owing to the diversity of topographic and climatic features of North-Eastern India, the aquatic resources are rich in endemic fish germplasm. The varied freshwater resources harbour 274 fish species, out of the 806 freshwater fish reported from India. Good growing, larger size fish like *Tor tor* (Hamilton), *Tor puttitora* (Hamilton), *Semiplotus semiplotus* (Mc Clelland) and *Neolissocheilus hexagonolepis* (Mc Clelland) are already identified as a major protein food item from the aquatic habitat and attention has been paid to develop their fishery. But most of the small food fish, which are treated as unwanted for conventional farming have good potency as ornamental fish and are popularly known as "Indian Aquarium Fish". Some of them are *Brachydanio rerio* (Hamilton), *Danio dangila* (Hamilton), *Botia dario* (Hamilton), *Schistura sikmaeinsis* (Hora), etc. Even some large food fish like *Channa marulius* (Hamilton), *Labeo calbasu* (Hamilton), *Neolissocheilus hexagonolepis* (Mc Clelland) and *Notopterus chitala* (Hamilton) are reared in aquaria in their young stage. Studies indicate that about 250 local fish species having ornamental value, have attracted hobbyists, both locally and globally. Out of the total 250 native ornamental fish species of the region, the highest number of ornamental fish species have been recorded from Assam (187), followed by Arunachal Pradesh (165), Meghalaya (159), Manipur (139), Tripura (103), Nagaland (71), Mizoram (46) and Sikkim (29). Presently, about 32% native ornamental fish species of the region, which are marketed overseas as Indian aquarium fish are included in the export list. These fish species are collected, by some unregistered traders from their natural habitat and supplied to the exporters based in Kolkata. Conservation status of native ornamental fishes have shown that out of the 250 species, 10 are Critically Endangered (CR), 28 are Endangered (EN), 49 are Vulnerable (VU), 45 are Lower Risk Near Threatened (LRnt), 08 are Lower Risk Least Concern (LRlc), 03 are Data Deficient (DD) and 107 are Not Evaluated (NE). At present, most of the species occur rarely. For sustainable growth of ornamental fisheries in the North Eastern States, the native ornamental fish can be reared and bred in captivity and this vocation will generate jobs in rural areas. Rearing and breeding of threatened fish species will also be conducive to conservation practice.

INTRODUCTION

North-Eastern India comprising of Arunachal Pradesh, Assam, Manipur, Meghalaya, Mizoram, Nagaland, Sikkim and Tripura is gifted with vast aquatic resources in the form of rivers, streams, lakes, reservoirs, floodplain wetlands, ponds and large areas under rice-fish culture system. The region which lies between 21.50° to 29.50° North latitudes and 85.50° to 97.50° East longitudes and is bounded on the north by China and Bhutan, on the west by Bangladesh and on the east by Myanmar has an elevation of 15 to 5000 m from the sea level. The rivers, Brahmaputra and Barak, with their numerous tributaries flowing through the different states along with myriads of rivulets and lentic water bodies, which harbour diversified fish fauna, form the principal drainage of NE India. Owing to the diversity of topographic and climatic features of North-Eastern India, the aquatic resources are rich in fish germplasm. The varied freshwater resources of the region harbour 274 fish species, out of the 806 freshwater fishes reported in India (Mahapatra *et al.*, 2003b). The larger sized fish species like the *Tor tor*, *Tor puttitora* and *Semiplotus semiplotus* have already been identified as good table fish (Ghosh and Lipton, 1982). Considering their commercial importance, their fishery has been developed to some extent. But there are many small fish species, which are not important for commercial farming due to their poor growth rate. These small food fish, treated as unwanted for conventional farming, have good potency as ornamental fish and are popularly known as "Indian Aquarium Fish". The fish species are popular among the hobbyists and ornamental fish traders due to their varied shapes, sizes, colour, finnage, etc. (Mahapatra, 1999; Dutta, 2000; Mahapatra, *et al.*, 2001, 2003c, 2003d; Vinod *et al.*, 2001; Radhakrishnan and Kurup, 2002; Ghosh *et al.*, 2002).

A. ORNAMENTAL FISHERY – NEW AREA OF FISHERIES

The varied forms and fascinating beauty of some fish have attracted people from times immemorial and these are named as 'ornamental fish'. In China and Japan, gold fish and koi carp have been used as an ornamental fish since long. Aquarium keeping of fish began in 1805 and the 'first public aquarium' was opened at Regent's Park in England in 1853. Thereafter, aquaria keeping picked up further and by 1928, 45 display aquaria were opened, with over 500 of public aquaria presently functioning worldwide. However, the global market for ornamental fish for public aquaria is less than 1%, at present and over 99% of the market for ornamental fish continues to be confined to hobbyists (Swain *et al.*, 2003). Ornamental fish keeping, which started as a hobby, has now turned out to be a commercial aqua-business and has taken the shape of an industry (Mahapatra *et al.*, 1999, 2000; Mahapatra, 2003; Ghosh *et al.*, 2003). The current value of the entire industry has been estimated at US $ 15,000 million (Bartley, 2000). The South-East Asian countries contribute about 69% of the total world production. Singapore is at the top with 35% of the total production of the South-East Asian countries, followed by Hong Kong, Malaysia, Thailand, Philippines, Taiwan and

Japan (Anon, 1995). India's share in ornamental fish trade is estimated to be US $ 0.1 million. Considering the enormous and diverse indigenous fish resources of the country in general, and the North-East in particular, there is immense scope for her to become a potential candidate and a strong competitor in the international trade (Ghosh and Mahapatra, 1999; Ghosh *et al.*, 2002). Indian domestic trade in this area is growing annually at the rate of 20%, and the demand at the domestic level is higher than the supply.

B. STATUS OF ORNAMENTAL FISH GERMPLASM IN NE INDIA

North-Eastern India, with its diversified freshwater resources, harbours diversified native ornamental fish species. These include both classified and non-classified types of aquarium fish (Mahapatra *et al.*, 2003a). The small fish like *Botia dario* (maximum size 7.5 cm), *Danio dangila* (maximum size 15 cm), *Puntius shalynius* (maximum size 6 cm) and *Schistura reticulofasciatus* (maximum size 5 cm) are classified types of ornamental fish, which can be reared in an aquarium throughout their life span. On the other hand, some larger food fish like *Neolissocheilus hexagonolepis* (maximum size 274.5 cm), *Labeo gonius* (maximum size 61 cm), *Channa marulius* (maximum size 120 cm) and *Rita rita* (maximum size 60 cm) are also now treated as ornamental fish in their juvenile stage and termed as non-classified ornamental fish.

1. Collection and Evaluation

The systematic effort to collect ornamental fish germplasm from the North-Eastern region started only very recently. The Indian Council of Agricultural Research (ICAR) launched a new project, "Evaluation of Ornamental Fishery Potential in NEH Region – Its Farming and Export Feasibility" under the World Bank aided, "National Agricultural Technology Project". The responsibility was given to the Fisheries Division of the ICAR Research Complex for the NEH Region, Barapani. Random rapid field surveys were conducted during 2001-2003 in different hill states of the region viz., Arunachal Pradesh, Assam, Manipur, Meghalaya, Mizoram, Nagaland, Sikkim and Tripura. The survey covered important river systems like the Brahmaputra, Barak, Kameng, Subansiri, Siang, Lohit, Tizu, Umiam, Jinjiram, Gumti and Teestha and other water bodies, including streams, lakes, reservoirs, beels, ponds and paddy fields. The representative fish species were collected from the landing centres as well as from the respective local markets for further identification and characterisation. Some fish were also collected live and were oxygen packed in polythene bags and transported to the laboratory for domestication and further studies. The occurrence and abundance of different fish species was ascertained through direct observation and from the information gathered from the local fishermen and fish vendors.

The Fish species, with respect to commercial utilisation as food, sports and ornamentation, were evaluated on the basis of specific criteria. The criteria adopted for food fish were on the basis of consumer acceptability. Of the food fish, those that showed adequate growth in unit time and attained maximum size were put under cultivable category. Fish have been identified as sport fish, based on the preference of anglers. The potential ornamental fish have been identified on the basis of their ornamental characters like colour, shape, size, banding pattern, adhesive sucker, hardiness, preying habit, ugliness and transparent body. A survey was also carried out on the wholesale ornamental fish market and exporters based in Kolkata, regarding species-wise export potentiality. To compile the present inventory of fish fauna of the NEH states, the species recorded by the present authors and also the reported valid species by others, have been incorporated.

2. Inventory of fish species and assessment of ornamental potential

The up to date inventory of the fish species of the NEH region showed 274 potential fish species which belonged to 114 genera under 37 families and 10 orders. This includes 265 indigenous and 9 exotic fish species. The diversified fish fauna of the NEH includes 272 food fish, 9 sport fish, 48 cultivable and 250 ornamental fish (Fig. 9.1).

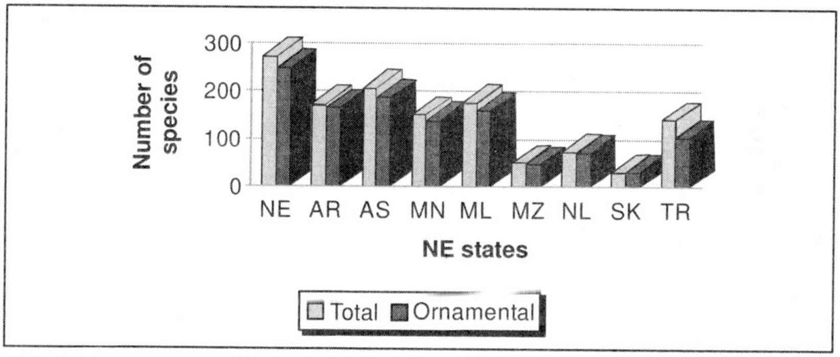

Fig. 9.1. Diversified fish fauna of the NEH region.

The study revealed that local people consume almost all the species (99.27%), except *Aplocheilus panchax* and *Chaca chaca.* A total of nine species (3.28%) were identified as major sport fish of the region. The important cold-water sport fishes of the region are mahseers (*Tor putitora, T. tor, T. chelynoides, Neolissocheilus hexagonolepis* and *N. hexastichus*) and Indian trout (*Barilius bola*). As many as 48 fish species (17.52%) of the region have been considered as cultivable fish, which can be farmed for the commercial production of flesh biomass from the aquatic habitat. A total of 250 fish species (91.24%) have been found to have potential ornamental value. The important ornamental fish like *Puntius shalynius, Danio dangila, Badis badis burmanicus, Acanthocobitis botia, Chaca chaca, Schistura beavani, Psilorhynchus sucatio and Hara hara* are largely unknown among the local masses. Out of the total 250 native ornamental fish species of the region, the highest numbers of ornamental fish species have been recorded from Assam (187), followed by Arunachal Pradesh (165), Meghalaya (159), Manipur (139), Tripura

(103), Nagaland (71), Mizoram (46) and Sikkim (29). The details of species-wise distribution in different North-Eastern States are presented in Table 9.1. But, in relation to the total number of fish species available and the number of fish species having ornamental value (Fig. 9.2), the highest percentage was recorded in Nagaland (97.26%), followed by Sikkim (96.67%), Arunachal Pradesh (95.93%), Mizoram (93.88%), Manipur (92.67%), Meghalaya (90.86%), Assam (90.78%) and Tripura (73.05%).

Table 9.1: Ornamental Fish Species of the North East and their distribution

Sl. No.	*Species*	*Distribution*							
		AR	*AS*	*MN*	*ML*	*MZ*	*NL*	*SK*	*TR*
1.	*Aborichthys elongatus* (Hora}	+	+	-	+	-	-	-	-
2.	*Aborichthys garoensis* (Hora)	-	-	-	+	-	-	-	-
3.	*Aborichthys kempi* (Chaudhuri)	+	-	-	+	-	+	-	-
4.	*Aborichthys tikaderi* Barman	+	-	-	-	-	-	-	-
5.	*Acanthocobitis botia* (Hamilton)	+	+	+	+	-	+	-	+
6.	*Acanthocobitis pavonaceus* (McClelland)	-	+	-	-	-	-	-	-
7.	*Acanthocobitis zonalternans* (Blyth)	+	-	+	-	-	+	-	-
8.	*Acanthopthalmus pangia* (Hamilton)	-	+	+	+	-	+	-	-
9.	*Acantopsis choirorhynchos* (Bleeker)	-	-	+	-	-	-	-	-
10.	*Ailia coila* (Hamilton)	+	+	-	+	-	-	-	+
11.	*Ailia punctata* (Day)	+	-	+	+	-	-	-	-
12.	*Amblyceps mangois* (Hamilton)	+	+	+	+	+	+	-	+
13.	*Amblyceps apangi* (Nath and Dey)	+	+	-	+	-	-	-	+
14.	*Amblyceps arunanchalensis* (Nath and Dey)	+	-	-	-	-	-	-	-
15.	*Amblypharyngodon mola* (Hamilton)	+	+	+	+	-	-	-	+
16.	*Anabas testudineus* (Bloch)	+	+	+	+	-	-	-	+
17.	*Anabas oligolepis* Bleeker	-	+	-	-	-	-	-	-
18.	*Anguilla bengalensis* (Gray and Hardwicke)	+	+	+	+	-	-	-	+
19.	*Aorichthys aor* (Hamilton)	+	+	+	-	+	-	-	+
20.	*Aorichthys seenghala* (Sykes)	+	+	+	+	-	-	-	+
21.	*Aplocheilus panchax* (Hamilton)	-	+	-	+	-	-	-	+
22.	*Apocryptes bato* (Hamilton)	-	+	-	-	-	-	-	-
23.	*Aspidoperia morar* (Hamilton)	+	+	+	-	+	-	-	+
24.	*Badis badis badis*(Hamilton)	+	+	+	+	+	+	-	+
25.	*Badis badis burmanicus* (Ahl)	-	-	-	+	-	-	-	-
26.	*Bagarius bagarius* (Hamilton)	+	+	+	+	-	-	+	+
27.	*Balitora brucei* (Gray)	-	+	+	+	-	-	-	-
28.	*Barilius barila* (Hamilton)	-	+	+	+	-	+	-	+
29.	*Barilius barna* (Hamilton)	+	+	+	+	+	-	+	+
30.	*Barilius bendelisis* (Hamilton)	+	+	+	+	+	+	+	+
31.	*Barilius bola* (Hamilton)	+	+	+	+	-	-	-	+
32.	*Barilius dogarsinghi* (Hora)	-	+	+	-	-	+	-	-
33.	*Barilius gatensis* (Valenciennes)	-	-	+	-	-	-	-	+
34.	*Barilius guttatus* (Day)	-	-	+	-	-	-	-	-
35.	*Barilius radiolatus* (Gunther)	-	+	-	-	-	-	-	-
36.	*Barilius shacra* (Hamilton)	+	+	+	-	-	-	+	+
37.	*Barilius tileo* (Hamilton)	+	+	+	+	-	-	-	+
38.	*Barilius vagra* (Hamilton)	+	+	+	-	+	-	+	+
39.	*Batasio batasio* (Hamilton)	+	+	+	+	+	-	+	+
40.	*Batasio tengana* (Hamilton)	+	+	+	+	-	-	-	-

41. *Bengala elanga* (Hamilton)	+	+	-	+	-	-	-	+
42. *Botia berdmorei* (Blyth)	-	+	+	-	-	-	-	-
43. *Botia dario* (Hamilton)	+	+	+	+	+	+	-	+
44. *Botia histrionica* (Blyth)	-	-	+	+	-	-	-	-
45. *Botia lohachata* (Chaudhuri)	+	+	-	+	-	-	-	+
46. *Botia rostrata* (Gunther)	+	+	-	+	-	-	-	+
47. *Brachydanio acuticephala* (Hora)	+	-	+	-	-	+	-	-
48. *Brachydanio rerio* (Hamilton)	+	+	+	+	+	+	+	+
49. *Catla catla* (Hamilton)	-	+	+	+	-	-	-	+
50. *Chaca chaca* (Hamilton)	-	+	-	+	-	-	-	+
51. *Chagunius chagunio* (Hamilton)	+	+	+	+	+	-	+	+
52. *Chagunius nicholsi* (Myers)	-	-	+	-	-	-	-	-
53. *Chanda nama* (Hamilton)	+	+	+	+	-	-	-	+
54. *Chandramara chandramara* (Hamilton)	+	+	-	+	-	-	-	-
55. *Channa barca* (Hamilton)	-	+		I	-	-	-	+
56. *Channa marulius*(Hamilton)	+	+	+	+	-	-	-	+
57. *Channa orientalis* (Bloch and Schneider)	+	+	+	+	+	+	-	+
58. *Channa punctatus* (Bloch)	+	+	+	+	-	+	-	+
59. *Channa stewartii* (Playfair)	+	+	-	+	-	-	-	+
60. *Channa striatus* (Bloch)	+	+	+	+	-	+	-	+
61. *Chaudhuria indica* (Yazdani)	-	-	-	+	-	-	-	-
62. *Chaudhuria khajuriai* (Talwar, Yazdani and Kundu)	-	+	-	+	-	-	-	-
63. *Chela cachius* (Hamilton)	+	+	+	+	-	-	-	+
64. *Chela laubuca* (Hamilton)	+	+	+	+	-	-	-	+
65. *Chitala chitala* (Hamilton)	-	+	-	+	-	-	-	+
66. *Cirrhinus mrigala* (Hamilton)	+	+	+	+	-	-	-	+
67. *Cirrhinus reba* (Hamilton)	+	+	+	+	+	-	-	+
68. *Clarias batrachus* (Linnaeus)	+	+	+	+	-	-	+	+
69. *Clupisoma montana* (Hora)	-	-	-	-	-	-	-	+
70. *Colisa fasciatus* (Schneider)	+	+	+	+	-	-	-	+
71. *Colisa labiosus* (Day)	+	+	-	-	-	-	-	-
72. *Colisa lalia* (Hamilton)	-	+	-	-	+	+	-	-
73. *Colisa sota* (Hamilton)	-	+	+	+	-	+	-	+
74. *Conta conta*(Hamilton)	+	+	+	+	-	+	-	-
75. *Crossocheilus burmanicus* (Hora)	+	+	-	-	-	-	-	-
76. *Ctenops nobilis* (McClelland)	-	+	-	-	-	-	+	-
77. *Danio aequipinnatus* (McClelland)	I	-	I	+	-	+	-	-
78. *Danio dangila* (Hamilton)	+	+	+	+	+	+	-	+
79. *Danio devario* (Hamilton)	+	+	+	+	-	-	-	+
80. *Danio naganensis* (Choudhuri)	-	-	+	-	-	+	-	-
81. *Danio regina* (Fowler)	-	+	-	-	-	-	-	-
82. *Erethestis montana montana* (Hora)	-	+	-	-	-	-	-	-
83. *Erethestis pussiles* (Mull. and Tros.)	+	+	+	-	-	+	-	+
84. *Esomus danricus* (Hamilton)	+	+	+	+	-	+	-	+
85. *Euchiloglanis kamengensis* Jayaram	+	-	-	+	-	-	-	-
86. *Eutropiichthys vacha* (Hamilton)	+	+	+	+	-	-	-	+
87. *Exostoma berdmorei* Blyth	+	-	-	-	-	-	-	-
88. *Exostoma stuartii* (Hora)	+	-	+	+	-	-	-	-
89. *Gagata cenia* (Hamilton)	+	+	+	+	+	+	-	+
90. *Gagata gagata* (Hamilton)	-	+	-	-	-	-	-	+
91. *Gagata sexualis* (Tilak)	-	+	-	+	-	-	-	+
92. *Ganiolosa manmina* (Hamilton)	-	+	-	-	-	-	-	+
93. *Garra annandalei* (Hora)	+	+	+	+	+	-	+	-
94. *Garra gotyla gotyla* (Gray)	+	+	+	+	+	+	+	+
95. *Garra gravelyi* (Annandalei)	-	-	+	-	-	-	-	-
96. *Garra kempi* (Hora)	+	+	+	+	-	+	-	-
97. *Garra lamta* (Hamilton)	+	+	+	+	+	+	+	+

98. *Garra lissorhynchus* (McClelland)	+	+	+	+	-	+	-	-
99. *Garra litanensis* (Viswanath)	-	-	+	-	-	-	-	-
100. *Garra* manipurensis (Viswanath and Sarojnalini)	-	-	+	-	-	-	-	-
101. *Garra mcClellandi* (Jerdon)	+	-	+	+	-	-	-	-
102. *Garra naganensis* (Hora)	+	+	+	+	+	+	-	-
103. *Garra nasuta* (McClelland)	+	+	+	+	+	-	-	-
104. *Garra rupecula* (McClelland)	+	+	+	+	-	+	-	-
105. *Glossogobius giuris* (Hamilton)	+	+	+	+	+	-	-	+
106. *Glyptothorax brevipinnis* Hora	+	-	-	-	-	-	-	-
107. *Glyptothorax manipurensis* (Menon)	-	-	+	-	-	-	-	-
108. *Glyptothorax pectinopterus* (McClelland)	+	-	+	-	-	-	-	-
109. *Glyptothorax platypogonoides* (Bleeker)	+	+	+	-	+	+	-	-
110. *Glyptothorax telchitta* (Hamilton)	+	-	-	+	+	-	-	+
111. *Glyptothorax annandelei* (Hora)	+	-	-	-	-	-	-	-
112. *Glyptothorax cavia* (Hamilton)	+	+	-	+	-	-	-	+
113. *Glyptothorax coheni* (Ganguly, Dutta and Sen)	+	-	-	-	-	-	-	-
114. *Glyptothorax conirostre conirostre* (Steindachner)	+	-	-	+	+	-	-	+
115. *Glyptothorax indicus* (Talwar)	+	-	-	-	-	-	-	-
116. *Glyptothorax striatus* (McClelland)	+	+	-	+	-	-	+	-
117. *Glyptothorax trilineatus* (Blyth)	+	-	+	-	-	-	-	-
118. *Gudusia chapra* (Hamilton)	+	+	+	+	-	-	-	+
119. *Hara hara* (Hamilton)	+	+	+	+	-	+	-	+
120. *Hara horai* (Misra)	-	+	-	-	-	-	-	-
121. *Hara jerdeni* (Day)	+	+	-	-	-	-	-	-
122. *Heteroneustes fossilis*(Bloch)	+	+	-	+	-	+	-	+
123. *Homaloptera modesta* (Vinciguera)	-	-	+	-	-	-	-	-
124. *Johnius coitor* (Hamilton)	-	-	-	-	-	-	-	+
125. *Kryptoterus indicus* (Datta, Burman and Jayaram)	+	-	-	-	-	-	-	-
126. *Labeo angra* (Hamilton)	-	+	+	-	-	-	-	-
127. *Labeo bata* (Hamilton)	+	+	+	+	+	-	-	+
128. *Labeo boga* (Hamilton)	+	+	+	+	+	-	-	+
129. *Labeo calbasu* (Hamilton)	+	+	+	+	+	-	-	+
130. *Labeo devdevi* (Hora)	+	+	-	-	-	-	-	-
131. *Labeo dyocheilus* (McClelland)	+	-	-	-	-	+	-	-
132. *Labeo gonius* (Hamilton)	+	+	+	+	-	-	-	+
133. *Labeo nandina* (Hamilton)	-	+	-	+	-	-	-	+
134. *Labeo pangusia* (Hamilton)	+	+	+	+	+	-	+	+
135. *Labeo rohita* (Hamilton)	+	+	+	+	-	-	-	+
136. *Laguvia ribeiroi* (Hora)	+	+	-	+	-	-	-	+
137. *Laguvia shawi* (Hora)	-	+	-	+	-	-	-	+
138. *Lepidocephalus annandalei* (Chaudhuri)	+	+	+	+	-	-	-	+
139. *Lepidocephalus berdmorei* (Blyth)	+	+	+	+	-	+	-	-
140. *Lepidocephalus guntea* (Hamilton)	+	+	+	+	+	+	+	+
141. *Lepidocephalus irrorata* (Hora)	-	+	+	+	-	+	-	-
142. *Lepidocephalus menoni* (Pillai and Yazdani)	+	+	-	+	-	-	-	-
143. *Macrognathus aral* (Bloch and Schneider)	+	+	-	+	-	+	-	+
144. *Macrognathus pancalus* (Hamilton)	+	-	+	+	-	+	-	+
145. *Maringua hodgati* (Hamilton)	-	+	-	-	-	-	-	-
146. *Mastocembelus armatus* (Lacepede)	+	+	+	+	+	+	-	+
147. *Mesonoemacheilus sijuencis*(Menon)	-	+	-	+	-	-	-	-
148. *Microphis deocata* (Hamilton)	+	+	-	-	-	-	-	-
149. *Monopterus cuchia* (Hamilton)	+	+	+	+	-	+	-	+
150. *Mystus bleekeri* (Day)	+	+	+	+	-	+	+	+

151. *Mystus cavasius* (Hamilton)	+	+	+	+	-	-	-	+
152. *Mystus menoda* (Hamilton)	-	+	-	-	-	-	-	-
153. *Mystus montanus* (Jerdon)	+	+	-	+	-	-	-	-
154. *Mystus tengara* (Hamilton)	-	+	-	-	-	-	-	-
155. *Mystus vittatus* (Bloch)	+	+	-	+	-	-	-	+
156. *Nandus nandus* (Hamilton)	+	+	+	+	-	-	-	+
157. *Nangra assamensis* (Sen and Biswas)	-	+	-	-	-	-	-	-
158. *Nangra nangra* (Hamilton)	-	+	-	-	-	-	-	+
159. *Nemacheilus corica* (Hamilton)	+	+	-	+	-	-	-	-
160. *Nemacheilus peguensis* (Hora)	-	-	+	-	-	-	-	-
161. *Neoeucirrhichthys maydelli* (Banerescu)	-	+	-	-	-	-	-	-
162. *Neolissocheilus hexastichus* (McClelland)	+	+	+	+	+	+	-	-
163. *Neolissocheilus hexagonolepis* (McClelland)	+	+	+	+	+	+	+	+
164. *Noenemacheilus labeosus* (Kottelat)	-	+	-	-	-	-	-	-
165. *Notopterus notopterus* (Pallas)	+	+	+	+	-	-	-	+
166. *Olyra burmanica* (Day)	-	-	-	-	-	+	-	-
167. *Olyra horai* (Prashad and Mukerji)	-	+	-	+	-	+	-	-
168. *Olyra longicaudata* (McClelland)	+	+	-	+	-	+	-	-
169. *Ompok bimaculatus* (Bloch)	+	+	+	+	-	+	-	+
170. *Ompok pabda* (Hamilton)	+	+	+	+	-	-	-	+
171. *Ompok pabo* (Hamilton)	+	+	-	+	-	-	-	-
172. *Ophisternon bengalense* McClelland	-	-	+	-	-	-	-	-
173. *Oreichthys casuatis* (Hamilton)	+	+	-	+	-	-	-	-
174. *Osphronemus goramy* Lacepede	-	-	-	-	-	-	-	+
175. *Osteobrama belangeri* (Valenciennes)	-	-	+	-	-	-	-	-
176. *Osteobrama cotico cunma* (Day)	-	-	+	-	-	-	-	+
177. *Osteobrama cotio cotio* (Hamilton)	+	+	+	+	-	-	-	+
178. *Osteochilus neilli* (Day)	+	-	-	-	-	-	-	-
179. *Pangasius pangasius (Hamilton)*	-	+	-	-	-	+	-	+
180. *Pisodonophis boro* (Hamilton)	-	+	-	-	-	-	-	-
181. *Pseudambasis baculis* (Hamilton)	+	+	+	+	+	-	-	+
182. *Pseudambasis lala* (Hamilton)	-	+	-	-	-	-	-	-
183. *Pseudambassis ranga* (Hamilton)	+	+	+	+	+	-	-	+
184. *Pseudechenies sulcatus* (Prashad and Mukerji)	+	+	+	+	-	+	+	-
185. *Pseudeutropius atherinoides* (Bloch)	+	+	-	+	-	-	-	+
186. *Psilorhynchus balitora* (Hamilton)	+	+	+	+	+	+	-	+
187. *Psilorhynchus gracilis* (Rainboth)	-	+	-	-	+	-	-	-
188. *Psilorhynchus homaloptera* (Hora and Mukherjee)	+	-	-	+	-	+	-	-
189. *Psilorhynchus sucatio* (Hamilton)	+	+	-	+	-	-	-	+
190. *Puntius chola* (Hamilton)	+	+	+	+	-	-	-	+
191. *Puntius clavatus* (McClelland)	+	+	+	+	-	+	+	+
192. *Puntius conchonius* (Hamilton)	+	+	+	+	+	+	-	+
193. *Puntius filamentosus* (Valenciennes)	-	+	-	+	-	+	-	+
194. *Puntius fraseri* (Hora and Misra)	-	+	-	-	-	-	-	-
195. *Puntius gelius* (Hamilton)	-	+	-	+	-	-	-	+
196. *Puntius guganio* (Hamilton)	+	+	-	+	-	-	-	-
197. *Puntius phutunio* (Hamilton)	-	+	+	-	-	-	-	-
198. *Puntius sarana orphoides* (Valenciennes)	-	-	+	-	-	+	-	-
199. *Puntius sarana sarana* (Hamilton)	+	+	+	+	-	-	+	+
200. *Puntius shalynius* Yazdani and Talukdar	+	+	+	+	-	-	-	-
201. *Puntius sophore* (Hamilton)	+	+	+	+	+	-	-	+
202. *Puntius terio* (Hamilton)	-	+	+	+	+	-	-	+
203. *Puntius ticto* (Hamilton)	+	+	+	+	+	+	-	+
204. *Rasbora daniconius* (Hamilton)	+	+	+	+	-	-	-	+
205. *Rasbora rasbora* (Hamilton)	+	+	+	+	-	+	-	+
206. *Rhinomugil corsula* (Hamilton)	-	+	-	+	-	-	-	+

207. *Rita rita* (Hamilton)	+	+	+	-	-	-	-	+
208. *Salmostoma bacaila* (Hamilton)	+	+	+	+	-	-	-	+
209. *Salmostoma clupeoides* (Bloch)	-	+	-	-	-	-	-	+
210. *Salmostoma phulo* (Hamilton)	+	+	-	+	-	-	-	-
211. *Schismatorhynchos nukta* (Sykes)	-	-	-	-	-	-	-	+
212. Schistura arunachalensis *(Menon)*	+	+	-	+	-	-	-	-
213. *Schistura beavani* (Gunther)	+	+	-	+	-	-	-	-
214. *Schistura cinticauda* (Blyth)	+	-	-	+	-	-	-	-
215. *Schistura denisoni dayi* (Hora)	-	-	-	+	-	-	-	-
216. *Schistura devdevi* (Hora)	+	-	-	+	-	-	+	-
217. *Schistura elongatus* (Sen and Nalbant)	-	-	-	+	-	-	-	-
218. *Schistura kangjupkhulensis* (Hora)	+	+	+	-	-	+	-	-
219. *Schistura manipurensis* (Chaudhuri)	+	+	+	-	+	+	-	-
220. *Schistura multifaciatus* (Day)	+	+	+	+	+	+	+	-
221. *Schistura nagaensis* (Menon)	+	-	-	-	-	+	-	-
222. *Schistura prashadi* (Hora)	-	-	+	+	-	+	+	-
223. *Schistura reticulofasciatus* (Singh and Banarescu)	-	+	-	+	-	-	-	-
224. *Schistura savona* (Hamilton)	+	+	+	+	-	-	-	-
225. *Schistura scaturigina* (McClelland)	+	+	-	+	+	+	-	+
226. *Schistura sikmainensis* (Hora)	+	+	+	+	-	+	-	-
227. *Schistura singhi* (Menon)	-	-	-	-	-	+	-	-
228. *Schistura vincigeurra* (Hora)	-	-	+	-	-	-	-	-
229. *Schizopygopsis esocinus* (Heckle)	+	-	-	-	-	-	-	-
230. *Schizopygopsis progastus* (McClelland)	+	+	-	-	-	-	+	-
231. *Schizopygopsis stoliczkae* (Steindachner)	+	-	-	-	-	-	-	-
232. *Schizothorax richardsonii* (Gray)	+	+	+	-	-	+	+	-
233. *Securicula gora* (Hamilton)	+	+	-	+	-	-	-	-
234. *Setipinna phasa* (Hamilton)	+	+	+	+	-	-	-	-
235. *Sicamugil cascasia* (Hamilton)	-	+	+	+	-	-	-	+
236. *Silurus afghana* (Gunther)	+	+	-	-	-	-	-	-
237. *Silurus berdmorei* (Blyth)	-	+	-	-	-	+	-	-
238. *Silurus torrentis* (Kobayakawa)	+	-	-	-	-	-	-	-
239. *Sisor rhabdophis* (Hamilton)	+	-	+	-	-	-	-	-
240. *Somileptes gongota* (Hamilton)	+	+	-	+	-	-	-	+
241. *Strongylura strongylura* (Van Hasselt)	-	-	-	-	-	+	-	+
242. *Tetradon cutcutia* (Hamilton)	+	+	+	+	-	-	-	+
243. *Tor chelynoides* (McClelland)	-	-	-	+	-	-	-	-
244. *Tor mosal* (Hamilton)	-	+	-	-	-	-	-	-
245. *Tor progeneius* (McClelland)	-	+	+	-	-	+	-	-
246. *Tor putitora* (Hamilton)	+	+	+	+	-	-	+	+
247. *Tor tor* (Hamilton)	+	+	+	+	+	+	+	+
248. *Triplophysa gracilis* (Day)	-	-	-	-	-	-	+	-
249. *Wallago attu* (Schneider)	+	+	+	+	-	-	-	+
250. *Xenentodon cancila* (Hamilton)	+	+	+	+	+	-	-	+

AR – Arunachal Pradesh, AS - Assam, MN – Manipur, ML – Meghalaya, MZ - Mizoram, NL – Nagaland, SK – Sikkim, TR – Tripura
+ Present; - Absent

Puntius shalynius

Botia dario

Pseudochenis sulcatus

Fig. 9.2. Distribution of fish fauna in NE states and number of species having ornamental value.

Diversified ornamental criteria were recorded in many species and an inventory is being developed, based on their ornamental value (Table 9.2).

Some of the potential native ornamental fish species have been shown in Fig 9.3.

Table 9.2: Diversified ornamental fish

Ornamental characters	*Examples*
Beautiful colour	*Pseudecheneies sulcatus, Tetraodon cutcutia, Colisa lalia, Botia berdmorei*
Stripes and banding pattern	*Botia rostrata, B. dario, Brachydanio rerio, Schistura reticulofasciatus*
Chameleonic habit	*Badis badis, Puntius shalynius*
Jumping behaviour	*Esomus danricus, Chela laubuca, Salmostoma bacaila*
Charming predatory habit	*Channa orientalis, C. striatus, Glossogobius giuris*
Calm behaviour	*Ctenops nobilis, Nandus nandus*
Transparent body	*Chanda nama, Pseudambassis baculis, P. ranga*
Small size	*Brachydanio rerio, Danio dangila*
Hardiness	*Anabas testudineus*
Suckers	*Garra gotyla gotyla, G. lissorhynchus, G. mcllendi, Glyptothorax striatus*

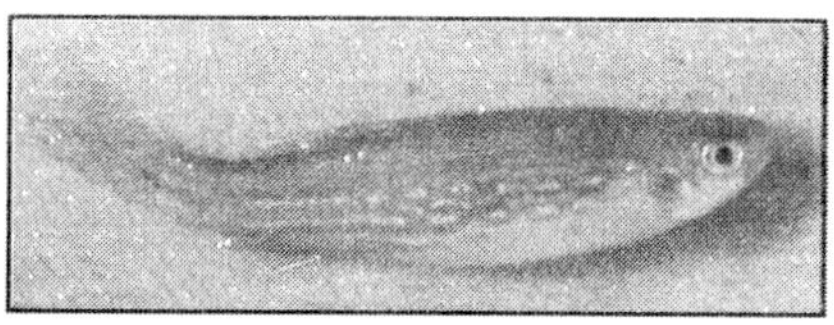

Danio dangila

Chaca chaca

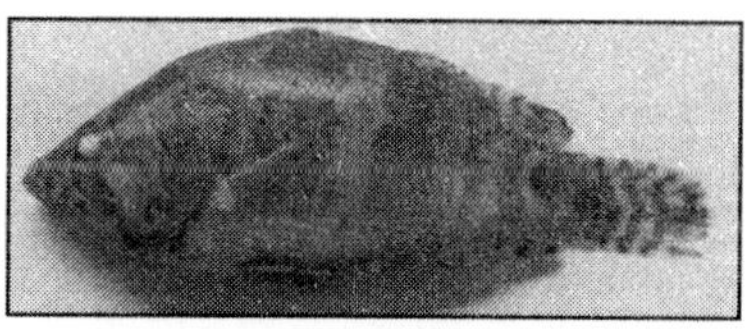

Nandus nandus

Badis badis

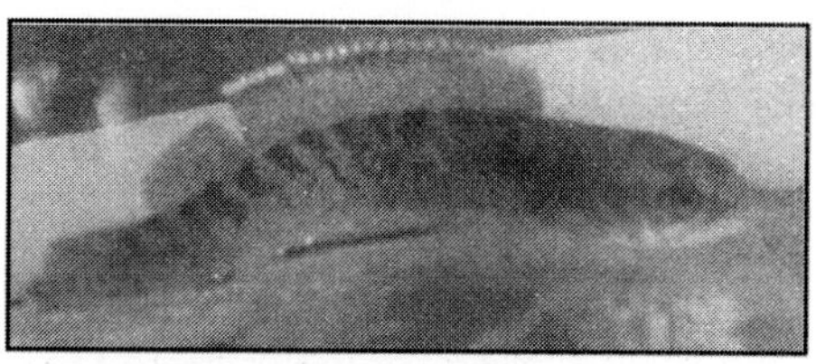

Channa stewartii

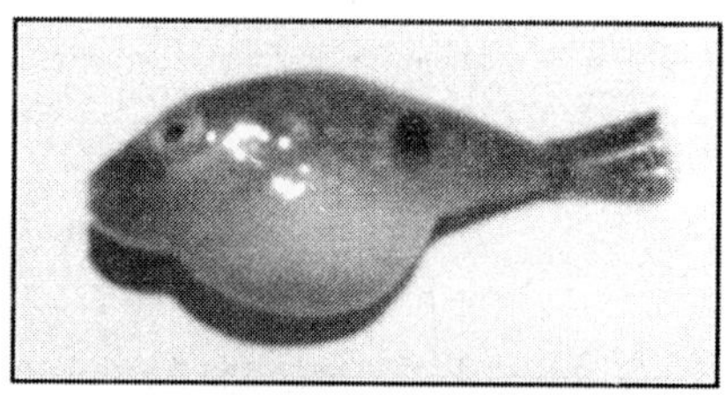

Tetraodon cutcutia

Fig. 9.3. Potential native ornamental fish species of the NE region

C. BREEDING AND LARVAL REARING OF NATIVE ORNAMENTAL FISH

The fishery division of the ICAR Research Complex for the NEH Region has already initiated work in this line and has successfully bred species like *B. rerio, P. shalynius, L. gonius, D. dangila, P. sophore and C. fasciata.*

1. Breeding and larval rearing of *Zebra danio, B. rerio*

The *Zebra danio* is a prolific breeder and their breeding season commences from April and continues till August. The eggs are non-adhesive and their breeding habit is of egg scatter type. The females are more silvery and larger in size as compared to the males. The belly is swollen in females, especially in the gravid ones; while in males, the body is generally slim. The male and female fish were separated, based on their morphological characters. An aquarium tank, 60 cm long, of about 50 L capacity provided with a perforated tank divider, was found to be suitable for the conditioning of the zebra danio. The males were introduced on one side and the females on the other side of the divider, in the same aquarium tank. The fish, when conditioned on a good variety of diet, including live feed for about two to three weeks, were ready for breeding. Successful spawning was recorded by introducing a shoal of 20 fully matured vigorous males and females in a ratio of 3 males: 2 females into the spawning tank. The spawning tank contained about 10 cm (4 inches) water with a steady temperature of 28-29 °C, and gentle aeration. A water pH of around 7.0 and hardness of normal to 250 ppm was maintained, which gave good results. The bottom of the aquarium was completely covered with pebbles of about 6 to 8 mm diameter. After spawning, the eggs fell in between the pebbles arranged at the bottom of the breeding tank. This prevented the parent fish from eating the eggs. A net fixed along the bottom of the tank (at a height of 2 inches from the bottom), instead of the pebbles, was also found to be successful in protecting the eggs from the parent fish. The eggs required a hatching time of 2 days, if the temperature was favourable. As soon as the tiny hatchlings were seen in the aquarium tank, the parents were removed. The hatchlings took 2 days to absorb their yolk sac. After 2 days, they were fed with infusorians for 4 days. Subsequently, rotifers, smaller daphnia were fed for a week. After that they were fed with larger plankton and powdered formulated feed.

The moustache danio *D. dangila* and the chameleon fish *B. badis,* which are also egg scatter type of fish, were bred almost in a similar way.

2. Breeding of giant gourami, *Colisa fasciatus* (Fig. 9.4)

For breeding purpose, the male and female fish were kept in separate tanks for a few weeks. When the abdomen of the female became remarkably swollen with eggs, they were transferred to a smaller breeding tank, with 5-6 inches water

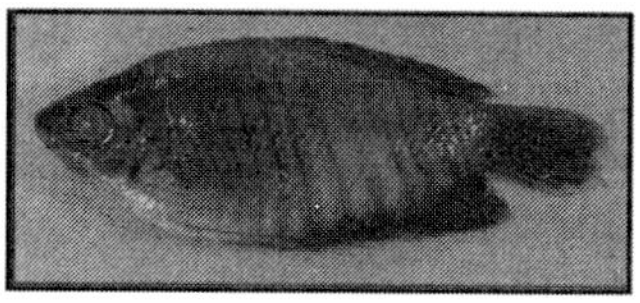

C. fasciata

Breeding nest

larval rearing

Fig. 9.4. Breeding process of C. fasciata

level. The tanks were provided with some floating plants. After one or two days, a good male was introduced into the breeding tank. The breeding tank was covered with a transparent plastic sheet or a glass sheet to keep the humidity and temperature high and to help maintain the bubble nest in a good condition. Then, the male soon began building a bubble nest. This is possible by taking a large gulp of air at the water surface and converting it into many smaller bubbles that are passed into the gill chamber and coated with an anti-burst agent before release.

During and after making the nest, the male displays it to the female and this usually ends with both the fish embracing the nest, resulting in the deposition of a large number of eggs in the nest. After breeding, the female was removed as the male guarded the eggs that remain attached to the floating bubble nest. Hatching took place within 24 hours. The moment the fry began leaving the nest, the male was also removed from the tank. After 36 hours, when the young ones remained in the free-swimming stage, they were provided with infusoria as a starter feed. After a week, the fry started taking newly hatched Artemia and small cladocerans. During this stage, the fry required vigorous feeding. Subsequently, when they had grown a little bigger, they were stocked in bigger cement tanks for further growth.

D. MARKETING CHANNELS AND SCOPE OF EXPORT

The collection of native ornamental fish and their marketing from the region is totally disorganised. The fish are collected by unregistered traders with the help of local fish collectors. These traders in turn, send these fish to West Bengal, either to the wholesalers or directly to the exporters (Fig. 9.5). In Kolkata and Howrah, there are twelve leading exporters who export these fish overseas. The local fish collectors get a very negligible price, while the exporters enjoy the lion's share (Fig. 9.6).

Among the wild-caught native ornamental fish species exported from the country, the North Eastern States alone contribute about 80%. The FOB price offered by exporters for native ornamental fish is significantly high and varies from about US$ 0.08 to US$ 9.33 (Table 9.3).

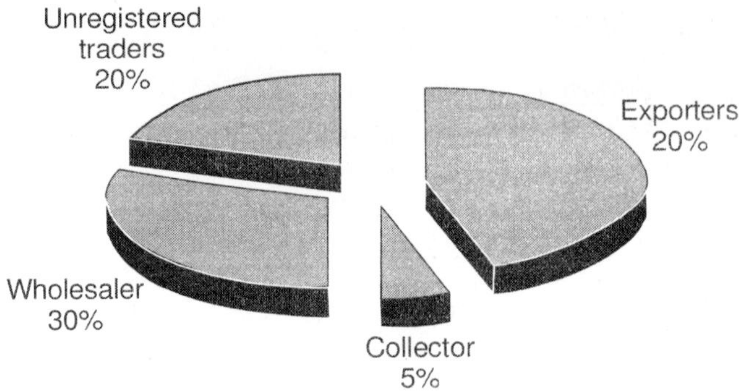

Fig. 9.5. Marketing Channel of native ornamental fishes.

COLLECTORS
(Get negligible Price)

↓

UNREGISTERED SMALL TRADERS
(Get Rs. 2-5 per piece)

↓

WHOLESALERS
(Get Rs. 3-10 per piece)

↓

EXPORTERS
(Get Rs. 5-20 (Even up to Rs. 100/- per piece)

Fig. 9.6. Profit percentage of different stakeholders

Table 9.3: Some of the native ornamental fish with their trade name and the price offered by the importer (FOB)

Sl. No.	*Species*	*Trade Name*	*Average price (In US $)*
1.	*Brachydanio rerio*	Zebra danio	0.08
2.	*Danio dangila*	Moustached danio	0.75
3.	*Puntius shalynius*	Shalyni barb	0.95
4.	*Channa stewartii*	Rainbow snake head	9.33
5.	*Chaca chaca*	Devil cat fish	1.25
6.	*Badis badis*	Chameleon fish	0.20
7.	*Channa punctatus*	Checkered snake fish	0.75
8.	*Nandus nandus*	Leaf fish	1.10
9.	*Botia dario*	Golden banded loach	0.20
10.	*Channa orientalis*	Special snake head	1.10
11.	*Garra gotyla gotyla*	Gotyla	2.25

E. CONSERVATION STATUS OF NATIVE ORNAMENTAL FISH

At present, most of the highly priced ornamental fish species are becoming vulnerable in the NEH region, due to the degradation of water bodies, pollution, poisoning, over fishing of broods and juveniles. Their population was abundant in their natural habitat, even one or two decades. But earlier these days, some of them are very rare in the NEH region and most of the species are being threatened. Conservation status of the native ornamental fish have shown that out of the 250 species, 10 are Critically Endangered (CR), 28 are Endangered (EN), 49 are Vulnerable (VU), 45 are Lower Risk Near Threatened (LRnt), 08 are Lower Risk Least Concern (LRlc), 03 are Data Deficient (DD) and 107 are Not Evaluated (NE) and these have been illustrated in Fig. 9.7.

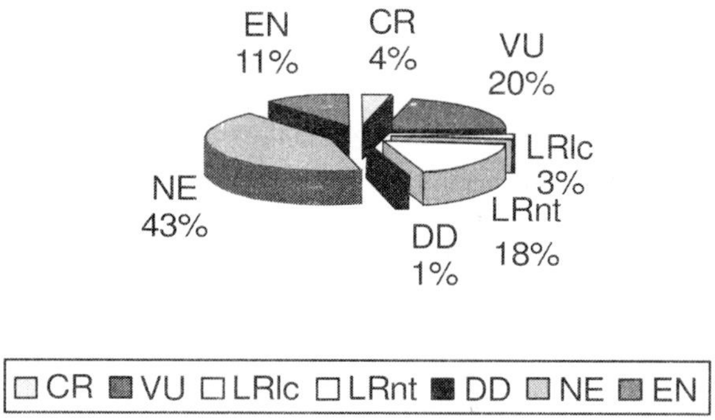

Fig. 9.7. Conservation status of native ornamental fish species in NEH Region

F. FUTURE STRATEGIES

North-East India, provided with 250 ornamental fish species, is one of the global hotspots for faunal diversity and has bright prospects in the ornamental fish trade, and contributes a lion's share on the part of India. To boost up the local economy and for self-employment, this native ornamental fish can play a key role. The value of the entire industry, like the indigenous germplasm resource of ornamental fish, ornamental plants and support activity in related supplies, can help the backward downtrodden people to maintain their livelihood.

To ensure fish welfare and environmental protection as well as sustainable growth of this industry, proper attention needs to be given, on the following lines.

1. Captive breeding and farming needs to be done for sustainable supply of the ornamental fish species. This will also help in the conservation, as well as sustainable use of biological diversity. Captive breeding will also help in the ranching of some species, whose stock is found to be declining, so as to revive their natural fishery.
2. Legal banning on the indiscriminate catch of brood and juveniles is important. This will ensure the protection of species that are or may become

endangered. However, provision should be made to sell endangered ornamental fish species, which are 'farm-raised', in which case, each farm-raised fish must be tagged, and the sale, should be monitored so as to ensure that the wild stocks of the same species are not marketed. Also, unnecessary banning of some species whose stocks are in plenty in the wild should be avoided to prevent natural wastage.

3. An improved information system and accurate reporting is crucial for fish that are caught from the wild and marketed, as well as for farm-raised fish. Species-wise detailed statistics, separately for capture and culture fisheries, is the need of the hour.
4. Government and leading NGO's should come forward for organise marketing channels for fair and equitable sharing of benefits derived from this trade.
5. As the ornamental fish culture is mostly indoors, there is immense scope for the involvement of women folk in this aqua-business.
6. Adequate transfer of technology to the local people on the ornamental value of uncustomary bioresources, their conservation status, their captive breeding and farming techniques, system design, proper handling, disease diagnosis and treatment and quarantine measures are important. In this context, it is essential to take some measures to educate all the stakeholders in this field for the sustainable growth of this industry.

REFERENCES

Anon (1995). Hand book on aquafarming. M.P.E.D.A. Cochin.

Bartley D. (2000). Responsible ornamental fisheries. FAO Aquaculture Newsletter 24: 10-14.

Dutta O.K. (2000). Ornamental icthyofauna of Assam. In: Workshop on Northeast Fish Germplasm Inventory and Conservation, February 10-11, 2000, organised jointly by the National Bureau of Fish Genetic Resources, Lucknow and the North Eastern Council, Shillong p 38-39.

Ghosh S.K. and Lipton A.P. (1982). Ichthyofauna of the NEH Region with special reference to their economic importance. ICAR Spl Bull, No 1 (ICAR Research Complex, Shillong), p 126.

Ghosh A. and Mahapatra B.K. (1999). Ornamental fishery resources of West Bengal with a note on their sustainable utilisation, exploitation and conservation. In: National Seminar on 'Eco-friendly Management of Resources for Doubling Fish Production–Strategies for 21st Century'. Inland Fish Soc India and Cent Inland Fish Res Inst, Barrackpore, 22-23 December, 1999.

Ghosh A., Mahapatra B.K. and Datta N.C. (2002). Studies on native ornamental fish of West Bengal with a note on their conservation. Environment and Ecology 20(4): 787-793.

Ghosh A., Mahapatra B.K. and Datta N.C. (2003). Ornamental fish farming – successful small-scale aqua business in India. Aquaculture Asia 8(3): 14-16.

Mahapatra B.K. (1999). Ornamental fish culture. Freshwater Fisheries Research Station, Kulia. Directorate of Fisheries, Govt of West Bengal, Publication No 10, p 16.

Mahapatra B.K. (2003). Studies on breeding and larval rearing of an exotic ornamental cichlid, *Pterophyllum scalare* (Lichenstein) with a note on its cultural prospects. J Natcon 15(2): 65-70.

Mahapatra B.K., Dutta A., Basu A., Dey U.K. and Sengupta K.K. (1999). Observation on the spawning and rearing of Angelfish, *Pterophyllum scalare* (Lichenstein). In: Eco-Friendly Management of Resources for Doubling Fish Production Strategies for 21st Century (Sinha M, Dhirendra Kumar and Katiha PK eds), Proceedings of the National Seminar, December 22-23, 1999. Inland Fisheries Society of India, Barrackpore, p 102-106.

Mahapatra B.K., Ghosh A. and Datta N.C. (2000). Breeding and rearing of ornamental fish, Guppy, *Poecilia reticulata*, Peter and Gold fish, *Carassius auratus* (Linnaeus) for prospective entrepreneurship development. Green Technology 3: 26-33.

Mahapatra B.K., Vinod K. and Mandal B.K. (2001). Prospects of puntius barbs in the NEH Region with reference to export potentiality in ornamental fish trade. In: Workshop on Integration of Fish Biodiversity Conservation and Development of Fisheries in North Eastern Region through Community Participation, Dec 12-13, 2001, organised jointly by the NBFGR, Lucknow and the North Eastern Council (NEC), Shillong.

Mahapatra B.K., Ghosh A. and Datta N.C. (2003a). Scope of indigenous ornamental fishery trade in Eastern and North Eastern India. In: Welfare Biology in the New Millennium (Kar D, Dey SC and Datta NC eds). Allied Publishers Pvt Ltd, India p 87-91.

Mahapatra B.K., Vinod K. and Mandal B.K. (2003b). Fish biodiversity of Meghalaya with a note on their sustainable utilisation. Aquacult 4(1): 1-10.

Mahapatra B.K., Vinod K. and Mandal B.K. (2003c). Studies on native ornamental fish of Meghalaya with a note on their cultural prospects. Aquacult 4(2): 171-180.

Mahapatra B.K., Vinod K. and Mandal B.K. (2003d). Scope of ornamental fishery in Meghalaya. In: Approaches for Increasing Agricultural Productivity in the Hill and Mountain Ecosystem. (Bhatt BP, Bujarbaruah KM, Sharma YP and Patiram eds). ICAR for the NEH Region, Umiam, Meghalaya, India p 397-401.

Radhakrishnan K.V. and Kurup B.M. (2002). Inland indigenous ornamental fish resources of Kerala. Fishing Chimes 22(4): 42-53.

Swain S.K., Jena J.K. and Ayyappan S. (2003). Prospects of export-oriented freshwater ornamental fish culture in India. Fishing Chimes 22(10 &11): 111-113.

Vinod K., Mahapatra B.K. and Mandal B.K. (2001). *Brachydanio rerio* (Hamilton) and *Danio dangila* (Hamilton) - Promising species for ornamental fisheries in Meghalaya and strategies for judicious exploitation. In: Workshop on Integration of Fish Biodiversity Conservation and Development of Fisheries in the North-Eastern Region through Community Participation, Dec 12-13, 2001, organised jointly by the NBFGR, Lucknow and the North Eastern Council (NEC), Shillong.

Chapter **10**

Medicinal Plants and Conservation: Some Issues and Concerns

D. Syiem and M.B. Syiem

Department of Biochemistry, North-Eastern Hill University, Shillong - 793 022, Meghalaya

Plants are a vital part of the environment. Medicinal and aromatic plants (MAPs) have been used for health and for flavouring food and drinks for centuries. Plants constitute a major remedy in the traditional medical system. Nowadays, the traditional knowledge of the use of plants for health care has provided initial leads towards drug development. Pharmaceutical companies can cut down costs and time for drug development through bioprospecting, using the initial knowledge of herbal practitioners, to narrow down their search for prospective molecules from plants. In the context of the North-East, medicinal plants provide a strong link between people and nature. Maintaining such links through the sustainable harvesting of MAPs can be a crucial factor for ensuring the continuation of the traditional cultural landscape in which MAPs thrive besides contributing to the preservation of traditions generally associated with the use of MAPs.

INTRODUCTION

Plants being an important part of the environment, they produce many essential products ranging from timber to medicines, to secondary metabolites of industrial importance, such as alkaloids, steroids, pigments, fragrances, and sweeteners, etc. Plants also provide mental and spiritual comfort to people all over the world.

Despite major breakthroughs in allopathic drugs, a large section of the rural population continues to use indigenous herbal medicines for their curative, preventive and prophylactic properties.

Besides being used for health, flavouring of food and drinks for centuries, plant extracts are considered to be the main source of herbal medicines. Since time immemorial, they have been used for curing various illnesses.

For many commercial purposes, medicinal plants are grouped in the broader category of 'medicinal and aromatic plants' (MAP), covering not only medicinally used plants, but also related and overlapping purposes like food, condiments and cosmetics (Hamilton, 2003).

PLANTS IN TRADITIONAL MEDICINE

Plants constitute a major remedy in the traditional medical system. The practice continues today because of their biomedical benefits, their easy access and place in cultural beliefs. Medicinal plant derivatives are non-narcotic. This has led to an upsurge for herbal medicine/herbal products in developed countries. In recent years, traditional medicine is seeing a revival at a fast pace. The use of MAPs has seen a sharp rise in the west, as well as in India and China. As per WHO, 80% of the population in developing countries relies on traditional medicines.

There are several traditional systems of medicine that have been broadly classified into 3 major categories. (Phytomedica@conserveafrica.org)

1. Traditional Medical Systems, e.g., Chinese, Tibetan, Ayurveda, Siddha, Unani, transmitted through the writer word
2. Traditional or Indigenous medical knowledge, including folk medicine, transmitted orally, and
3. Shamanistic medicine, with a strong spiritual element.

World Health Organization (WHO) has reported that 21,000 medicinal plants are utilised by approximately a billion people in different parts of the world, of which at least 100 species are used regularly. Raw materials for medicine preparation are extracted from approximately 2000 species of plants that are well documented for their curative properties, and over 1300 species of plants are known for their aroma and flavour. (phytomedica@conserveafrica.org)

MEDICINAL PLANTS AS A SOURCE OF PHARMACEUTICAL DRUGS

In the present day scenario, medicinal plants are used as a source of drugs and as a potential source of remedy for diseases that are currently difficult to treat/cure. Drugs such as taxol from *Taxus baccata,* vincristin from *Vinca rosa,* Colchicine from C. *automnale* are a few examples of anti-cancer drugs from plant sources.

GLOBAL SCENARIO ON MEDICINAL PLANTS TRADE

Herbal medicine is growing 10-20% annually, in Europe and North America, over the recent years (Hamilton, 2003). Trade in herbal medicines includes related botanicals sold as health food, food supplements, herbal teas and other purposes related to health. Medicinal plants form an important part of the world's economy. As per WHO, the global market of MAPs is 62 billion dollars, of which India's share is very small, despite its huge resources both in terms of medicinal plants as well as a rich heritage of traditional knowledge of their use, providing a huge potential for bioprospecting. While China's earning is Rs. 18,000 crores, Indias earnings on export is only Rs. 500 crores, despite the presence of a strong industrial component comprising more than 2000 odd pharmaceutical firms. (www.medherb.com)

WILD HARVESTING OF MEDICINAL PLANTS

Harvesting of MAPs from the wild is generally a seasonal activity providing a supplementary income for the rural people. On the other hand, small-scale cultivation of MAPs often in multicropping systems, can be a mean by which low-income families can boost their earnings (Lange, 1998).

Wild harvesting of MAPs provides cash income for the poorest sectors of society, especially in the North-East. It is however, unfortunate that prices paid for the material collected are generally low. In wild harvesting, problems of uncertainty of the market, such as regular purchase and prices, etc are quite frequent. Fairer prices for collectors and incentives for better management are needed, which could be promoted and regulated by the respective State Governments.

In the context of the North-East, medicinal plants provide a strong link between people and nature. Maintaining such links, through the sustainable harvesting of MAPs, can be a crucial factor for ensuring the continuation of the traditional cultural landscape in which MAPs thrive, besides contributing to the preservation of traditions generally associated with the use of MAPs. Conservation of habitats, vulnerable or otherwise, such as 'sacred groves' can attract tourists and thus, help in creating employment for the local people, specially where employment opportunities are scarce.

While many MAPs are common in the wild and can be freely harvested, some of the most popular and most traded are under severe threat from over-harvesting. They face commercial or local extinction, if not, global extinction (Lange, 1998). As reported, in Europe, between 20,000 to 30,000 tons of wild plant materials are collected annually and more than 500 MAPs were harvested from the wild in France, during 1988-89 (Kuipers, 1997; Lange, 1998).

Trade in MAPs is largely unmonitored and harvesting practices are frequently unsustainable. In most developing countries, effective regulatory mechanisms and controls over the amounts harvested are not in place, but policies and framework are being framed and legislated, at least in India, a case in point is the newly constituted National Biodiversity Act and the State Medicinal Plants Board. In the context of Meghalaya and other North-Eastern States, where there is a mixed bag of land-holding system, as well as overlaps in administration between the government and the respective District Councils, there are complications and serious managerial problems.

Earlier we have reported (Syiem *et al.*, 1999) the use of at least 56 medicinal plants in Meghalaya, all of them extracted from the wild. Many more are being sold to middle men from other states at throwaway prices. Species such as Lycopodium, lichens, and orchids, including pitcher plants are being rampantly extracted and sold. The State Forest Department and the local autonomous District Council have little to say on the matter. In general, there is little knowledge of the ecology and regeneration abilities/agrotechniques of many MAPs and accordingly, little understanding of the steps necessary to ensure the maintenance of populations and their species.

An advantage of maintaining wild populations through *in situ* means is that reserves of genetic variability remain. The sum total of all the wild populations of a species, across its geographical range, constitutes an invaluable resource. Wild populations frequently contain genes of value in plant-breeding, for instance, to increase the levels of active principles or to convey resistance to disease. Countries can protect their wild gene pools by working with users to introduce policies that link controlled harvesting with habitat management and monitoring. In the context of Meghalaya, working with the managers/ owners of the 'sacred grove', which contains rich gene pool of MAPs, is a case in point.

ISSUES

As this sector is largely unorganised and hidden, the absence of a regulatory mechanism, coupled with the practice of collection, which is mostly from the wild has lead to species depletion or threat of extinction for certain habitats. In addition, the conflicts open arise between the statutory and traditional or customary laws, and the Government, District Councils and the local traditional institutions, "Durbars" and heads, "Syiems", has resulted in the weakening of customary laws that had traditionally regulated the use of natural resources, resulting in over exploitation of medicinal plants. Terms such as biopiracy are frequently used to describe the prevailing situation. On the whole, these pose a serious threat to conservation efforts, decline in local knowledge in the use of medicinal plants and also on cultural survival.

In the North-East, addressing conservation issue has become complicated with the need to strike a balance between conservation on the one hand, and rural livelihood on the other. It is therefore, important to address the medicinal plants and related activities in a holistic manner, touching on the many dimensions that medicinal plants have on people.

Therefore, an effective strategy of conservation is one that requires conservationists to have an appreciation of the culture, economiy and social structure and dynamics of the local societies, in addition to the knowledge that they need about the biology and ecology of the plants, themselves.

MEDICINAL PLANTS AND CONSERVATION

The special significance of medicinal plants in conservation stems from their cultural, livelihood or economic roles. Several recommendations have been made and compiled, relating to the conservation of medicinal plants, beginning with the historic Chiang Mai Declaration, 'saving lives by saving plants' in Thailand (1988), at the Bangalore conference, in 1998, and the recently concluded 3rd world congress at Chiang Mai, on the theme, "medicinal plants for human welfare", in 2003. They include the need for co-ordinated conservation action, based on both *in situ* and *ex situ* strategies, community participation and gender perspectives in the development of policies and programmes, the need for more information on the medicinal plant trade, inventorisation, development of sustainable harvesting practices, encouragement of micro-enterprises by the rural communities and the protection of traditional resources and intellectual property rights.

A recent meeting, organised by WWF-UK and WWF/TRAFFIC, Germany, representatives of WHO, IUCN, TRAFFIC and WWF, discussed the need to revise the 1993 guidelines on the Conservation of Medicinal Plants, in light of significant new developments in the field of medicinal plant conservation, such as community involvement, gender (apart from Governments, NGO's), incentive-based approaches and certification, GACP and GMP. A new sector included in the revised guidelines is the commercial sector (Source: phytomedica list server). These are global guidelines that were published by WHO, IUCN and WWF. (www.phytomedica@conserveafrica.org)

It has been emphasised by conservationists that conservation of medicinal plants should be with those people, who own, manage or make use of these species. This has been the approach by the FRLHT, Bangalore vis-à-vis it's medicinal plant conservation initiative (under the UNEP), which has highlighted the need of involving the major stakeholders. Medicinal plant conservation touches on many dimensions, probably, the most important being the role of medicinal plants, in biological and ecological conservation which stems from the foundation that they can provide for the involvement of people in the conservation of natural habitats (Hamilton, 2003).

CULTIVATION OF MEDICINAL PLANTS

Out of about 2,000 MAPs traded in Europe, between 1200 to 1300 are native to the continent, with only about 130-140 species derived from cultivated stock. An estimated 70,000 hectares of land are devoted to the cultivation of MAPs in the European union, France, Hungary and Spain (Lange, 1998).

Cultivation is commercially attractive to companies because it gives them greater control over quality and supply. Its feasibility depends on the ability of the species to thrive as a monocrop, while its economic viability depends on the volumes required and the market price (www. bellanet.org). MAPs likely to be cultivated are those that are high yielding and responsive to the economics of sale. Slow growing, space demanding or low-yielding species are less likely to be economical and attractive to commercial growers. Cultivation is particularly advantageous for growers where there are long-standing partnerships and contractual arrangements to supply manufacturers (Kuipers, 1997; Lange 1998).

Domestication of a species can be difficult and expensive. As a result, wild harvesting is likely to remain the only option for many species. Wild harvesting is generally much cheaper than expenses incurred in the establishment of cultivation, a process that may take 20 years (Kuipers, 1997; Lange 1998).

In the Indian context, the recent Biodiversity Act and the Medicinal Plants Board will provide a policy framework on medicinal plants *per se*, but it is yet to be seen how this will be operationalised at the ground level. Cultivation of MAPs, however, is a promising means of meeting an expanding market, while reducing the pressure on the population of wild plants (Lange, 1998). Cultivation can have the potential of saving the wild population of MAPs, thus, contributing towards the conservation of their genetic diversity. Harvesting could be made more sustainable by just cutting branches. Increased cultivation reduces the need for wild harvesting. However, it may be noted that collections from the wild may be unavoidable or even preferrable for those many MAPs that are slow growing (e.g., *Taxus baccata*) and difficult to domesticate, or for which the requirements are low, since the cost from wild-collection is typically much less than that of cultivation. Some of the risks associated with collections from the wilds are:

1. Over-harvesting of endemic MAPs species, with very restricted geographical distributions, can make them vulnerable to extinction; loss of genetic diversity through the reduction or elimination of local plants population with unique genetic characteristics; and the unnecessary destruction of plants resulting from careless/ destructive harvesting practices (Harnischfeger, 2000).
2. Some practitioners and patients prefer to use wild stock in the belief that wild stock have greater efficacy than cultivated ones. Wild harvested material can be much higher priced and sought after.

However, with cultivated stock, it is easier to ensure that standards of quality are maintained, including given levels of active compounds. Further it is easier to control post harvest treatment, by providing buyers with consistent products.

CONCLUSION

What should be the approach vis-à-vis MAPs, cultivation or wild harvesting? This is not only a either/or a situation, rather it is a question of developing structures that support the existing sustainable practices of wild harvest, supporting institutions that contribute to the preservation of wild gene pools, but at the same time, where feasible, to introduce cultivation practices. Wild-growing plants are sources of genetic diversity needed by the MAP industry for developing new cultivars. Furthermore, as reported, many MAPs are unsuited to high-volume mono cropping, or the market for them is too small to involve large-scale cultivation. Both of these have social dimensions that cannot be ignored. Taken together, both can be an effective means of alleviating problems associated with the rural poor, and at the same time providing solutions to issues related to medicinal plants, in particular, and biodiversity as a whole.

REFERENCES

Hamilton A. (2003). Discussion paper for Medplant. www.medplant.net

Harnischfeger G. (2000). Proposed guideline for the commercial collection of medicinal plant material. Journal of Herbs, Species and Medicinal Plants 7: 43-50.

Kuipers S.E. (1997). Trade in medicinal plants. In: Medicinal Plants for Forest Conservation and Healthcare, vol. II. Non-wood forest products (Bodeker G, Bhatt KKS, Burley J and Vantomme P eds), Food and Agricultural Organisation of the United Nations, Rome, Italy, p 45-59.

Lange D. (1998). Europe's medicinal and aromatic plant; their use, trade and conservation: An overview. TRAFFIC International, Cambridge, U K.

Syiem D., Kharbuli B. and Kayang H. (eds) (1999). Biodiversity North East India Perspectives. Published by North Eastern Biodiversity Research Cell, Shillong, India.

Chapter 11

Biodiversity of North-East India and its Conservation

Pramod Tandon and Suman Kumaria
Plant Biotechnology Laboratory, Department of Botany, North-Eastern Hill University, Shillong - 793 022, Meghalaya

Plant genetic resources are the basis of human survival and economic well-being, as they provide food, clothing, shelter, medicines, biomass energy and industrial raw materials, besides offering a potential for providing many more, yet unknown, benefits to the future generations. The Indian region having diverse geoclimatic conditions, boasts of a rich diversity of plants, animals and microorganisms. Indian flora accounts for 10.78% of the global flora. The endemism of Indian biodiversity is about 33% of the country's recorded flora, which is mainly concentrated in the North-East, the Western Ghats, North-West Himalayas, and the Andaman and Nicobar Islands. North-East India represents a distinct biogeographic zone, rich in bioresources, ethnic cultures and folklore traditions. The plant resources of NE India are enormous and represent about 80% (ca. 8000 spp.) of the floristic wealth of India. The forests are very rich in timber-yielding trees. The region contains floristic elements, some of which are not found anywhere else (endemic species), while some others are reported from regions of India, which are geographically separated from other countries by great distances. The plant genetic resources are getting depleted at an alarming rate. One quarter of the plant species are estimated to be at risk of extinction within the next generation. The conservation of plant genetic resources has for long been realised as an integral part of plant genetic improvement programmes. Conservation may be *in situ* or *ex situu*, either in the natural or semi-natural habitats, or in some purposely-built environment.

INTRODUCTION

Biological diversity or biodiversity, the study of all organisms, species, populations and the genetic variations among these, includes the complex assemblage of communities and ecosystems. It refers to the interrelatedness of genes, species, and ecosystems and their interactions with the environment. Three levels of biodiversity, namely genetic, species and ecosystem diversity exist. Biodiversity provides the basic essentialities, namely, food, fodder, shelter, biomass, etc. for the survival of mankind. As one of the resource–ecosystems of the environment, it provides the basic raw materials, which are the primary infrastructure and the basic requirements for any economic or other income generating activities. It executes the job of recycling, filtering and absorbing, besides detecting and indicating hazard levels within the environment. Biodiversity helps to maintain the biophysical systems, chemical balance of the earth and climate stabilisation. It is above all, a life supporting system. Biodiversity deserves to be conserved in its total form and its complete existence within the environment is essential.

Over millions of years, new biological species have evolved. The ever-changing new ecological conditions have caused many species to disappear. The changes in the climatic conditions of a particular habitat have led to changes in the inhabitants' population due to their ability or inability to adapt to the changes. Those that can adapt, continue to grow and prosper, while others dwindle and die out in course of time. Habitat loss and destruction, usually as a direct result of human activity and population growth, is a major force in the loss of species, populations, and ecosystems. Alterations in the ecosystem composition, such as the loss or decline of a species, can lead to a loss of biodiversity. Technological interventions and human activities have not only changed the set up and status of evolution, but have also accelerated the depletion, steep decline and deep extinction of biodiversity reserves. The ever-expanding population of humans on earth has created havoc on earth and on its species. The changing ecosystems, soils, waters and atmosphere have led to a crisis in biodiversity. Although, extinction being a rare and an obscure event, is a natural part of evolution and accounts for the extinction of 1-10 species a year. In face, scientists have estimated that the rates of extinction have drastically increased in this century i.e., about 1000 species per year.

The knowledge of biodiversity, which has increased in recent years, is due to the importance of biodiversity with its utilitarian values (medicine and agriculture), ecosystem services (continued production of atmospheric oxygen), and moral, ethical and aesthetic values (Eldredge, 1998). Plant biodiversity is the single greatest resource that humankind has treasured from nature. Scientists have described more than 2,50,000 species of mosses, ferns, conifers, and flowering plants, and have estimated that there may be more than 50,000 plant species yet to be documented, primarily in the tropical forests of the world (Tuxill, 2000).

According to Udvardy (1975), the Indian region falls in two biogeogaphical realms, viz., Palaearctic (which includes the whole Europe, Northern China, Japan, Soviet Russia, North Africa, Persia, etc.) and Indo-Malayan (includes the Malayan peninsula and islands of the Malayan Archipelago). The wide variety in physical features and climatic situations have resulted in a diversity of ecological habitats, like forests, grasslands, wetlands, coastal and marine ecosystems and desert ecosystems, which harbour and sustain the immense biodiversity. With only 2.4% of the total land area of the world, the known biological diversity of India contributes 8% to the known global biological diversity. According to the latest assessment, India is placed as 10th among the plant rich nations of the world and 4th among the Asian countries. Some habitats, particularly among terrestrial systems, possess a greater number of species than others. Indian flora accounts for 10.78% of the global flora. The endemism of Indian biodiversity is about 33% of the country's recorded flora, which is mainly concentrated in the North-East, Western Ghats, North-West Himalayas, and the Andaman and Nicobar Islands. According to the Red List of Threatened Species (IUCN, 2000), 44 plant species are critically endangered, 113 are endangered and 143 are vulnerable (Table 11.1). The floral species found in India have been categories into different divisions (Table 11.2).

Table 11.1: Threatened Plants of India by Status Category

Ex	*EW*	*CR*	*EN*	*VU*	*LR/cd*	*LR/nt*	*DD*
7	2	44	113	87	1	72	14

Ex-Extinct; EW-Extinct in the Wild; CR-Critically Endangered; VU-Vulnerable; LR/cd-Lower Risk conservation dependent; LR/nT-Lower Risk near threatened; DD-Data Deficient
Source: IUCN, 2000

Table 11.2: Floral species found in India

Division	*No. of species*
Bacteria	850
Viruses	Unknown
Algae	6500
Fungi	14500
Lichens	2000
Bryophytes	2850
Pteridophytes	1100
Gymnosperms	64
Angiosperms	17500

Source: MoEF 1999

With around 48,000 plant species and 81,000 animal species, India is one of the world's 12 mega-biodiversity centres. The flowering plants of India comprise about 17,500 species, which represent 6% of the world's known flowering plants. About 315 families and 2250 genera of flowering plants are known to exist in India in different ecosystems (Nayar, 1997). However, although the ethos of conservation is ingrained in India's cultural heritage, development activity is increasingly threatening this rich biodiversity.

BIODIVERSITY OF NORTH-EAST INDIA

North-East India and the whole of the Eastern Himalayas are one of the hotspots of the globe (Myers *et al.*, 2000). The plant resources of NE India are enormous and they represent the rich floristic wealth of India. The lower groups of plants like liverworts, lichens, ferns, etc. have not been explored to a great extent. The information on these groups of plants is based only on a few randomly collected species. There are many more species belonging to both flowering and non-flowering plants, that need to be discovered in the whole North-Eastern region. NE India represents a distinct biogeographic zone, rich in biodiversity, ethnic cultures and folklore traditions. About half of the total geographical area of the North-Eastern region is covered by forests.

The vegetation of NE India has been broadly classified into 3 major types viz., (i) the tropical, (ii) the temperate and (iii) the alpine vegetation (Murti and Joseph, 1984). The tropical forests comprise moist and dry deciduous, evergreen and semi-evergreen forests and savannah type of vegetation. The tropical evergreen forests are found in the Assam valley, the foot-hills of the Eastern Himalayas, the lower parts of the Naga Hills and Manipur and the lower elevations of the Khasi Hills. Tropical moist deciduous and dry deciduous forests are found in Tripura, Goalpara, Kamrup, Nowgong and Darrang in Assam, and parts of Meghalaya. Tropical grasslands in the riparian flats are distinct from those at higher altitudes of the Meghalaya Plateau and the lower parts of North Cachar and Mikir Hills. Subtropical mixed forests occur in Western Kameng and the inner valleys of Sing, Tirap and Lohit districts in Arunachal Pradesh. Subtropical pine forests occur within the areas of Khasi and Jaintia hills of Meghalaya and in Rupa valley of Kameng district in Arunachal Pradesh. The temperate forests are confined to elevations from 1700 to 3000m. Based on different elevations, these can be further categorised into lauraceous, high level oak-hemlock, coniferous and birch-rhododendron forests. The alpine and sub-alpine type of vegetation is found in the higher ranges at an elevation of 3000 to 4500 m. This vegetation occurs in the Aka hills of Kameng district of Arunachal Pradesh, Naga Hills and in Manipur.

The occurrence of many primitive plants such as *Altinglia, Euptelea, Tetracentron, Pycnarrhena, Haematocarpus, Aspidocarya, Magnolia griffithii, M. pealiana, M. gustavii, Stanntonia, Paryatia, Exbucklandia, Decaisnea, Holboellia, Distyllum, Houttuynia, Myrica esculenta, Alnus* and *Betula* have rendered North-East India as the 'cradle of flowering plants'. In spite of the rich vegetation, the flora of this region remains largely unexplored, which hinders the full utilisation of the plant resources.

The plant resources of NE India have been grouped under two major categories viz. agricultural and forest plant resources. The agricultural resources are confined mostly to Assam, where they contribute significantly to the state income. Important crops are rice, tea and jute, maize, cotton, jute, potato, pine-apples, oranges, banana, ginger and betel leaves etc. The various forest resources of the region are shown in Table 11.3.

Table 11.3: Forest resources of North-East India

Sl. No.	Forest Resources	Plant Species
1.	**Timber**	*Shorea robusta, Tectona grandis, Michelia champaka, Bombax ceiba, Shorea assamica, Dipterocarpus macrocarpus, Terminalia myriocarpa, Albizzia leebek, Cinnamomum glanduliferum, Alseodaphne dumicola, Phoeba attenuata, P. goalparensis, Litsaea cebifera, Juglans regia, Engelhardtia specata, Podocarpus neriifolia, Taxus baccata, Pinus insularis, P. wallichiana, P. excelsa, Cedrus deodara, Engelhardtia acerifolia, Quercus semicarpifolia, Q. griffithii, Q. semiserrata, Q. lanceaefolia, Castanopsis hysterix*, etc.
2.	**Bamboos**	*Bambusa tulda, B. pallida, B. vulgaris, Dendrocalamus strictus, D. hamiltonii, Melocanna bambusoides, Chimenobambusa khasiana, Arundinaria calloea, Phyllostachya mannii, Dendrocalamus gigenteus, Bambusa khasiana, B. balcooa, Arundinaria hirsuta, A. suberecta, Chimenobambusa polystachya, Bambusa elegans.*
3.	**Orchids**	*Aerides fieldingii, A. multiflorum, A. odoratum, Anoectochilus sikkimensis* (Jewel Orchid), *Arundina graminifolia* (Bamboo Orchid), *Calanthe mosuca, Cirrhopetalum ornatissimum, Coelogyne spp., Cryptochilus sanguinea, Cymbidium spp., Dendrobium spp., Paphiopedilum* spp., *Phaius tankervilliae* (Nun Orchid), *Pholidota articulata, P. imbricata, Pleione praecox, P. humilis, P. maculata, Renanthera imoschotiana, Rhynchostylis retusa, Thunia alba, T. marshalliana, Trias pusilla, Vanda coerulea, V. teres* etc.
4.	**Medicinal plants**	*Ageratum conyzoides, Aristolochia saccata, Buettneria platyphylla, Blumea balsamifera, Centella asiatica, Clerodendron colebrookianum, Lucas linifolia, Mahonia nepalensis, Nepenthes khasiana, Naravelia zeylanica, Nicotiana tabacum, Oxalis corniculata, Piper betele, P. diffusum, Dioscorea prazeri* etc.
5.	**Essential oils, gums dyes, resins and tannins**	*Cymbopogon martini, Acacia arabica, Pinus wallichiana, Indigofera tinctoria, Rubia cordifolia, Lawsonia alba, Terminalia catappa, T. citrine, T. tomentosa, Canarium benghalensis, C. sikkimensis, Acacthopanax trifoliatus, Aporusa diocia, Baccaurea sapida, Indigofera dosua, Smilax acreata, Turpinia nepalensis*, etc.
6.	**Paper and Match Industry**	*Albizzia lebbek, Alstonis scholaria, Bombax ceiba.*

DEPLETION OF THE PLANT RESOURCES

Forests are a part of the primeval environment and have a very important role in preserving environmental values. As long as the essential structure of forests is maintained, they can be manipulated or 'managed' in such a way so as to optimise their environmental benefits.

It is very unfortunate that plant resources are getting depleted at an alarming rate. A great number of plant species, including several unique and irreplaceable varieties, are becoming extinct and many more are awaiting a similar fate. Extinction of plant species means loss of opportunity to discover more useful forms. This genetic erosion has alarmingly reduced the diversity and variation within the species. It is a known truth that plant genetic resources have limitations regarding their rates of growth, reproduction and adaptation. As such, additional loss of plant genetic resources, on account of the destructive activities of man, will greatly affect the welfare of future generations. The rate at which extinction of plant resources is occurring, needs immediate concerted efforts because, in the face of accelerating losses, our greatest enemy is time.

The loss of biodiversity may be natural or man made. The natural factors responsible for posing a threat to plant species include changes in the biotic parameters in the environment, such as cyclones, long unfavourable weather spells, earthquakes, etc., and biotic parameters, such as natural competition and biological imparities of the species. The prominent man-made causes, are loss or fragmentation of habitat due to clearance of forests for various developmental activities, expansion of agriculture and aquaculture in prime forest lands and natural wetlands, over-exploitation of forest resources beyond their carrying capacity, without any consideration for their its sustainability, and the introduction of alien-species. The exponential growth, in both human as well as cattle population over the last century, has resulted in an increasing gap in the demand and supply of natural resources. Human population in India has grown over four times during the period 1901-2001, and today it represents about 17 per cent of the world's population. Similarly, the cattle population in India today constitutes about 18 per cent of that of the world. The resulting demand on the forest resources has lead to almost 80 per cent loss in the natural habitats (Mackinnon and Mackinnon, 1986; UNEP, 1995), including the degradation of water bodies. Apart from that, the advent of the green revolution in the early 1960's, relying on a handful of crop varieties, also led to an enormous loss in agro-biodiversity.

Shifting agriculture is the characteristic feature of the North-Eastern region. This land use system involves slashing the vegetation, burning the dried slash, raising a mixture of crops on a temporarily nutrient enriched soil for a year or two, fallowing the plot for the re-growth of natural vegetation, and eventual return to the same plot for another cropping phase after a few years (Ramakrishnan, 1992). Due to shifting cultivation and large-scale deforestation, there has been a continuous degradation of the land, leading to ecological imbalance including soil and water loss, which has ultimately led to problems in agricultural production (Verma and Bhatt, 2001). There are also a large number of problems in the management of agricultural production, which mainly includes gap in extension, poor motivation and awareness, inadequate training and lack of viable improved technology for *Jhum* cultivation. It used to be an appropriate and sustainable land use practice in diverse socio-economic set-ups where the dependent human population was within the carrying capacity of a 10-15 year *Jhum* cycle. However, in recent times, shifting cultivation is regarded as environmentally destructive as it has a very low input-output ratio. Shifting cultivation has become unsustainable, primarily due to the increase in population that has led to an increase in food demand. In order to meet the growing food demand, the *Jhum* cycle (the intervening period between two successive slashes) has shortened, leading to an overall decrease of crop yield (Anonymous, 1997). As a result, more virgin forest areas have been covered under shifting cultivation. Thus, the vicious cycle continued and more forest areas were converted to wastelands as a result of repeated *Jhum* having very short (often 2-3 years) cycles. Table 11.4 shows the areas covered by shifting cultivation in North-East India.

Table 11.4: Areas affected by shifting cultivation in NE India

State	*Geographical area (sq km)*	*Cumulative area of shifting cultivation (1987-1997) (million ha)*
Arunachal Pradesh	83,743	0.23
Manipur	22,327	0.36
Meghalaya	22,429	0.18
Mizoram	21,081	0.38
Nagaland	16,579	0.39
Tripura	10,486	0.06
Total		1.60

(Source: Anonymous, 1999)

CONSERVATION OF PLANT BIODIVERSITY

Loss of plant species is occurring at a rapid rate. There is an urgent need for sustaining biodiversity due to the following reasons:

- Forests render the climate more equable, prevent soil erosion and landslides and help in flood controlling floods.
- Most of today's food crops have been domesticated from wild tropical plants.
- About 80% of the world's population relies on plants or plant extracts for medicines.
- Pollination and seed dispersal by birds, insects and animals is essential to increase the diversity of genetic recombinations.
- Food security and the development of new crop strains for tomorrow.
- Survival of humans and other species is dependent on the producer.
- Wildlife serves as a gene library; premature extinction of species leads to irreversible loss of genetic information that influences the future evolution of life on earth.
- Aesthetic value.

Conservation of plant genetic resources is the only way to guarantee food supplies for the future. The conservation and subsequent use of such resources are complicated by cultural, economical, technical and political issues. Conservation may be *in situ* or *ex situ*, either in the natural or semi-natural habitat, or in some purposely-built environment. The choice of one or the other technique, or a combination of both, will depend on the particular case. *In situ* conservation involves the protection of genetic resources in the natural environment through the protection of the environment itself. It is an ideal and dynamic approach that allows plants to interact and co-evolve with other components of the ecosystem, including insects, animals and microbes. *In situ* conservation is costly to maintain and is highly susceptible to natural calamities like forest fires, extreme weather conditions, and damage by diseases and animals. The dominant approach of conserving plant genetic resources is by removing them from their natural habitats and protecting them at specialised institutions, such as botanical gardens, nurseries and gene banks. During the last few years, there has been an increase in the number of plant collections and in accessions in *ex situ* storage. Botanical Gardens have played an important role in *ex situ* conservation programmes, particularly in acclimatisation, rehabilitation, multiplication and judicious exploitation of plant resources. These gardens must have coordinated efforts through an international conservation network which would ensure that the rarest plants receive priority for propagation and ultimately, reintroduction. Gene banks focus exclusively on storing seeds of crop varieties and their immediate wild relatives. It is considered essential that all wild related species and varieties of an economic plant should be collected and grown at one place so that they can be easily assessed and utilised in crop improvement and hybridisation.

Biotechnology can play an important role in plant conservation programmes. It represents an interface of basic and applied sciences, involving gradual transformation of science into technology. Many plant species of agricultural, forestry and medicinal importance are vegetatively propagated and maintained in germplasm collections. The use of plant tissue culture technology for the conservation of germplasm through *in vitro* propagation and re-establishment has great potential in micropropagation and embryo rescue of plants (Tandon *et al.*, 1990; Tandon and Rathore, 1992, 1994; Kumaria and Tandon, 1994; Prance, 1997; Seeni and Sabu, 1997; Szendark *et al.*, 1997; Feijoo and Iglesias, 1998; Lynch, 1999; Nobre *et al.*, 2000; Mohammud *et al.*, 2002). The molecular marker technology and molecular diagnostics, *in vitro* technologies and cryopreservation techniques have also been applied for germplasm conservation (Tandon and Kumaria, 1998; Tandon, 2000).

There is a need for conservation strategies to be developed for the protection of species and ecosystems, involving a mixture of both *in situ* and *ex situ* strategies. A few important areas for such a strategy should include the following:

1. Population surveys and the creation of databases, based on adequate data available on the diversity of species and changes over time, so that a proper strategy for conservation could be designed.

2. For facilitating biodiversity conservation, it is important to generate maps of the protected areas, showing their continuity with the existing reserve and protected forests.
3. Improved protection efforts and a landscape approach to conservation, where protected and non-protected areas should be integrated through significant protection measures, initiated at both the state and community levels.
4. *In situ* conservation through management of the populations distantly related in the wild which would be essential for their survival.
5. Computer models which would provide estimates for the survival of the regular population in the face of newly identified threats, population status and other relevant research, for the next 100 years, should be generated. These models should also include different steps needed for improvement in the overall conservation status of the species.
6. The preservation plots and the protected areas in a number of states are the important means for conserving and protecting important plant species. These should be properly demarcated and actively maintained.
7. Since a large proportion of biodiversity occurs outside the network of protected areas, conservation programmes outside protected areas must become an integral part of forest management.

REFERENCES

Anonymous (1997). Agenda notes for the meeting on WDPCA Project. Ministry of Agriculture (SWC Wing) New Delhi.

Anonymous (1999). State of Forest Report. Forest Survey of India. Ministry of Environment and Forests, Govt of India p 113.

Eldredge N. (1998). Life in the Balance: Humanity and the Biodiversity Crisis. Princeton: Nevraumont Publishing Co.

Feijoo M.C. and Iglesias I. (1998). Multiplication of an endangered plant: *Gentiana lutea* L subsp *aurantiaca* Lainz using *in vitro* culture. Plant Tissue Culture and Biotechnology 4: 87-94.

IUCN (2000). Red list of threatened species. Gland, Switzerland: The World Conservation Union.

Kumaria S. and Tandon P. (1994). Clonal propagation and establishment of *Dendrobium fimbriatum* var. *oculatum* Hk f. In: Advances in Plant Tissue Culture in India (Tandon P ed), Pragati Prakashan, India, pp 218-231.

Lynch P.T. (1999). Tissue culture techniques in *in vitro* plant conservation. In: Plant Conservation Biotechnology (Bensen EE ed), Taylor and Francis, London p 41-62.

Mackinnon J. and Mackinnon K. (1986). Review of protected areas system in the Indo-Malaysian realm. IUCN, Gland.

MoEF (1999). National policy and macrolevel action strategy on biodiversity. Ministry of Environment and Forests, Government of India, New Delhi p 74.

Mohammud N. Amin., Nahar K., Ahmed F. and Ahmad S. (2002). Micropropagation of *Annona squamosa* Linn using explants (shoot tip and node) of field grown mature plants. Pakistan Journal of Biological Sciences 5(4): 394-397.

Murti S.K. and Joseph J. (1984). Plant resources of Northeast India. Meghalaya Science Society, Shillong Vol. II: 28-39.

Myers N., Mittermeier R.A., Mittermeier C.G., da Fonseca G.A.B. and Kent J. (2000). Biodiversity hotspots for conservation priorities. Nature 403: 853-858.

Nayar M.P. (1997). Changing pattern of Indian flora. Bull Bot Surv India 19: 145-154.

Nobre J., Santos C. and Romano A. (2000). Micropropagation of the Mediterranean species *Viburnum tinus*. Plant Cell Tiss Org Cult 60(1): 75-78.

Prance G.T. (1997). The conservation of botanic diversity. In: Plant Genetic Conservation (Maxted N, Ford-Lloyd BV and Hawkes JG eds), Chapman and Hall, London.

Ramakrishnan P.S. (1992). Ecology of shifting cultivation and ecosystem restoration. In: Ecosystem Rehabilitation, Vol 2, Ecosystem Analysis and Synthesis (Wali MK ed), SPB Academic Publishing Bv, The Netherlands, pp 19-35.

Seeni S. and Sabu K.K. (1997). Conservation and economic utilisation of plant resources through biotechnological means. In: Conservation and Economic Evaluation of Biodiversity Vol. 1. (Pushpangadan P, Ravi K and Santosh V eds), Oxford IBH Publishing Co, New Delhi, p 239-249.

Szendark E., Read P.E. and Miller V.I. (1997). *In vitro* culture initiation and multiplication of the hybrids of the once-dominant Eastern forest tree, the American chestnut (*Castanea* spp). In – Vitro 33(3): Pt 2, 51A.

Tandon P. (2000). Role of Biotechnology in the Conservation of Plant Genetic Reources in the 21st century - An Indian Perspective. Platinum Jubilee Lecture delivered at the 87th session of Indian Science Congress, Pune, January 4, 2000.

Tandon P. and Rathore T.S. (1992). Regeneration of plantlets from hypocotyle derived callus of *Coptis teeta*. Plant Cell Tissue and Organ Culture 28: 115-117.

Tandon P. and Rathore T.S. (1994). Mass clonal multiplication of the threatened Indian insectivorous plant (*Nepenthes khasiana* Hk F) through shoot bud culture. In: Advances in Plant Tissue Culture in India (Tandon P ed), Pragati Prakashan, India p 185-192.

Tandon P. and Kumaria S. (1998). Threats to plant diversity in high altitude areas of North-East India and conservation of rare and endangered plants using biotechnological approaches. In: Science at High Altitude, Proceedings of the National Symposium (Saha S, Ray PK and Sinha B eds), Allied Publishers Ltd, India p 140-147.

Tandon P., Rathore T.S. and Dang J.C. (1990). Mass multiplication and conservation of some threatened plant species of Northeast India through tissue culture. In: Abstracts VIIth International Congress on Plant Tissue and Culture, IAPTC, Amsterdam, Netherlands p 136.

Tuxill J (2000). The biodiversity that people made—and are now losing. World Watch 13 (3): 24-35.

Udvarday MDF (1975). A Classification of the Biogeographical Provinces of the World. IUCN occasional paper 18, Morges, Switzerland.

UNEP (1995). Global Biodiversity Assessment. Cambridge University Press, Cambridge.

Verma, N.D. and Bhatt B.P. (2001). (eds) Steps towards modernisation of agriculture in the NEH region. ICAR Research Complex for NEH Region, Umiam, Meghalaya, India p 536.

Chapter 12

Diversity in Agri-Horticultural Plants: An Experience with North-East India

D.K. Hore
National Bureau of Plant Genetic Resources
New Kench's Trace, Shillong - 793 013, Meghalaya

The North-East region of India is one of the twelve (12) mega-biodiversity centres of the world. The various plant /crop genetic resources found in this region have got paramount importance for future crop improvement programmes. The extent of crop genetic diversities that exist in the region have been enumerated according to the crop groups. The information furnished is based entirely on field explorations and collections made during various trips. The role of NBPGR in the region and its future approach for safeguarding the valuable genetic wealth of crops has been summarised. The present status of various germplasm conservation strategies undertaken in the region has been discussed.

INTRODUCTION

There are three levels of diversity in the plant kingdom. Firstly, ecosystem diversity refers to independent communities of species and their physical environment with characteristic assemblage of plants. Secondly, species diversity refers to species or cultivars of species within a given area. Thirdly, genetic diversity, refers to the variations in genes and genotypes between and within species, which permit an organism to adapt to changes in the environment. The diversity in agricultural plants supports our social and economic system in order to sustain by way of providing food and nutrition. In nature, the diversities comprise traditional landraces, primitive cultivars, wild and weedy relatives of cultivated plants, with variations among them being due to continuous evolution over time and space. High yielding varieties resulting from research programmes and natural outcrossing, also constitute a part of crop diversity. The present day plant breeders are dependent and in search of such original genes or genepools which can serve them in developing a desired crop variety.

Physiographically, the Indian subcontinent has been divided strictly into eight regions (excluding Myanmar) by Chatterjee (1939). Similarly, based on physiography, climate and cultural characteristics, the ICAR has recognised the divisions of subcontinent into 8 agroclimatic regions (Murthy and Pandey, 1978). The region is termed as the 'Hindustani Centre' (Zeven and Zhukovsky, 1975), where 152 cultivated species are enumerated and considered as their centres of diversity. About 35 agri-horticultural species were domesticated in the Indian gene centre. Approximately, 160 species, combining the exotic and indigenous plants, being are cultivated presently (Chandel and Rana, 1995). Over 325 wild relatives of various crop plants are native to this subcontinent and hold a natural reservoir of genes (Arora and Nayar, 1984). Floristically, the Indian gene centre is extremely rich (over 15000 species of higher plants, of which 4900 species are endemic). Of these, 800 species are of ethnobotanical interests, and well over 1200 species are of medicinal importance.

NORTH-EAST INDIA

Seven constituent states in the Eastern Himalayan ranges form the North-East region of India. These are Arunachal Pradesh, Assam, Manipur, Meghalaya, Mizoram, Nagaland and Tripura. The region lies in a strategic area, bordering China, Burma and Bangladesh on three sides and is geographically located between 20° to 29° N latitude and 89.40° to 97.25° W longitude. The total area of the region is 256 thousand sq km, of which 212 thousand sq km form hilly areas. The region has a total population of 33 million. Sixty-seven major ethnic communities are the inhabitants of this region.

Diverse physiography ranges from plains, plateaus, and hills to high mountains characterise the region. Accordingly, the temperature, rainfall, soil and vegetation also vary. The annual precipitation in the region is due to the South-West monsoons and it continues from May to October. The average rainfall in the region is recorded at a minimum of 1648 mm and a maximum of 6320 mm.

The land may be classified into five categories on the basis of land use, i.e., forest area, net area available for cultivation, uncultivated land excluding fallow land, fallow land and net area sown. The summarised Table 12.1 gives a picture of land resource utilisation against the total geographical area of each state of the North-East region.

VEGETATION

The natural vegetation of the region is characterised by the composition of the Eastern Himalayan and Burmese (Myanmar) floral elements. According to altitude and topography, they occupy either diversified mixed community or pure community. This is further determined by higher rainfall, scanty snowfall, high humidity and temperature. Dicotyledonous species are present almost three times more than monocotyledonous species. The dominant families are Orchidaceae, Poaceae, Cyperaceae, Rosaceae, Leguminosaceae, Asteraceae, Fagaceae, Magnoliaceae, Dipterocarpaceae, Urticaceae, Moraceae, Melastomaceae and Euphorbiaceae.

Table 12.1: Land use classification in the North-Eastern Region - 1989-90 and 1990-91 (Source – Basic statistics of the North Eastern Council, 1995) (in '000 ha)

State	*Geographical Area*	*Year*	*Reporting Area For land Utilization*	*Forest Area*	*Area put to non-Agricultural uses*	*Barren and un-cultivable land*	*Total of col. (6-7)*	*Permanent pastures and other grazing land*	*Cultivable waste-land*	*Others*	*Current fallow land*	*Fallow Land other than current follow land*	*Net Area Sown*
1	2	3	4	5	6	7	8	9	10	11	12	13	14
Arunachal Pradesh	8374	1989-90	5544	5200	29	48	77	-	-	44	25	49	149
		1990-91	5544	5200	29	48	77	-	-	44	25	49	149
Assam (c)	78444	1989-90	7852	1984	914	1541	245 5	184	104	247	88	84	2706
		1990-91	7852	1984	914	1541	245 5	184	104	247	88	84	2706
Manipur (j)	2233	1989-90	2211	602	26	1419	144 5	(n)	(n)	24	(a)	-	140
		1990-91	2211	602	26	1419	144 5	(n)	(n)	24	(a)	-	140
Meghalaya	2243	1989-90	2239 (f)	940	84	142	226	-	492	151	60	167	203
		1990-91	2239 (f)	939	84	142	226	-	493	153	59	167	202
Mizoram	2108	1989-90	2102	1303	10	201	211	4	74	3	183	259	65
		1990-91	2102	1303	10	201	211	4	74	3	183	259	65
Nagaland	1658	1989-90	1514	863	28	-	28	-	96	124	113	109	181
		1990-91	1532	862	28	-	28	-	99	125	113	110	190
Tripura	1049	1989-90	1049	606	131	(1)	131	(n)	1	39	1	1	270
		1990-91	1049	606	131	(1)	131	(n)	1	39	1	1	270

** Note: (a) Below 500 ha (c) Relates to the year 1981-82 (f) Excludes area under the Municipality and Town Committee (j) Ad-hoc estimates (n) Includes land under miscellaneous trees, crops and groves, etc.

There are about 1500 threatened forest species in India, out of which 650 plant species are said to be from the North-Eastern region, alone. This number may increase when the different Reserve forests, Sanctuaries, National parks and declared Biosphere areas are explored.

AGRICULTURAL CROP DIVERSITY

The Indian subcontinent is rich in the diversity of agricultural crops. There are a large number of crop species, which originated in this region and were subsequently domesticated. In the course of time, they were cultivated. Table 12.2 lists crop species, which are said to have originated from this subcontinent.

Table 12.2: Agri-horticultural crop species native to India

	Crop group	*Crop*	*Name*	*Family*
I	CEREALS	Rice	*Oryza sativa* L. (Various *indica* types)	Poaceae
		Dwarf wheat	*Triticum aestivum* (L.) Thell. Spp. *Sphaerococum* (Perc.) MK.	Poaceae
II	PSEUDO-CEREALS/ MILLETS	Kodo millet	*Paspalum scrobiculatum* L.	Poaceae
		Ragi	*Eleusine coracana* Gaertn.	Poaceae
		Raishan	*Digitaria cruciata* (Nees) A. Camus var. *esculenta* Bor.	Poaceae
		Job's tear	*Coix lacryma jobi* Gaertn. var. *ma-yuen* (Romanet) Stapf and var. *stencarpa* Stapf.	Poaceae
III	PULSES	Black gram	*Vigna mungo* (L.) Hepper	Papilionaceae
		Moth bean	*Vigna aconitifolia* Jacq.	Papilionaceae
		Pigeon pea	*Cajanus cajan* Milsp.	Papilionaceae
		Horse gram	*Dolichos uniflorus* Lam.	Papilionaceae
		Velvet bean	*Mucana utilis* Wall. ex Wight.	Papilionaceae
IV	OILSEEDS	Rai	*Brassica campestris* L.	Brassicaceae
		Sarson	*B. campestris* L. var. *sarson* Prain.	Brassicaceae
		Toria	*B. campestris* L. var. *toria* Duthie and Fuller	Brassicaceae
V	VEGETABLE	Brinjal	*Solanum melongena* L.	Solanaceae
		Cucurbits	*Luffa acutangula* (L.) Roxb.	Cucurbitaceae
			L. cylimdrica (L.) M. Roem.	Cucurbitaceae
			Trichosanthes dioica Roxb.	Cucurbitaceae
			Citrullus lanatus (Thunb.) Mansf.	
			Cucumis sativus L.	Cucurbitaceae
			Coccinia indica (L.) Voigt.	Cucurbitaceae
				Cucurbitaceae
		Radish	*Raphnus sativus* L.	Brassicaceae
VI	TUBER CROPS	Taro	*Colocasia esculenta* (L.) Schott.	Araceae
		Yam	Dioscorea spp.	Dioscoreaceae
		Elephant's Foot yam	*Amorphophallus campanulatus* (Roxb.) Blume	Araceae

	Crop group	Crop	Name	Family
VII	FIBRE CROPS	Jute	*Corchorus capsularis* L.	Tiliaceae
		Tree cotton	Gossypium arboreum L.	Malvacee
		Sunhemp	*Crotalaria juncea* L.	Papilionaceae
VIII	SPICES AND CONDIMENTS	Turmeric	*Curcuma longa L.*	Zingiberaceae
		Ginger	*Zingiber officinale Rosc.*	Zingiberaceae
		Bengal Cardamom	*Amomu aromaticum* Roxb.	Zingiberaceae
		Large cardamom	*Amomum subulatum* Roxb.	Zingiberaceae
		Black pepper	*Piper nigram* L.	Piperaceae
IX	BEVERAGE	Tea	*Camellia sinensis* (L.) O. Kuntzd	Ternstroemia-ceae
X	FRUITS	Mango	*Magnifera indica* L.	Anacardiaceae
		Orange, Lime and Lemons	*Citrus* spp.	Rutacae
		Phalsa	*Grewia asiatica* L.	Tiliaceae
		Bael	*Aegle marmelos* (L.) Corr.	Rutaceare
		Lotka	*Baccaurea sapida* Muell. –Arg.	Euphorbiaceae
		Banana	*Musa* spp.	Musaceae
		Jack Fruit	*Artocarpus heterophyllus* Lamk.	Moraceae
XI	SUGAR YIELDING CROP	Sugarcane	*Saccharum officinarum* L.	Poaceae
			Saccharum sinense Roxb.	Poaceae
XII	BAMBOOS	Bambusae	*Bambusa arundinaceae* (Retz.) Willd.	Poaceae
			B. polymorpha Munro	Poaceae
			B. strictus Nees	Poaceae
			B. tulda Roxb.	Poaceae
			Cephalostachyum capitatum Munro	Poaceae
			Dendrocalamus hamiltonii Nees and Arn.	Poaceae
			D. longispathus Kurz.	Poaceae
			Melocanna baccifera (Roxb.) Kurz.	Poaceae
			Neohouzeaua dullooa (Munro) A. Camus	Poaceae
			Sinocalamus giganteus (Walp.) Keng	Poaceae

EXTENT OF CROP DIVERSITY

Based on agroecological situations and phytogeographical regions of India, Arora (1991) has given a picture of crop plant diversity, their wild relatives and diversity in economic plants. In a nutshell, the present author will, however, give an account of the distribution and extent of crop diversity in the North-East region of the country. A floristic account of the region was depicted by Hooker (1872-97), Kanjilal (1934-40), Bor (1960) and subsequently, by the scientists from the Botanical Survey of India and the Arunachal Pradesh Forest Department, who enumerated and described many economic plant species.

Factors responsible for the occurrence of much genetic diversity are (i) Ecogeographical and physiographical diversity, (ii) Heavy precipitation, (iii) Ethnic communities and their complete dependence on agriculture, (iv) Crop introduction from the adjacent bordering countries, (v) Natural migration, (vi) Remoteness and inaccessibility of the area, leading to the maintenance of primitive cultivars.

With regard to cultivated species, the total gene pool for each crop and their diversity in the region in not yet fully understood. Tribal people, whose basic profession is agriculture, dominate the region. Diverse topography and ecological conditions also prevail in this region. Crop diversities are dispersed over the region and they have been preserved through the ages by the local farmers who have, inherited them from their previous generations. The occurrence of such indigenous/native diversities is stated crop group/crop wise.

I. Cereals

Rice (*Oryza sativa* L.) is the principal crop of this region. Hardly, 5% the area is cultivated out of the total rice growing area of the country in this region and thus it contributes to about 8% of the total rice production in India. More than 30 HYV are cultivated under different rice ecosystems. Based on the exploration and collection of germplasms (Hore and Sharma, 1991), there are more than 8000 rice diversities occurring in the North-Eastern region of the country. At least 9 taxa of wild relatives of rice (Hore and Sharma, 1993) can be found in localised pockets. Attention is still to be paid to various deep-water varieties of rice, particularly in Assam state. Black kerneled rice is a novelty of Manipur, while red kerneled is of Mizoram. Autumn rice is cultivated in every state of the region. Summer rice is cultivated in Assam, Meghalaya and Tripura (Borthakur, 1992). Both winter and summer rice cultivation is absent in Arunachal Pradesh.

Maize (*Zea mays* L.) is a secondary crop, especially in the hilly areas of this region. Abundant variability has been observed and the NBPGR Regional Station has collected at least 800 diversities. So far dent and flint corn are commonly found, while pop and waxy type are rarely cultivated. Maize has been classified into 15 distinct races and 3 sub-races on the basis of the occurrence of different maize varieties (Singh, 1975). The states of Nagaland, Manipur and Sikkim are potential areas of maize crop diversities. Its production may be enhanced and encouraged if the farmers get proper storage, communication and marketing facilities. Except naturally occurring hybrids of corn, invasion of HYV is very less in the region. The introduction of *Euchlaena mexicana* Schrad (Teosinte) and *Tripsacum dactyloides* (L.) L. (Gama grass) is a recent one in some localised pockets and it is used as a forage crop. These two genera are closely related to the corm crop and have played a dominant role in understanding the evolution of maize.

Barley (*Hordeum vulgare* L.) is one of the oldest cultivated cereals. The crop has limited cultivation in the high altitudes of Arunachal Pradesh and Sikkim. Perhaps, due to less consumption and preference, its cultivation is not popular. A few varieties, however, were collected by the NBPGR from Arunachal Pradesh and Sikkim states. Mostly they are the naked types, awned and cold-adaptable genotypes. Similarly, wheat (*Triticum aestivum* L.) is cultivated in the high altitudes of these two states and these are cold-tolerant genotypes. HYV like *Kalyan Sona* and *Sonalika* which have been introduced in Assam are performing well due to their wide adaptability and resistantce to loose smut.

II a. Pseudocereals

Buckwheat (*Fagopyrum esculentum* Moench) is one of the minor grain crops, said to be a native of Central Asia. The distribution/cultivation of this crop is done in the higher altitude areas of the Himalayas. In the North-Eastern region, it is cultivated in Sikkim, Meghalaya and Manipur. Dwarf varieties have been introduced in the plains of Assam to a limited scale. *Fagopyrum cumosum* Meissn which is a wild relative of Buckwheat is a perennial type, found in Meghalaya (Joshi and Paroda, 1991).

Amaranth is also another crop, which is cultivated to a limited scale in this region. The vegetable types which belong to the species *Amaranthus tricolor, A. caudatus* and *A. spinosus* are used in Meghalaya, Assam, Manipur, Nagaland, Tripura and Arunachal Pradesh, while the grain amaranth, which belongs to the species *A. hypochondriacus,* is cultivated in the high altitudes areas of Arunachal Pradesh. The importance of this future food crop has been well-documented (Joshi and Rana, 1991).

Chenopods (*Chenopodium album* L.) are also an important pseudocereal for the high hills of the NE region. The species is cultivated in *jhum* land along with other crops. The tall type variety bears a heavy inflorescence and the apical stalk bends due to its weight. This variety is usually cultivated for grains in Phek and Tuensang district of Nagaland. Variability is pronounced in plant height, leaf size, days to flower, maturity, size of inflorescence and grain yield. The promising potentialities of this crop in the Western Himalayan region have been studied (Pratap and Kapoor, 1987).

II b. Millets

These are small-seeded cereal and forage grasses, which embrace the 10 (ten) genera of family, Poaceae. The species includes *Pennisetum americanum, Setaria italica, Panicum miliare, Eleusine coracana, Paspalum scrobiculatum, Echinochloa frumentaceae, Brachiaria ramosa, Eragrostis tef, Digitaria* sp. and *Coix lacryma-jobi.* Among the millet growing countries, India occupies the first position. In the North-Eastern region, the major cultivated millets are foxtail millet (*Setaria italica*), Finger millet (*Eleusine coracana*), kodo millet (*Paspalum scrobiculatum*), pearl millet (*Pennisetum americanum*), raishan (*Digitaria cruciata* var. *esculenta*) and Job's tear

(*Coix lacryma-jobi*). Within these species, again there are a good number of variabilities existing, especially in foxtail (Panicle and caryopsis characteristics) and finger millet (arrangement of spikes). Tribal pockets of Arunachal Pradesh and Meghalaya (Garo Hills) are rich in diversities of *Eleusine* and *Setaria*. Raishan (*Digitaria cruciata* var. *esculenta*), an obsolete crop of the West Khasi Hills of Meghalaya is gradually disappearing due to changes in food habits (Singh and Arora, 1972).

Jowar (*Sorghum bicolor*) is another important food and fodder crop of dry land agriculture. Cultivation of *Sorghum* and their variability is available in Mikir and the NC Hills of the region. It is cultivated for the grains, which are palatable, the seeds differ in size, colour and shape of inflorescence.

The distribution of various cultivated millets and their wild relatives in the region is given below (Bor, 1960):

Species	*Distribution*
Eleusine indica Gaertn.	Wild in plains and hills up to 5000 ft.
Eleusine coracana Gaertn.	Cultivated in hills in between the range of 2000-5000 ft.
Avena sativa L.	Cultivated in Arunachal Pradesh.
Avena fatua L.	Introduced and naturalised in Meghalaya
Setaria italica Beauv.	Cultivated throughout the hilly region up to 6000 ft.
Setaria glauca Beauv.	Wild in hills and plains.
Pennisetum americanum (L.) K Schum.	Upper Assam.
Pennisetum alopecuroides (L.) Spreng.	Naga hills in between 4000-5000 ft. altitude.
Sorghum nitidum Pers.	Common in hill grasslands.
Sorghum halepense (L.) Pers.	Plains of Assam.
Coix lacryma-jobi L.	Meghalaya, Nagaland and Arunachal Pradesh
Coix gigantean Roxb.	Khasi Hills of Meghalaya.

Cultivation of other millets, such as Kodo (*Paspalum scrobiculatum* L.), Prose (*Panicum miliaceum* L) Barnyard (*Echinochloa frumentacea* (Roxb) Link), Small millets (*Panicum miliare* Lamk) and Teff [*Eragrostis teff* (Zuoc) Trot] are unknown to this region.

III. Pulse crops or grain /pored legumes

Pulses are important crops, which produce root nodules, fix atmospheric nitrogen and thereby, improve soil fertility. They contribute sufficient protein in the diet, besides serving as excellent forage and cattle feed. The cultivated area is about 20% of the total area under food grain crops in the country. The major cultivated species of the region and their wild relative(s) are given below:

A. Grain legumes

	Cultivated species	*No. of available diversities*	*Wild related Species*
1.	Rice bean – *Vigna umbellata* (Thunb.) Ohwi and Ohashi.	40 covering 5 Botanical vars. var. *major* var. *rymbaija* var. *macrocarpa*	V. *carpensis* V. cantiang V. *vexillata* V. *pilosa*
2.	Adzuki bean – *Vigna angularis* (Willd.) Ohwi and Ohashi.	Introduced	–
3.	Green gram Wilgczek– *Vigna radiata* (L.)		*Vigna radiata* var. *sublobatus*
4.	Black gram–*Vigna mungo* (L.) Hepper		
5.	Khesari–*Lathyrus sativus* L.	-	-
6.	Soya bean – *Glycine max* (L.) Merril	6	-
7.	Cowpea–*Vigna unguiculata* (L.) Walp.	8	-
8.	Pigeon pea–*Cajanus cajan* Milsp.	90	*Atylosia barbata* Baker *A. elongata* Benth. Threat
9.	Lentil – *Lens culinaris* Medic	-	-
10.	Pea – *Pisum sativum* L.	6 var. *hortense* var. *arvense*	-

B. Vegetable legumes

	Cultivated species	*Diversity in Cultivars*	*Wild related species*
1.	Sem – *Lablab purpureus* (L.) Sweet.	10	*Dolichos falcatus* Klein *D. biflorus* L.
2.	French bean – *Phaseolus vulgaris* L.	4	-
3.	Jack bean – *Canavalia ensiformis* DC.	1	-
4.	Tree bean – *Parkia roxburghii* G. Don (Sharma *et al.*, 1993)	2	-

C. *Tuberous legumes*

	Species	*No. of diversity*	*Remarks*
1.	*Moghania vestita* (Benth ex Baker) O. Kuntze	1	Confined mostly to the Khasi and Jaintia hills district of Meghalaya (Singh and Arora, 1973)
2.	*Eriosema chinense* Vogel	1	An underexploited and under-utilised tuberous herb, grows in hilly slopes, among grasslands.
3.	*Pachyrhizus erosus* (L.) Urb.	2	Cultivated in semi-wild condition in the plains of Assam and the foothills of Meghalaya.

IV. Oilseeds

Oil, which is an ingredient of various food preparations is extracted from oilseeds. In fact, oil can be extracted from various plant species. While edible oil is abundantly used in the plains of the region, the use of vegetable oil is lesser. There are 203 varieties of oilseeds, which are identified, released and introduced in cultivation: 44 varieties have been developed and introduced directly after selection and screening from the germplasm.

Crop	*Species*	*Area of concentration of diversities*	*Wild relatives*
Sesame	*Sesamum indicum* (L.)	Mikir, Cachar Hills, Mizoram, Nagaland.	-
Groundnut	*Arachis hypogaea* (L.)	Introduced in Hills	-
Niger	*Guizotia oleifera* (DC.)	Introduced in Assam plains	-
Castor	*Ricinus communis* (L.F. Br.)	Assam and Meghalaya	-
Rape seed			
a. Yellow	*Brassica campestris* var. yellow *sarson* (L.)		-
b. Brown	*B. campestris* var.brown *sarson* (L.)		-
c. Toria	*B. campestris* var. *toria* (L.)		-
Indian Mustard	*Brassica juncea* (Czern and Coss)	Assam Plains	
Black Mustard	*Brassica nigra* (L.) Koch.)	Mikir and Cachar Hill, including Assam plains	-
Bhanjira	*Perilla frutescens* (L.) Britt.	Subtropical hilly areas of the NE Region (Sharma *et. al.*, 1989)	-
Linseed	*Linum usitatissimum* (L.)	Introduced in Assam plains	-

V. Vegetables

Plenty of vegetable crops are grown in the region. Cole crops and leafy vegetables are grown more in the hills due to their growth and adaptation, with respect to temperature and altitude. Remarkable polymorphic variabilities are observed in cucurbits (Hore and Sharma, 1990a), which include the following varieties and land races:

(a) Cucurbits

	Cultivated species	*Area of concentration for diversities*	*Range of diversity*
1.	*Cucurbita maxima* Duch ex. Lam	Throughout the region	Extensive
2.	*Cucurbita moschata* (Duch ex. Lam) Duch ex. Poir	Hilly areas	Moderate
3.	*Cucurbita ficifolia* Bouche	Meghalaya	Introduced and naturalized
4.	*Cucurbita pepo* L.	Meghalaya	Limited
5.	*Coccinia grandis* (L.) Voigt	Assam plains	Limited
6.	*Cucumis sativus* L.	Throughout the region	Wide
7.	*Cucumis callosus* (Rottl) Cogn	Foothills of Meghalaya and Mizoram	Confined to limited pockets
8.	*Luffa acutangula* (L.) Roxb	Tropical areas of Assam	Wide
9.	*Luffa cylindrica* (L.) M Roem	Tropical and Subtropical areas of Assam, Meghalaya and Manipur	Moderate
10.	*Momordica charantia* L.	Assam, Meghalaya, Arunachal, Tripura.	Moderate
11.	*Momordica cochinchinensis* (Lour) Spreng	Assam, Meghalaya, Manipur	Limited
12.	*Momordica dioica* Roxb. ex Willd	Garo Hills	Meagre
13.	*Trichosanthes anguina* L.	Meghalaya, Tripura, Assam	Limited
14.	*Trichosanthes dioica* Roxb	Tropical areas of Assam and Tripura	Limited
15.	*Cyclanthera pedata* (L) Schard	Hills of Meghalaya, Manipur, Nagaland	Meagre
16.	*Benincasa hispida* (Thunb) Cogn	Assam, Nagaland and Meghalaya	Wide
17.	*Lagenaria siceraria* (Molina) Standl.	Throughout the region	Wide
18.	*Sechium edule* (Jacq) Sw	High Hills of Meghallaya, Manipur, Mizoram and Nagaland	Moderate

(b) Wild species

Wild economic species or relatives of cultivated cucurbitis are an invaluable material for breeding programmes. Examples of such species are – *Cucumis hystrix, hodgsonia macrocarpa, Luffa aegyptica, cucumis callosus, gymnopetalum cochinchinensis, Bryonopsis laciniosa* and *Trichosanthes cordata.*

Similarly, Okra *Abelmoschus esculentus* (L) Moench is cultivated in the tropical and subtropical altituds of the region. Wide range of variability exists in shape, size and longitudinal fissures of immature fruits. Perkin's long green cultivar is recommended for hilly areas. The sparsely distributed *A. manihot* and *A. moschatus* are two wild relative species of the crop, available in the region.

The primary centre of origin of brinjal (*Solanum melogena* L.) is India, extensive variability in the region. Brinjal can be classified into two groups according to its longevity, i.e., annual and perennials. Three botanical varieties such as *esculentum, surpentinum* and *depressum* are widely cultivated. At least, 100 different cultivars exhibit a vast range of morphological variability, with respect to size, shape, colour, and weight. The other related and widely occurring species of *Solanum* growing in the region are:

Species	*Remarks*
Solanum macrocarpon L.	Introduced in the region.
S. xanthocarpum Schrad & Wendl	Used as vegetable and for medicinal purposes.
S. indicum L.	Domesticated, used as vegetable and medicine.
S. mammossum L.	Possibly introduced, ornamental with high solasadine percentage.
S. khasianum Cl.	Wild and cultivated for solasadine alkaloid.
S. torvum Swartz.	Wild, sold in the market of West Java.
S. barbisetum Nees.	Ripe fruits are eaten.
S. ferox L.	Wild, leaves are used medicinally.
S. spirale Roxb.	Wild, but domesticated for medicinal use.
S. sisymbrifolium Lam.	Native of America, wildly grown.
S. Kurzii Br.	Endemic to the Garo Hills, Meghalaya, edible.
S. gille *Raddi.*	Introduced in the region as a vegetable.

Species of chillies such as *Capsicum annum* L. and *Capsicum frutescens* are very much used as a vegetable, and as a spice for flavouring dishes. Based on the survey in the region, the NBPGR, Regional Station has already identified more than 50 chilli diversities, which are already characterised and conserved in the National Gene Bank.

VI. Tuber crops

Tuber crops play a dominant role in the life of tribals. They are commonly used as food and require little care for cultivation. The major crop genera of this group include Taros (*Colocasia, Alocasia, Xanthosoma*), yams (*Dioscorea* spp), potato (*Solanum tuberosum*) and sweet potato (*Ipomoea batatas*). Cassava (*Manihot esculenta)* is an introduced crop in this region. The major species of this group, available in this region (Pandey, 1992), are:

(a) ***Taros***: *Colocasia esculenta* (L.) Schott; *C. affinis* Schott; *C. fallax* Schott; *C. manni* Hook f; *Alocasia cucullata* (Lour) Schott; *A. fallax* Schott; *A. macrorrhiza* (L.) Schott; *Xanthosoma sagittifolium* (L.) Schott.

(b) ***Elephant foot yam***: *Amorphophallus campanulatus* (Roxb.) Blume.

(c) ***Yam***: The species belonging to the genus *Dioscorea* are known as yam. It is a very large genus comprising over 600 species. The plants are rhizomatous or tuberous and are climbers by habit. A few species such as *Dioscorea alata, D. esculenta, D. bulbifera, D. hamiltonii* and *D. pentaphylla* are important as food crops. Morphologically, twining of stem characters distinguishes the species. Bulbils are the storage organs in some species. Yams are rich in carbohydrates, starch grains and vitamin C. Calcium, iron and phosphorus are among the components of the mineral fraction of the tuber. Certain yam species contain quantities of alkaloids (Dioscorine) and steroid derivatives (Diosgenin). The edible fresh species *Dioscorea* can be utilised in boiled, pounded, mashed and fried forms. Industrially, they can be used in the preparation of corticosteriod drugs, poisonous alkaloids for pharmaceutical purposes and oral contraceptives (from Diosgenin).

The occurrence of species and their corresponding varieties with their distribution in the region is listed in Table 12.3.

Table 12.3: Species and varieties of *Dioscorea* and their distribution in the North Eastern states

Specific Name	*Varieties*	*Distribution*
D. alata L.	var. *alata* L. var. *tarri* P and B var. *vera* P and B	Assam (N. Lakhimpur), Meghalaya, Mizoram (Aizwal, Sairang, Banglang)
D. arachnida Pand B	-	Assam, Meghalaya
D. belophylla (Prain) Haines	-	Arunachal Pradesh, Meghalaya, Nepal, Bhutan.
D. bulbifera L.	var. *Crispata* (Roxb.) Prain var. *Pulchella* (Roxb.) Prain	Arunachal Pradesh, Assam, Meghalaya, Mizoram (300 m to 1800 m).
D. cumingii P. and B.	var. *inaequifolia* (Elmer ex. P and B.) Burkill.	Assam (N. Lakhimpur, Karbi-Anglong).
D. decipiens Hook f.	-	
D. deltoidea Wall. ex. Griseb	-	Kameng district in Arunachal Pradesh.
D. esculenta (Lour.) Burk	-	Meghalaya (Garo Hills)
D. glabra Roxb.	-	Assam (Goalpara, N. Lakhimpur, Dibrugarh, Kamrup, Arunachal Pradesh (Siang , Lohit, Khonsa), Meghalaya (Garo Hills, Nongpoh), Nagaland (Kohima) – Alt. 225 to 1035 m.
D. hamiltonii Hook f.	-	Tripura (Damcherra 35 m.), Arunachal Pradesh (Siang, Khonsa), Meghalaya (Garo Hills, Ri-Bhoi area –Umtrew), Assam (Kaziranga, Kamrup), - Alt. 20. m to 1130 m.
D. hispida Dennst.	var. *hispida* var. *daemona* (Roxb.) P and B	Assam (Kamrup, Darrang dists.), Meghalaya (Dist. Ri-bhoi, Garo Hills), Mizoram (Lungleh) – Alt. 240 m. to 600 m
D. japonica Thunb.	-	Nagaland, Mizoram
D. kumaonensis Kunth.	var. *kumaonensis* var. *stramena*	Meghalaya (Jowai, Jarain) –Alt. 1450 m., Nagaland.
D. lepcharum P. and B.	-	Assam, Arunachal Pradesh.
D. listeri P and B.	-	Arunachal Pradesh, Assam, Nagaland
D. melanophyma P and B.	-	Arunachal Pradesh, Meghalaya, Sikkim
D. oppositifolia L.	var. *oppositifolia*	Assam (Nagaon, Karbi-Anglong, N-C Hills), Meghalaya (Jowai), Arunachal Pradesh (Lohit, Subansiri and Khonsa)
D. pentaphylla L.	var. *pentaphylla*	Assam (kaziranga), Meghalaya (Mawphlang, Nartiang – 1350 m, Nongstoin, Jowai, Rongrem in Garo Hills, 170 m), Arunachal Pradesh (Khonsa, Alt. 450 – 1066 m, Siang, Subansiri), Nagaland (Kohima).

Specific Name	Varieties	Distribution
	var. *communis* P. and B.	Assam Assam
	var. *hortorum* P. and B.	Assam, Mizoram (Banglang).
	var. *kussok* P. and B.	Arunachal Pradesh (Siang)
	var. *suli* P. and B.	
D. prazeri P and B	-	Sikkim, Assam, Nagaland
D. pubera Bl	-	Meghalaya (Khasi Hills)
D. pyrifolia Kunth	var. *ferruginea* P. and B.	Arunachal Pradesh, Assam.
D. scortechinii P and B	-	Assam (Sibsagar).
D. trinervia Roxb. ex P and B	-	Meghalaya (Sohra to Dwaki – shaded forests), Arunachal Pradesh (Khonsa), Alt. 250 m.
D. wallichii Hook f	-	Arunachal Pradesh (Aka hills)
D. wattii P and B	-	Meghalaya, Assam (N. Lakhimpur, Kakoch F.), Arunachal Pradesh (Lohit), Tripura (Chandrapur), Sikkim

VII. Sugar-yielding crop

The main sources of sugar are the plants named *Saccharum officinarum*. Though the cultivation and production of sugar is very less in the region, it is bestowed with rich genetic resources and the wild relatives of closely allied genera (Hore and Sharma, 1995). The major related wild genera of *Saccharum* are *Erianthus, Miscanthu, Sclerostachya* and *Narenga*. Their occurrence and population are frequent and abundant in their respective habitats. The states of Meghalaya, Manipur and Arunachal Pradesh are rich in the diversity of sugarcane, however, their cultivation is very limited.

VIII. Beverage

India is considered as a secondary centre of the origin for tea. Tea crop is cultivated mainly in Assam and Tripura. *Camellia sinensis* (L.) O. Ktze and its various clones, contributing high yield, are maintained by tea gardens. Besides, *C. kissi, C. caudate, C. japonica* L. are 3 known and related species of tea occurring in this region (Singh and Bezbaruah, 1988), *Camellia sasanqua* and *C. irrawadiensis* are introduced species. The major wild relative species of Tea are *Gardenia excelsa, Eurya japonica, Eurya acuminata* and *Schima wallichi* . The primary centres of tea are the mountainous areas of China and North of North-East India. The secondary centre is Assam, where the *C. sinensis* var. *assamica* (Masters) Wight, is cultivated.

IX. Fibre crops

A wide range of various fibre yielding crop diversities including their wild relatives occur in this region. These are as follows:

Crop(s)	*Species*	*Wild relatives*	*Distribution*
Cotton	*Gossypium arboreum* L.	-	Assam
	G. herbaceum L.	-	Garo Hills, Meghalaya
	Bombax ceiba L.	-	Assam
Jute	*Corchorus capsularis* L.	-	Assam, Meghalaya
	C. olitorius L.	-	
Mesta	*Hibiscus cannabinus*	var. *viridis* *rubber* *simplex* *vulgaris* *purpureus*	Assam, Tripura
Roselle	*H. sabdariffa* L.	-	Garo Hills and Hills of Assam.
Ramie	*Boehmeria nivea* (L.) Guad	*B.utilis*	Assam, Meghalaya
		B. rcgulosa	Assam
Matcane	*Schumannianthus dichotomus* (Roxb.) Gagnep and North Bengal	-	Lower Assam (Plains)
Hemp	*Cannabis sativa* L.	-	Wildly grown in Assam and Meghalaya as secondary vegetation

X. Spices

Ginger (*Zingiber officinale* Rosc.) and Turmeric (*Curcuma longa* L.) are well-adapted crops in this region. Variability in these crops was noticed in respect of their fibre content and curcumin content, respectively. Besides these, Black pepper (*Piper nigrum* L.), True cardamom [*Elettaria cardamomum* (L.) Maton], Cinnamon (*Cinnamomum zeylanicum* Bl.) Tejpat (*C. tamala* Nees and Eberm), Bengal cardamom (*Amomum aromaticum* Roxb.) and Nepal cardamom (*Amomum subulatum* Roxb.) are cultivated in Assam and Sikkim areas.

Promising cultivars of ginger, namely *'Thingalaidon'* and *'Thingpui'* were originally derived from their wild relatives in the Lushai Hills of Mizoram. These varieties are well distributed in Meghalaya. Similarly, *'Lakadong'* variety of turmeric is well known for its high curcumin content. Its home is Jowai, Meghalaya. *Curcuma zeodoaria* Rosc., from which 'Shoti' (Barley) food is prepared is grown wildly in Meghalaya and the plains of Assam.

Leaves of Coriander (*Coriander sativum* L., *Eryngium foetidum* L.) and Pipperming (*Mentha piperata* L.) are used for flavouring and eaten as salad with cooked food.

XI. Fumitories and masticatories

Two species of tobacco (*Nicotiana tabacum* L. and *N. rustica* L.) are cultivated extensively for the purpose of smoking and chewing. The former species is grown in Assam, while the latter needs a cooler climate, hence it is grown in a limited scale in most of the hill districts of the region. The occasionally found wild tobacco plants are escapes of cultivation. Inter-specific crosses have been made to introduce male sterilising cytoplasm and genes conditioning resistance to disease.

Table 12.4: Area-wise important fruits in the NE Region

(in thousand hectares)

State	*Pineapple*	*Orange*	*Other Citrus*	*Banana*	*Apple*	*Pear*	*Stone fruits*	*Papaya*	*Mango*	*Guava*	*Coconut*	*Total*
Arunachal Pradesh	0.26	1.67	-	-	2.312	0.125	0.305	-	-	-	-	4.67
Assam	3.40	2.00	1.10	20.00	-	-	-	2.48	-	-	5.0	25.98
Manipur	6.02	4.00	2.00	8.50	-	1.60	0.40	0.2	0.1	1.0	-	23.82
Meghalaya	7.30	4.80	1.60	2.88	-	0.30	0.31	-	-	-	-	17.85
Mizoram	0.21	0.38	-	0.9	-	-	-	0.69	-	-	-	2.18
Nagaland	0.11	0.36	-	-	-	-	-	-	-	-	-	-
Tripura	2.83	1.21	-	8.50	-	-	-	0.30	-	-	0.8	7.14
Total	20.13	14.42	4.70	35.08	2.31	2.03	1.02	3.67	0.1	1.0	5.8	84.44

XII. Fruits

The cultivation of fruits in India dates back to ancient times. Among the indigenous fruits, mango, banana, phalsa, jackfruit, bael, amla, lemons, citrons seem to be of Indian origin. Peach, sweet orange, litchi might have come from China and some common fruits, such as guava, papaya, sapota, custard apple, and pineapple are from continental America though introduced in the country by the Portuguese settlers. Temperate fruits and nuts like apple, pear, plum, cherry, apricot, walnut, pomegranate, pistachio-nut, grapes were introduced from the Central Asiatic region.

The state-wise area distribution under different major fruit cultivations in the region, is given in Table 12.4. It indicates that pineapple, orange, banana and papaya are popular and more suitable for the local climatic conditions. Subtropical fruits like apple are well acclimatised to the high altitude areas of Arunachal Pradesh.

A list of major and minor fruits that are available (both in cultivation and wild) in this region, along with their wide variety of wild relatives and their distribution, have been furnished in Table 12.5.

Table 12.5: List of Major Fruit Diversities in the NE Region

Common Name	*Species*	*No. of Cultivars in the region*	*No. of wild relatives*	*Distribution*
TROPICAL				
Mango	*Magnifera indica* L.	6	2	Tropical areas of Assam, Meghalaya, Mizoram, Tripura.
Litchi	*Litchi chinensis* Senn.	2	-	Introduced in Assam and Tripura.
Jackfruit	*Artocarpus heterpophyllus* Lamk.	3	1	Foothills and plains of the entire region.
Jamun	*Syzygium cumini* (L.) Skeels	1	2	Cultivated plains of Assam
Water Rose Apple	*Syzygium aquemn* (B.f.) Alston	1	-	Tropical and Subtropical areas of Assam, Meghalaya, Tripura.
Sugar apple	*Annona squamosa* L.	1	1	-
Elephant apple	*Dillenia indiaca* L.	-	2	Assam, Foothills of Pasighat, Tripura.
Wood apple	*Aegle marmelos* (L.) Corr.	-	2	Tropical areas of Assam, Meghalaya and Tripura.
Ber	*Zizyphus mauritiana* Lamk.	1	5	Plains and Hills up to 500 m. alt.
Carambola	*Averrhoa carambola* L.	1	-	Plains of Assam, Meghalaya and Tripura.
Cashewnut	*Anacardium occidentale* L.	1	-	Introduced in the lower altitude / foothills of Garo and Mikir Hills.
Coconut	*Cocos nucifera* L.	2	-	Lower Assam, Tripura.

Common Name	Species	No. of Cultivars in the region	No. of wild relatives	Distribution
Pineapple	*Ananas comasus* L.	2	-	Introduced and naturalised.
Papaya	*Carica papaya* L.	3	1	Introduced in the region, adapted nicely in Mizo Hills.
Pomegranate	*Punica granatum* L.	1	-	Sporadically introduced in home gardens.
Watermelon	*Citrullus lanatus* (Thunb.) Mats. and Nakai.	1	-	Assam plains, riverbed areas.
Cucumber	*Cucumis sativus* L.	8	2	Throughout NE region
Aonla	*Emblica officinalis* Gaertn.	1	2	Plains and Hills of Meghalaya.
Jalpai	*Elaeocarpus floribundus* Bl.	-	1	Lower Hills of Meghalaya.
Guava	*Psidium guajava* L.	6	1	Tropical and Subtropical (upto 1000 m.) zone of NE region.
Mangosteen	*Garcinia mangostena* L.	-	1	Mid-altitude areas of Meghalaya.
Banana	*Musa acuminata* Colla. *Musa balbisiana*	18	14	Throughout the tropical and subtropical zones.
SUBTROPICAL				
Lime, lemon and oranges	*Citrus* spp.	17 plus 52 vars.	-	Lime and Lemon in both tropical and subtropical while oranges in the subtropical zone only (Kaul, 1981).
Passion fruit	*Passiflora edulis* Sims.	2	-	Meghalaya, Nagaland.
Mulberry	*Morus alba* L.	. 1	1	Assam, Meghalaya.
Peach	*Prunus persica* Benth. and Hook.f.	4	5	Meghalaya.
Pear	*Pyrus communis* L.	4	3	Meghalaya, Nagaland and Arunachal Pradesh.
Plum	*Prunus domestica* L. spp. *insitita* (L.) Schneid	9	-	Meghalaya and Mizoram
Straw berry	*Fragaria vesca* L.	3	-	Hills of NE region
Apple	*Malas sylvestris* (L.) Mill	24	2	Arunachal Pradesh and introduced in Nagaland.
Docynia	*Docynia hookeriana* Dcne	1	1	Meghalaya
Cotoneaster	*Cotoneaster rotundifolia* Wall. ex Hook. f.	1	7	Meghalaya
Sorbus	*Sorbus lanata* Wenzig	1	8	Meghalaya (mostly apomictics)
Rubus	*Rubus ellipticus* Smith	-	16	Meghalaya. Nagalnd, Arunachal Pradesh

There are a few dozen economically important minor fruits, which are necessary to be exploited appropriately for large-scale cultivation (Pandey, 1993). They can help farmers to earn cash in difficult times. The potential species under this category include: *Elaegnus latifolia, Elaegnus pyriformis, Rubus ellipticus, Rubus lasiocarpus, Myrica esculenta, Myrica sapida, Pyrus pashia, Prunus nepalensis, Baccaurea sapida, Castanopsis indica, Viburnum foetidum, Calamus erectus, Gynocordia odorata, Emblica officinalis, Artocarpus lakoocha, Garcinia laceafolia, Magnifera sylvatica, Citrus aurantium, Citrus medica, Musa balbisiana, Phoenix humilis* and *Elaeocarpus floribundus.*

XIII. Bamboo

Bamboos are the predominant species of the region and can be termed as 'Green Gold'. Approximately, 63 different species covering 19 genera of Bamboo are found in this region (Sharma *et al.*, 1993). The utilisation of various bamboo species is associated with the daily life of the tribal communities. Besides, 4 paper mills of the region solely depend on bamboo as a raw material for manufacturing paper pulp. The majority of bamboo vegetation is concentrated in the North Cachar and Mikir Hills. Sporadic distribution of various species extends up to 1500 m or beyond the altitude hills, right from the plain areas of the region. The genera occurring in the zone are – *Arundinaria, Bambusa, Cephalostachyum, Chimnobambusa, Dendrocalamus, Dinochloa, Gigantochloa, Melocalamus, Melocanna, Neohouzeaua, Oxytenanthera, Phyllostachys, Thamnocalamus, Semiarundinaria, Pseudostachyum, Pleioblastus, Teinostachyum, Sinobambusa* and *Thyrsostachys*. The diversity of species available in the states of Arunachal Pradesh, Meghalaya and Assam are 41, 38 and 33 species, respectively, while some species are common to all these states.

XIV. Cane

This economic plant species has enough potentialities because of its widespread occurrence in the region. Large-scale extractions have resulted in diminishing the cane population. However, a status paper to this effect (Hore, 1994) was prepared for this region, which included 15 species of *Calamus,* 1 species of *Daemonorops* and 3 species of *Plectocomia,* found to be occurring in this region.

XV. Orchids

This group of plants has got immense horticulture importance. The long lasting, beautiful flowers can earn foreign exchange through cut flower trade. There are about 700 species, covering 132 genera existing in this region. The orchids in this region contribute to about 50% of country's orchid species wealth, out of which 157 species are endemic. Many genera/species have migrated from the Indo-Myanmar-Malayam region. Attempts to conserve various orchid species by the efforts of the Orchid Research Centre, Tipi, Arunachal Pradesh are praiseworthy. On the basis of occurrence and importance of the species, a new project, Directorate on Orchids, has been initiated by the ICAR in Sikkim. The occurrence of North-East Indian orchids has been analysed, elucidated and reviewed (Hore and Sharma, 1990b).

Apart from agricultural plants, there is a vast diversity in other economic plants, which includes (a) medicinal and aromatic plants (b) wild forage plants (c) native ornamental plants, and (d) forest trees. Commercial cultivation of economically important species is not done on a large scale. In fact, they are extracted from nature randomly. In case of forest tree species, leaving aside some monoculture cultivations (*Tectona grandis, Shorea robusta, Michelia champaca, Lagerstroemia flos-reginae, Gmelina arborea, Anthocephalus chinensis, Pinus khasiana*), most of the diversified tree species are subjected to felling without any regeneration/afforestation. In recent years, the most sought after trees are–*Taxus baccata* and *Aquillaria agallocha.* Those are being exploited illegally and randomly from their habitats for commercial purposes.

RESPONSIBILITIES, ACTION OF NBPGR AND THEIR FUTURE APPROACHES

Against this rich crop diversity scenario in the region, NBPGR, Regional Station, Shillong has been entrusted with the responsibility of collection, evaluation and *ex situ* conservation of mandate crop genetic wealth of the region. Since 1985, the scientists of the station have collected more than ten thousand accessions of diversities belonging to several crops (Table 12.6). About 30% area of the region, is still to be explored (which is unapproachable or inaccessible). There are several Biospheres reserves, National parks, Game sanctuaries and Reserve forests situated in the region. Assessment of plant genetic wealth in such reserves is not yet fully understood. The Nokrek reserve of the Garo Hills has been declared as a gene sanctuary for *Citrus indica* (Singh, 1981). Habitat preservation of *Nepenthes khasiana* (Pitcher plant) in Jarain by the Forest department is an ideal approach for conservation. A similar approach is needed for the conservation of *Lilium mackliniae* in the Sirohi hills of Ukhrul, Manipur.

Table 12.6. Statewise collection of crop group accessions since establishment of the station

Crop group	*Assam*	*Meghalaya*	*Manipur*	*Tripura*	*Nagaland*	*Arunachal*	*Mizoram*	*Sikkim*	*Total*
Cereal	870	801	21	230	463	548	210	52	3195
Pseudo-cereals	82	202	-	50	117	54	21	-	526
Grain legumes	210	522	45	101	260	67	69	14	1288
Oilseeds	315	152	-	57	117	14	53	-	708
Tuber crops	180	454	1	93	53	75	23	52	931
Vegetable	426	646	151	352	161	61	114	06	1917
Spices and Condiments	83	418	13	81	76	22	40	34	767
Fibre crops	131	50	-	27	02	-	5	-	215
Fruit crops	25	116	16	39	01	135	6	44	382
Sugar crops	-	4	115	-	-	-	2	-	121
Miscellaneous	89	266	57	62	77	73	37	12	673
	2411	**3631**	**419**	**1092**	**1327**	**1049**	**580**	**214**	**10723**

The station distributed about 3500 accessions of various crop germplasm to different research institutes, during the period 1987-1995.

Attention has also been given to popularize under-utilised crops and to increase their yield potentialities, so that they can serve as alternate food plants. Experiments are being undertaken on such crops, through the All India Coordinated Research Project on under-utilised and under-exploited plants.

Introduction of exotic crop plants, which can adapt to the local environmental condition, is another function of the station. The desert type exotic guava and passion fruit are all well acclaimatised by now. Cultivation of forgotten crops (*Digitaria cruciata* var. *esculenta, Passiflora quadrangularis, Tetragonia expansa, Vigna umbellata* var. *rambaija, Moghania vestita, Perilla frutescens,* wild races of *Allium sativum*) needs to be continued, in order to avoid their further genetic erosion.

One of the NBPGR Regional Stations is located in this region at Shillong, Meghalaya. The responsibilities entrusted to them are to explore, collect and conserve the gene pool of valuable crop germplasm that are present in this region. A small gene bank is going to be established very soon, so that suitable crop germplasm can be maintained in *ex situ* conditions. Efforts are underway to develop human resources, as well as infra-structural facilities.

STATUS OF CONSERVATION

Conservation of plant genetic resources diversity is an important aspect, which is carried out in two ways, i.e., *in situ* and *ex situ* methods. The former strategy was barely initiated in the early seventies, when India became a member of the MAB programme launched by the UNESCO. The object is to conserve, for present and future use, the diversity and integrity of biotic communities within natural ecosystems and to safeguard the genetic diversity of species on which their continuous evolution depends.

In situ conservation includes the protection of large areas under different categories, such as in the form of Biosphere Reserves, Wild Life Sanctuaries, Nature Reserves or habitat / plot preservation. This way, entire biomes are safeguarded and the extinction of species is deterred, but there is little impact on useful plants, unless protection laws are strictly enforced. The system allows the wild species to remain in their natural communities. For semi-wild or semi-cultivated species, it helps to domesticate economic plants and safeguard landraces in their respective areas of cultivation. On the other hand, *ex-situ* conservation leads to the domestication of in mass reservoirs or genetic reserves. It can be seed storage in farmers' level or in the seed gene bank under controlled systems. Further, the plants parts of vegetatively propagated material can be maintained through tissue/meristem cultures. Another approach of *ex situ* conservation is the maintenance of genetic diversity in plants, in Botanical Gardens or in recreational parks, as a last effort.

In the context of the North-Eastern region, there are a total of 3 biospheres reserves, 4 national parks, 31 wild life sanctuaries and 8 game reserves (Table 12.7), which cover about 10 % of the total geographical area. In India, the diversity of wild relatives of crop plants amounts to about 320 species, of which about 75 are endemic, and 132 species belong to the wild genetic diversity occurring in the NE region.

Table 12.7. Number of different nature reserves in North-Eastern states

State (s)	*Biosphere Reserves*	*National Parks*	*Wild life Sanctuaries/ Reserved Forests*	*Game Reserve*	*Total area (Sq kms)*
Assam	1(Manas)	1(Kaziranga)	8	2	4384
Arunachal Pradesh	1(Namdapha)	1(Mouling)	12	-	19325
Meghalaya	1(Nokrek)	1(Balphakram)	3	-	302
Mizoram	-	-	1	4	941
Manipur	-	1(Keibul)	1(Yangoupokpi)	-	225
Nagaland	-	-	2	2	222
Tripura	-	-	4	-	*
Total	3	4	31	8	-

* Not available.

The information about the availability of genetic diversity in economic plants is still scanty. Nokrek Reserve in Meghalaya and its adjacent areas possess at least, 10 diversities of *Citrus* and 6 diversities of foxtail millets. The sacred groves of the Khasi and Jaintia Hills of Meghalaya are unique examples of the prevalence of traditional values dating back to the Paleolithic times of human history. These are forbidden forests and no one is allowed to cut trees or kill animals inside these forests. The forest which is a natural storehouse of at least 650 species of orchids, 450 species of grass, 28 species of bamboos and 65 species of rhododendrons is being conserved, through traditional beliefs, religion and culture.

The habitat reserve of *Nepenthes khasiana* at Jarain in the Jaintia Hills and Baghmara Hills of the South Garo Hills are to be protected more carefully.

The Manas wild life sanctuary in Assam is rich in *Dalbergia sisso, Elaeocarpus ganitrus, Terminalia myriocarpa* and *Mesua ferra.* Similarly, the ecosystem of Kaziranga ranges from marshy grassland to mixed hilly deciduous forest, with emphasis being laid on the animal species, which inhabit it.

The Namdapha Biosphere reserve in Arunachal Pradesh is dominated by a variety of tree species of forestries importance. The Hollung tree (*Dipterocarpus retusus* Bl.) is a novelty of this reserve, in addition to 50 timber yielding species. The forest is tropical and subtropical evergreen type and the vegetation can be classified in three categories, which encompass the floristic elements of species belonging to trees, shrubs, herbs, lianas, climbers, stragglers, epiphytes, saprophytes, lichens, palms, banana and orchid groups. The surveyed area and the floristic analysis reveal that there is an occurrence of 1150 plant species (Chauhan, 1994) in this reserve. The richness of additional plant species diversities in Arunachal Pradesh and the preparation of their inventory (Haridasan and Deori, 1991) provides valuable information after the account of "Flora of British India" by J.D. Hooker. This includes 683 plant species belonging to 104 families. The establishment of an orchid sanctuary at Sessa, Arunachal Pradesh, is a noble attempt towards the conservation of native orchid species.

The state of Mizoram is rich in indigenous bamboo and banana. A total of 941 sq km area is under reserve forests. Blue mountain area is rich in orchids and Rhododendrons. Submontane forests are rich in the populaton of *Magnifera sylvatica, Parkia roxburghii,* and *Aquillaria agallocha.* The natural forests of higher elevation, in Mamit-Tuidam area, are being conserved. Multiplication of cane species in such a habitat is a good attempt by the Forest Department. The state needs to establish a *Bambusetum,* where the representative species of bamboos can be conserved. The state is also rich in the occurrence of *Citrus* species diversity.

Two wild life sanctuaries and two game reserves have been declared in Nagaland state so far. The regions adjacent to Burma are rich in various important economic plants. *Panax pseudoginseng* has recently been discovered in the state. The state has established a Botanical Garden at Kohima. A good number of variabilities in hill paddy, maize, soyabean, rice bean, Job's tear, tuber crops, cucurbitis are available in the state. The distribution of *Citrus indica* is also found in the Intangki Reserve. *Vanda coerulea, Heliconia dasyantha, Ensete superba* and *Citrus ichangensis* are the novelties of the State.

Manipur has hardly 225 sq km area of protected forests, which include one national park and one wild life sanctuary. The state is proud of various *Magnifera* and wild relatives of the *Saccharum* species diversity. *Zizania latifolia,* an aquatic species is considered as a wild relatives of cultivated rice; it is found in Loktak Lake. The state is rich in cultivated paddy diversities. The black kerneled rice is available in the high hills of Ukhrul district. Siroy Hills are famous for 'Siroy lily' (*Lillium mackliniae*), which is considered an endangered species.

The state of Tripura is the smallest among the region, where only 4 (four) wild life sanctuaries exist. The Jampui hills area, adjacent to Mizoram state is rich in *Citrus, Zizyphus* species and indigenous hill paddy diversities. The wild relative of Guava (*Psidium guianensis*) is dispersed all over the hillocks of West Tripura district. The state is also rich in paddy, jute, and bamboo germplasm diversity. The bamboo-based cottage industry of the state, which exports various curie items abroad, is the pride of the country.

Ex situ conservation of crop genetic resources is the task of NBPGR, by way of storing different crop germplasm material in National Gene Bank. For vegetatively propagated material, the germplasm is conserved in *in vitro* repositories and clonal field Gene Bank. In the North-Eastern region such facilities are very limited. The NBPGR Regional Station, Shillong does not conserve such material here. In fact, they have sent 2470 accessions of germplasm of rice, maize, rice bean, chilli and soyabean, belonging to this region, to the National Gene Bank in New Delhi. 66 accs. of ginger and 20 accs. of *Musa* germplasms were also sent to the NFOTCR, for *in vitro* culture and storage. The activities for *in vitro* culture and storage of citrus, orchids, economically important plants and endangered plants were initiated by the ICAR Research Complex for the NEH Region, Barapani, Orchid Research and Development Centre, Tipi, North-Eastern Hill University, Shillong and Botanical Survey of India, Shillong, respectively.

REFERENCES

Arora R.K. (1991). Plant diversity in the Indian gene centre. In: Plant Genetic Resources Conservation and Management – Concepts and Approaches (Paroda RS and Arora RK eds) IBPGR regional office for South and Southeast Asia, New Delhi p 25-54.

Arora R.K. and Nayar E.R. (1984). Wild relatives of crop plants in India. NBPGR Sci Monogr 7: 90.

Bor N.L. (1960). The grasses of Burma, Ceylon, India and Pakistan (excluding *Bambuseae*). Pergamon Press, Oxford-London-New York–Paris.

Borthakur D.N. (1992). Genetic resources of the Northeast region. In: Agriculture in the Northeastern Region with Special Reference to Hill agriculture, Beecee Prakashan, Guwahati p 108-126.

Chandel K.P.S. and Rana R.S. (1995). Plant genetic resources system–an Indian perspective. In: Proceedings of the South Asia National Coordinators meeting on Plant Genetic Resources (Arora RK and Ramanathan Rao V eds), IPGRI office for South Asia, New Delhi, India p 25-43.

Chatterjee D. (1939). Studies on the endemic flora of India and Burma. J Royal Asiat Soc Bengal NS (Science) 5: 19-67.

Chauhan A.S. (1994). Resource potential of proposed Namdapha Biosphere Reserve of the Northeastern region. Indian J Hill Farmg *7*(1): 22-29.

Haridasan K. and Deori D. (1991). A list of plants not recorded in the flora of British India that are recorded from or found in Arunachal Pradesh. Arunachal For News 9(2): 21-34.

Hooker J.D. (1872-97). The Flora of British India, Vol. 1-7, L. Reeve and Co, London.

Hore D.K. (1994). Rattan of northeast India and strategies for their conservation. In: Management of Minor Forest Products Sustainability Oxford and IBH Publishing Co Pvt Ltd, New Delhi, Calcutta p 560-566.

Hore D.K. and Sharma B.D. (1990a). Cultivated cucurbitis and their wild relatives in India. India J Pl Genet Resources 3(1): 81-90.

Hore D.K. and Sharma B.D. (1990b). Orchids of northeastern India and their conservation. J. Orchid Soc India 4(1 and 2): 113-122.

Hore D.K. and Sharma B.D. (1991). The collection, conservation, evaluation and uses of rices of South Asia with special reference to Northeast India. India J Pl Genet Resources 4(1): 61-72.

Hore D.K. and Sharma B.D. (1993). Wild rice genetic resources of Northeast India. Indian J Pl Gene Resources 6(1): 27-32.

Hore D.K. and Sharma B.D. (1995). Diversity and distribution of sugarcane complex in NE India. Indian J Forestry 18(1): 38-42.

Joshi B.D. and Rana R.S. (1991). Grain Amaranthus: The Future Food Crop. NBPGR, Shimla Sci Mongr 3: 1-152.

Joshi B.D. and Paroda R.S. (1991). Buckwheat in India. NBPGR Shimla Sci Monogr 2: 117.

Kanjilal U.N. (1934-1940). Flora of Assam, Vol I-IV, Government of Assam.

Kaul G.I. (1981). Development of horticulture in the Northeastern region – a case study of North Cachar Hills District. NEC Secretariat, Shillong.

Murthy S.R. and Pandey S. (1978). Delineation of agro-ecological region of India. 11th Congress, International Society of Soil Science, Edmonton, Canada, 17-27 June 1978.

Pandey G. (1992). Collecting aroids diversity in Northeastern India. Indian J Pl Genet Resource 5(2): 97-102.

Pandey G. (1993). Indigenous minor fruit genetic resources and their marketing status in the Northeastern hills of India. J Hill Res 6(1): 1-4.

Pratap T. and Kapoor P. (1987). The Himalayan grain chenopods-III. An underexploited food plant with promosing potential. Agric Ecosystems Environ 19: 71-79.

Sharma B.D., Hore D.K. and Mandal S. (1989). *Perilla*: An oil and protein rich underexploited crop of the Northeastern hills. J Oilseeds Res 6: 386-389.

Sharma B.D., Hore D.K. and Salam J.S. (1993). *Parkia roxburghii* – a useful tree of Northeastern India. Indian J Pl Genet Resources 6(2): 171-173.

Singh B. (1975). Races of maize in India. ICAR, New Delhi, p 106.

Singh B. (1981). Establishment of the First Gene Sanctury in India for *Citrus* in the Garo Hills. Concept Publishing Company, New Delhi.

Singh H.B. and Arora R.K. (1972). Raishan, *Digitaria* sp.–minor millet of the Khasi Hills, India Econ Bot 26(4): 376-380.

Singh H.B. and Arora R.K. (1973). Soh-phlang, *Moghania vestita*- a leguminous root crop of India. Eco Bot 27(3): 332-338.

Singh I.D. and Bezbaruah H.P. (1988). Genetic resources in Tea in plant genetic resources – Indian perspective (Paroda RS, Arora RK and Chandel KPS eds) p 435-442.

Zeven A.C. and Zhukovsky P.M. (1975). Dictionary of Cultivated Plants and their Centres of Diversity. Pudoc, Centre for Agricultural Publishing and Documentation, Wageningen.

Chapter 13

Microbial Diversity: A Review

M. S. Dkhar, H. Kayang and M. Dkhar
Department of Botany, School of Life Sciences, North-Eastern Hill University, Permanent Campus, Umshing, Mawkynroh, Shillong – 793 022, Meghalaya

Microorganisms, which plays a vital role in the functioning of the ecosystem, occur virtually in every habitat in the biosphere. Microorganisms, including fungi and bacteria related to forests, agriculture, jhum, grasslands, protected and degraded forests of India and other parts of the world, have been reviewed. Comprehensive accounts of saprophytes or symbionts, including mycorrhiza, which are concerned with soil fertility, have been given and discussed.

INTRODUCTION

Microbial diversity is an unseen national, as well as international resource that deserves greater attention than it has bean receiving. It is the largest untapped resource, both for understanding how biological systems function as well as for new biotechnologies. It encompasses the spectrum of variability among all types of microorganisms in the natural world and as altered by human intervention. Microorganisms occur virtually in every habitat in the biosphere. In fact, recent discoveries have indicated their existence even on other planets too. They even occur in extreme environments even in abiotic conditions, unfavourable for most forms of life. Microorganisms represent by far, the richest repertoire of molecular and chemical diversity in nature. They underlie basic ecosystem processes, such as biogeochemical cycles and food chains as well as maintain vital and often elegant relationships between themselves and higher organisms. Microorganisms provide the fundamental base to all types of ecosystem. Without microorganisms, all life on earth would cease.

Focussing on microbial diversity is timely. Diverse microorganisms are essential for a sustainable biosphere. They are able to recycle nutrients, produce and consume gases that affect the global climate, destroy pollutants, treat wastes, and they can even be used for biological control of plant and animal pests. The study of microbial diversity is also important to treat new and emerging diseases and to advance biotechnology. New technologies, particularly in nucleic acid analysis, computer science, analytical chemistry and habitat sampling and characterization, place the study of microbial diversity at the cutting edge of science. Over the ages, humans have been highly successful in applying processes carried out by microorganisms to solve problem in agriculture, food production, human health, environmental quality and industry. They have played a pivotal role in the evolution of life on earth. Recent technologies developed in molecular biology and genetics offer great promise to develop the potential of microbial diversity. The world of microorganisms and their activities in nature are vast and therefore, it is not possible to deal with the entire microbial activity in nature. However, in the present study, we have tried to present an overview of the functional aspects of microorganisms in relation to environmental factors.

UNDERSTANDING FACTORS AFFECTING MICROBIAL DIVERSITY

Studies on microbial diversity are important in order to understand microbial ecology in soil and other ecosystems (Atlas, 1984; Reid, 1994). The information on microbial diversity from North-East India is sporadic, as compared to other parts of India and no comprehensive study has so far been made to explore and conserve the microbial diversity of North-East India as a whole. North-East India is known to possess diverse forest types, from tropical evergreen forests to moist alpine shrub forests which occupy 59.9% of its geographical area, which is much higher than the all India average of 28%. Its forests constitute a powerful ecological unit having vital environmental significance.

Studies on physico-chemical characteristics have generally been conducted in forests, agricultural and grassland soils. Various physico-chemical characteristics which affect microbial diversity are temperature, pH, moisture content, climate, vegetation, organic matter constituents, altitude, etc. (Singh, 1968; Mishra and Kanaujia, 1972; Das 1980; Dkhar 1983; Das and Mishra, 1986).

Soil microbial diversity in general, is influenced by the nature of the crop grown and the physico-chemical characteristics of soil viz., nature of the particles of which the soil is composed of, i.e., nutrients, aeration, temperature, precipitation, organic matter content, moisture, pH and depth (Waksman, 1927; Mishra and Kanaujia, 1972; Deka, 1981; Baruah 1983; Campbell, 1983; Behera and Mukherji, 1985; Tiwari *et al.*, 1986; Tiwari, *et al.*, 1987 a, b; Tiwari, 1988; Arunachalam *et al.*, 1997; Berg *et al.*, 1998; Coyne, 1999; Kayang *et al.*, 2000). The physico-chemical characteristics of soil regulate the microbial populations and their activities; the state of litter decomposition governs microbial diversity in the soil (Mishra, 1966; Tiwari *et al.*, 1987a). Therefore, due to their inseparable relationships, these parameters were widely investigated by various workers (Waksman, 1927; Thakur and Morris, 1928; Warcup, 1950).

Saksena (1955) stated that the high moisture content of soil was favourable for the growth of fungi, as long as there was no water logging. Brown (1958), Cook and Lawrence (1959), Pugh (1962) and Wohlrab *et al.*, (1963) reported that there is a shift in the species composition of soil mycoflora which parallels pioneer vegetational succession.

It is widely accepted that each of the vegetational communities harbours a characteristic soil fungal population, particularly the assemblage of more predominant species. However, there is no evidence that microfungi form discrete communities, rather they constitute a continuum, the species composition gradually changing with differences in the vegetational cover and soil characteristics (Wicklow and Whittingham, 1974). The changing climatic conditions at different altitudes are accompanied by differences in the composition of the vegetation. Thus, a microorganism that subsequently decomposes a little of this vegetation may require a different range of metabolic reactions than those utilised in other environments. Furthermore, a decrease in the diversity of fungal species at higher altitudes is paralleled by a decrease in plant species. As the decomposition of litter is also controlled by differences in the chemical composition of litter, so the spectrum of fungal species concerned with litter decomposition is likely to differ in the same way (Schinner, 1978). The pattern of distribution of fungi is also affected by nutritional conditions (Rehder and Schafr, 1978) and micro-climatic factors which have a decisive influence on the distribution of individual species.

Microbes are important organisms that drive the turnover of organic matter in soil and aquatic environments. They are the focal point of analysis when dealing with soil respiration, mineralisation, gas emissions or primary steps of the food web, known as the microbial loop (Azam *et al.*, 1983).

Soil microorganisms are largely hidden under ground populations; frequently have a greater mass than the plants and animals above ground (Jenkinson and Ladd, 1981; Sparling, 1985). It is also suggested that the total functional diversity of microorganisms in different soils may be similar and that a combination of environmental and plant factors influence microorganisms, which are active, and become culturable and proliferate under different conditions (Colwell *et al.*, 1985). Climatic regimes, seasonal variations in the year, type of vegetation, soil management practices and deforestation are the other important factors which influence the distribution and diversity of microorganisms in soil (Kauri, 1982; Lundgren, 1982; Behera and Mukherji, 1985; Bossio and Scow, 1995; Acea and Caraballas, 1996; Gorres *et al.*, 1998; Lupiwayi *et al.*, 1998; Zeller *et al.*, 2001).

The major underlying principle of diversity studies is probably the assumption that interactions between populations in a habitat lead to an organised and stable community (Atlas, 1984). Diversity can vary with a number of factors, such as disturbances and stress, in addition to ecological interactions. Microbial communities are inherently stable, but at the same time are also dynamic structures (Campbell, 1983; Bossio and Scow, 1995). The changes in microbial communities, resulting from agricultural practices, ecosystem management and global climate changes, can have profound impacts on ecosystem dynamics (Shukla and Mishra, 1992).

Behera *et al.*, (1991), while conducting an ecological investigation of some micro fungi in a tropical forest soil of Orissa, could isolate 36 fungal species and one sterile mycelium from the soil. They observed seasonal variations of fungal population to be pronounced in the upper layers of the soil. They also observed a positive correlation between fungal population, soil moisture and organic matter contents.

Microbial activity is an important key to understanding biological processes in soil. It is influenced by soil degradation, transformation of organic matter and soil structure. Changes in soil microbial activity, in response to stress, may eventually affect plant productivity and ecosystem functioning (Garcia *et al.*, 1994). Elliot and Lynch (1994), Pankhurst *et al.*, (1996) and Giller (1997) suggested that broad functional diversity might be additionally important in influencing the resilience residents of soil. Being the most important component of the soil environment, microorganisms exert considerable influence on soil fertility and plant growth (Tiwari *et al.*, 1991). The microbial population, as observed by Jha *et al.*, (1992), showed a positive and significant correlation with organic C in more degraded forest stands, and the disturbance to the soil and vegetation adversely affected the microbial populations and endogonaceous spores.

Berg *et al.*, (1998) investigated the abundance and micro stratification of bacteria and fungi inhabiting the organic layers of a scots pine forest. They suggested that the abundance of bacteria was influenced by water and that of fungi by water and temperature.

Tiwari and Sharma (1998) reported that the fungal and bacterial populations in highland soil increased with an increase in altitude up to 1100m, but thereafter, the population declined sharply. They also observed a positive correlation between fungal and bacterial population with organic matter content of the soil.

Soil microbial communities exist within a highly heterogeneous physical structure, which affects the spatial distribution of solutes, nutrients and of the communities, themselves (Young and Ritz, 1998). As a result, soil processes related to microbial activity often exhibit very high spatial variability (Parkin and Shelton, 1992; Robertson *et al.*, 1997; Rover and Kaiser, 1999; Stoyan *et al.*, 2000).

Generally, microbial populations and their activities in no tillage soil are greater than in conventional tillage soil. But environmental disturbances like clear cutting of forests affects the microbial populations and their activities. When the natural forest is cut completely the roots die, root decomposition takes place and the debris from leaves and branches on the soil surface enter into the soil, serving as food for microbial growth (Coyne, 1999). Thus, there is an increase in microbial growth and their activities in a short term. Soil disturbance activities, such as shifting cultivation and others related to agriculture and forestry have a considerable impact on the biological health of soil systems.

Sami *et al.*, (2000) stated that soil fungi and bacteria are the major organisms responsible for nutrient cycling and for controlling the amount of nutrients available to plants. Liu *et al.*, (2000) suggested that substrate availability was the most important factor affecting the diversity and activity of soil microorganisms within a season. Among the stress treatments soil moisture was not the factor causing differences in microbial diversity and activity, but it was a predictor for some microbial responses under particular kinds of stress.

Singh *et al.*, (2001) studied the effects of varying degrees of forest degradation on microbial communities, their activities and the various physico-chemical characteristics of the soil. Physico-chemical characteristics of the soil, viz., moisture content, water holding capacity, pH, organic carbon, total and available nitrogen and phosphorus contents declined with the increasing severity of forest degradation. Biological characteristics of the soil, such as soil respiration, bacterial and fungal populations, spore distribution and diversity of VAM fungi, microbial biomass and C and N contents also declined due to site degradation, suggesting the detrimental effect of *jhum* cultivation on soil fertility. Analysis of VAM fungal population revealed that degraded and moderately degraded sites exhibited lower indices of species diversity of VAM fungi, as compared to non-degraded sites suggesting higher levels of VAM fungi diversity at non-degraded sites. Microbial diversity at different sites analysed at molecular levels, using standard methodology, revealed that, in general, degradation caused a decline in microbial diversity.

Priha *et al.*, (1999) and Graystone *et al.*, (2001) pointed out that the structure and functional diversity of microbial communities in the soil is highly related to plant species composition above-ground, thus providing an important link between above and below-ground processes in terrestrial ecosystems. Bakermans and Madsen (2002) observed that the composition of microbial communities reflects the physical, chemical, geological and biological characteristics of their habitats.

Kourtev *et al.*, (2001) hypothesised that microbial communities in the initial soil will, in time, differentiate along with the different plant species and this will in turn lead to dangers in microbial functional capabilities. They further hypothesised that the various measures of community structure and function could be correlated with each other. Changes in microbial structure and function were accompanied by changes in ecosystem level, soil characteristics (nitrogen concentrations and pH) and processes (nitrogen mineralisation).

SIGNIFICANCE OF MICROBIAL DIVERSITY IN NATURE CONSERVATION

Only in the last decade mycologists and microbiologists started to pay attention to microorganisms as objects of nature conservation. Nowadays, knowledge of biodiversity and its conservation is receiving much attention all over the globe due to the politicisation of issue. On the other hand, microbiologists have not paid serious attention to microbes even though they exhibit far greater diversity than any other living organisms of the world. For planning the application of microbial diversity to biotechnology and for acquiring a better understanding of taxonomic approaches, it is necessary to study the ecology and distribution of the species in detail. Increased understanding of microbial species should lead to better conservation and management of whole ecosystem.

Knowledge of microbial diversity is important because of their role in regulating the population of other organisms and ecosystem processes, and of equal importance to microbial diversity itself, are variations in their own population that affect host-plant diversity and cycling. Microorganisms, however, are important, especially due to the fact that all life is dependent on microbial processes. The last decade has produced an extraordinary new awareness on microbial diversity (Dykhuisen, 1997).

Microbial communities, particularly bacteria and micro-fungi, constitute an essential component of biological characteristics in soil ecosystems. Directly or indirectly, animal wastes and the dead bodies of animals and tissues of plants eventually enter the soil. In time, they all disappear and are transformed into substances that enrich the fertility of the soil. Microorganisms are responsible for such transformations, converting dead plants and animal matter into simple organic substances that nourish the plants. Thus, by serving as a link amongst various components of energy, they play a key role in maintaining life on earth. In these processes, the microorganisms reform a variety of functions, e.g., biochemical transformation of nitrogen and nitrogenous compounds (nitrogen fixation, proteolysis, ammonification, nitrification, assimilatory nitrate reduction, denitrification), biochemical transformation of carbon dioxide and other carbon compounds (CO_2 fixation, degradation of organic carbon compounds), biochemical transformation of sulphur and sulphur compounds (oxidation and reduction of sulphur compounds), and the biochemical transformation of phosphorus, iron and essential elements, etc. Thus, such soil microbial processes play a key role in maintaining the atmospheric balance of various elements vital for life and also in mediating many natural functions in the ecosystem.

Degradation of natural ecosystems and consequent decline in vegetation cover is a major ecological concern. Soil physio-chemical properties can be improved by the manipulation of soil microbiological activities and the manipulation of microbial associations like mycorrhizae, *Rhizobia* and plant growth-promoting microorganisms, microbe residue association and biofertilisers, etc. Biofertilisers have been shown to possess a significant role in the restoration of soil fertility. Role of *Azotobacter, Azospirillium, Rhizobium, Frankia,* blue green algae, mycorrhizae in the restoration of degraded soils is well documented. These organisms have been exploited for the reclamation of degraded lands like coastal dunes, saline soils, eroded lands and mine spoils. In recent years, emphasis has been laid on the role of mycorrhizae in the revegetation of coal mine spoils. Mycorrhizae play a significant role in plant succession, thus, maintaining diversity of plant communities in such highly degraded ecosystems.

In response to the increased concern for environmental quality, innovative approaches need to be incorporated into agricultural systems. Management of VAM fungi is one important aspect of such an approach. However, the obligate symbiotic nature of VAM fungi continues to impede research on improving our understanding of these beneficial fungi. Furthermore, limited supply of VAM inocula has restricted field-testing. High production costs, as well as problems associated with soil-based pot cultures, have dampened the enthusiasm for commercial development. Nonetheless, the inability to presently use fermentation processes to produce VAM inocula should not deter further development of current technologies to produce these fungi in large quantities. Several areas of VAM inoculum production and application technology merit further investigation. Development of nutrient film and aeroponic culture technologies, along with sheared root processing of colonised roots produced in these systems, should make it possible for low cost, high quality inocula to be widely available for nursery inoculation and for moderate size field testing in the near future.

Much is unknown about the storage of VAM inocula. Further research on the storage properties of VAM inocula will definitely spur commercial production and the development of new formulations. Innovative inoculation technologies are needed to provide the most efficient application of available inocula in the wide array of crop production systems. Co-inoculation of VAM fungi with other beneficial organisms has great merit and should allow for a more holistic approach to rhizosphere health in the future. As inoculum production and inoculation technologies of VAM fungi undergo further development, involvement of all those concerned with crop production and plant health will ensure VAM inocula of consistent high quality and availability.

Microorganisms are essential components of biocoenoses by their functions as decomposers of organic matter, parasites of other organisms and mutualistic symbionts in mycorrhizae. Microorganisms can be excellent bio-indicators in view of their functions, niche differentiation and species diversity. For instance, many species of ectomycorrhizal fungi are indicators of the degree and kind of air pollution, wood-inhabiting fungi for the intensity of forestry practice, and grassland fungi for soil conditions, type of management and degree of disturbance.

Microorganisms possess many qualities that can be exploited for monitoring the presence of unwanted substances in the environment. The range of environmental extremes tolerated by microorganisms is much broader than that of other forms of life. Cells of bioluminescent bacterium (*Photobacterium phosphoreum*) have been used to detect air pollution.

Microorganisms are important (potential) sources of food and medicines, and they can be used for selective delignification of wood and straw, as well as for the degradation of xenobiotics. Their economic value can be counted in billions of US dollars, on account of their commercial uses, which include the production of amino acids, antibiotics, beer, biocontrol agents, cheese, enzymes, fermented foods, food (mushrooms), flavours, food colourants, fuel (ethanol, biogas), organic acids, preservatives, single cell protein, vitamins, herbicides, bio-pesticides and waste biodegradation.

Soil fertility is related to the soil parent material, litter types and the presence of basic microbial species richness. Therefore, within an ecosystem, species composition certainly matters, and changes in species diversity can therefore, lead to changes in soil fertility. Substantial alterations in soil fertility can be driven by changes in plant species composition, both within and among ecosystems. Disturbing the fungal diversity by several means, however, disturbs the microbial balance in nature. These include indiscriminate use of inorganic fertilisers and pesticides in agriculture to kill or suppress non-target microbes, air, water and soil pollutants, biotic activities, and collection for their use as food, medicines, etc. Therefore, useful fungal species are adversely affected, along with the harmful fungi affecting the whole biodiversity of ecosystems.

In natural habitats, microorganisms are an integral part of the biotic community, which interacts dynamically with the abiotic components to form ecosystems. Nowadays, particularly attention is given to soil functionality, largely to microbial activity, which is measured by determining both microbiological parameters and biochemical activities simultaneously (Bonmati *et al.*, 1991). However, in order to maximise the beneficial effects of microbial activity, there is a need for greater understanding of the factors influencing microbial diversity and activity. The abundance and activity of soil microorganisms are influenced by various environmental (e.g., soil type, nutrient status, pH and moisture) as well as plant (e.g., species and age) factors. They are not distributed uniformly in the environment, rather the abundance and activity of the organisms changes along environmental gradients. Biotic processes often result in the aggregation of organisms, even within homogeneous environments (Legendre and Fortin, 1989). Therefore, in order to maximise the beneficial effects of microbial activity, greater understanding of the factors influencing microbial diversity and activity is needed. Microbial growth in soil is carbon limited and therefore, the presence of organic matter has an influence on microbial populations (Lynch and Whipps, 1990; Wardle, 1992).

With increasing emphasis on environmentally friendly, low-input agricultural practices, there is growing interest in the management of soil microbial communities to enhance plant growth. Soil is rarely manipulated for microbial growth. However, if a soil is altered to affect plant growth, which is the essence of agriculture, these manipulations affect the distribution and types of microbial population. Thus, the microorganisms in turn, affect plant growth (Coyne, 1999).

Wardle and Giller (1996) pointed out that despite their comparable mass, we know much less about microorganisms than we do about higher plants and animals. Microorganisms have an enormous impact and role in our daily lives, including everything, right from maintaining the biosphere to improving our lifestyle (Hunter- Cevera, 1998).

The important functional attributes of soil microorganisms include population number, biomass, respiratory efficiency and diversity patterns. These microorganisms are responsible for the breakdown of organic matter, nutrient transformation (Das, 1980; Baruah, 1983; Tiwari 1988; Dkhar and Mishra, 1992; Kayang, 2001), influencing plants and are considered crucial to the functioning of soil ecosystems. Apart from their indispensable role in the biogeochemical cycling of nutrients, they are reported to be useful indicators of the fertility and quality of soil, since they are sensitive to changes occurring in the soil environment. In fact, soil microorganisms play a very important role in soil fertility, not only because of their ability to carry out biochemical transformation, but also due to their importance as a source and sink for mineral nutrients (Jenkinson and Ladd, 1981).

The analysis of the functional diversity of soil population is a very useful way for characterising and comparing the microbial communities. Little is known about the importance of the functional diversity of soil microbial communities for the sustained functioning of terrestrial ecosystems (Beare *et al.*, 1995, Moore *et al.*, 1995 and Pankhurst *et al.*, 1996; Giller *et al.*, 1997). It has generally been hypothesised that the reduction in soil microbial diversity will result in the reduction of functional capability of soils (Giller *et al.*, 1997). A high degree of functional redundancy in soil microbial communities may result in there being no effect of the changes in microbial diversity on the function of communities (Andren *et al.*, 1995; Giller *et al.*, 1997). However, broad functional diversity may be additionally important in influencing the resilience of soil (Elliot and Lynch, 1994; Pankhurst *et al.*, 1996; Giller *et al.*, 1997).

The decomposition of dead plant material is an important process responsible for the release of nutrients from the litter and soil. It plays a major role in the structure and dynamics of ecosystems. It also helps in the transfer of elements and energy and in control mechanisms and feedback processes. It is a key process in all-terrestrial ecosystems, because it controls nutrient availability and hence, primary production. In addition, interactions of phylloplane, rhizosphere and rhizoplane microorganisms with plants are of much significance. In some cases, microorganisms may be keynote species which if lost, would lead to major changes in the ecosystem.

The relationships among microbial diversity, microbial activity, plant quality and ecosystem sustainability of disturbed and range lands are still poorly understood. Ecosystem functioning, before and after disturbance, can be governed by soil microbial population dynamics (Kennedy and Smith, 1995). The microbial communities in forest soils, rich in organic matter, may differ considerably from those found in grasslands and in arable soils, with lower organic matter content (Frostegard and Baath, 1996). Therefore, research on microbial diversity and activity may provide some advance evidence of ecosystem degradation. In managed ecosystems, the early detection of soil biological activity may allow for changes in the management system before the functioning of the entire ecosystem is impaired.

FUTURE PROJECTION

Microorganisms have enormous potential to make our life better on this earth. They are an integral part of our lives and possess a vast range of activities to support and improve the quality of our life; many of their properties are still unexplored and in future, we will largely depend on microorganisms to solve our natural as well as man-made problems. Conservation of the gene pools of microorganisms is needed in order to extend our understanding of evolutionary processes and the resulting diversity in taxa, morphological structures and ecological strategies.

Our knowledge of the relation between microorganisms and different management practices in forests and nature reserves is still very scanty. Therefore, such studies from different parts of North-East India are needed to explore, not only new fungal and bacterial diversity, but also to gather knowledge on their substrate relationship, culture and physiology, seasonal variations, quantitative estimations, decomposition biology, etc. Finally, the consequences of changes in microorganisms for the functioning of ecosystems are still poorly understood. Increased insight into these relationships will without doubt, contribute to a better appraisal of microorganisms as essential components of ecosystems.

REFERENCES

Acea M.J. and Caraballas T. (1996). Changes in physiological groups of microorganisms in soil, following wild fires. FEMS Microbiol Ecol 20: 33-39.

Andren O., Bengtsson J. and Clarholm M. (1995). Biodiversity and species redundancy in decomposed litter. In: The significant and regulation of soil biodiversity (Collins HP, Robertson GP and. Klug MJ eds), Kluver Acadamic Publishers, Dordrecht p 141-151.

Arunachallam A., Maithani K., Pandey H.N. and Tripathi R.S. (1997). The impact of disturbances on detrital dynamics and soil microbial biomass of *Pinus kesiya* forest in Northeastern India. Forest Ecology and Management 88: 273-282.

Atlas R.M. (1984). Diversity of microbial communities. In: Advances in Microbial Ecology (Marshall KC eds), Vol 7, Plenium Press, New York p 1-47.

Azam F., Fenchel T., Field J.G., Gray J.S., Meyer-Reil L.A., Thingstad F. (1983). The ecological role of water-column microbes in the sea. Mar Ecological Prog Ser 10: 257-263

Bakermans C. and Madsen E.L. (2002). Diversity of 16S rDNA and Naphthalene Deoxygenase genes from coal tar waste contaminated aquifer waters. Microb Ecol 44: 95-106.

Baruah M. (1983). Studies on population dynamics and activity measurement of microbial communities of paddy field soils. Ph.D. Thesis, North Eastern Hill University, Shillong, India p 175.

Beare M.H., Coleman D.C., Crossley D.A.Jr., Henrix P.F. and Odum E.P. (1995). A hierarchical approach for evaluating the significance of soil biodiversity to biogeochemical cycling. Plant and Soil 170: 5-22.

Behera N. and Mukherji K.G. (1985). Seasonal variations and distribution of microfungi in the forest soils of Delhi (India). Folia Geobotanica Phytotaxton 20: 291-311.

Behera N., Pati D.P. and Basu S. (1991). Ecological studies of microfungi in a tropical forest soil of Orissa, India. Tropical Ecology 32(1): 136-143.

Berg M.P., Kniese J.P. and Verhoof H.A. (1998). Dynamics and stratification of bacteria and fungi in the organic layers of pine forest soil. Biology and Fertility of Soils 26: 313-322.

Bonmati M., Ceccanti B. and Nannipieri P. (1991). Spatial variability of phosphatase, urease, protease, organic carbon and total nitrogen in soil. Soil Biology and Biochemistry 23: 391-396.

Bossio D.A. and Scow K.M. (1995). Impact of carbon and flooding on the diversity of microbial communities in soils. Applied Environmental Microbiology 61: 4043-4056.

Brown J.C. (1958). Soil fungi of some British sand dunes in relation to soil type and succession. J Ecol 46: 641-664.

Cooke W.B. and Lawrence D.B. (1959). Soil mould fungi isolated from recently glaciated soils in South Eastern Alaska. J Ecol 47: 529-549.

Campbell R. (1983). Microbial ecology. In: Basic Microbiology (Wilkinson JF ed), Vol 5, Second Edition, Blackwell Scientific Publications, Oxford.

Colwell R.R., Brayton P.R., Grimes D.J., Roszak D.B., Huq S.A. and Palmer L.M. (1985). Viable but non-culturable *Vibrio cholera* and related pathogens in the environment: Implications for the release of genetically engineered microorganisms. Biotechnol 3: 817-820.

Coyne M. (1999). Soil microbiology. An explanatory approach. Delmer Publisher, New York p 462.

Das P.K. (1980). Microbial decomposition of pine (*Pinus kesiya* L) litter: An Ecological Study. Ph.D Thesis, North Eastern Hill University, Shillong, India p 233.

Das P.K. and Mishra R.R. (1986). Dehydrogenase activity and Co_2 evolution from *Pinus kesiya* Ex Gordon forest soils. Geobios 13: 12-15.

Deka H.K. (1981). Study on soil microbiological population of *jhum* fallows of different ages. Ph.D. Thesis, North Eastern Hill University, Shillong, India.

Dkhar M.S. (1983). Studies on the ecology of edaphic microbial populations and their activities in maize fields. Ph.D Thesis, North Eastern Hill University, Shillong, India p 158.

Dkhar M.S. and Mishra R.R. (1992). Changes in chemistry during the decomposition of maize residues. Proc Nat Acad Sci India 62(B): 437-440.

Dykhuisen D.E. (1997). Santa Rosalia revisited: Why are there so many species of bacteria? Antonie Van Leewenhoek 73: 25-33.

Elliot L.F. and Lynch J.M. (1994). Biodiversity and soil resilience. In: Soil Resilience and Sustainable Land Use (Greenland DJ and Sza Boles I eds), CAB International, London p 353-364.

Frostegard A. and Baath E. (1996). The use of phospholipid fatty acid analysis to estimate bacterial and fungal biomass in soil. Biology and Fertility of Soil 22: 59-65.

Garcia C., Hemandez T. and Costa F. (1994). Microbial activity in soil under Mediterranean environmental conditions. Soil Biol Biochem 26: 1185-1191.

Graystone S., Griffith G., Mawdsley J., Campbell C. and Bardgett R.D. (2001). Accounting for variability in soil microbial communities of temperate upland grassland ecosystems. Soil Biol Biochem 33: 533-551.

Giller K.E., Beare M.H., Lavelle R., Izac A.M.N. and Swift M.J. (1997). Agricultural intensification, soil biodiversity and agro system functions. Applied soil Ecology 6: 3-16.

Gorres J.H., Dichia M.J., Lyons J.B. and Amadon J.A. (1998). Spatial and temporal patterns of soil biological activity in a forest and or an oil field. Soil Biol Biochem 30: 219-230.

Hunter-Cevera J.C. (1998). The value of microbial diversity. Curr Opinion Microbial 1: 279-285.

Jenkinson D.S. and Ladd J.N. (1981). Microbial biomass in soil measurement and turnover. In: Soil Biochemistry (Paul and Ladd JN eds), Vol 5, Marcel Dekker Inc, New York and Basel p 415-471.

Jha D.K., Sharma G.D. and Mishra R.R. (1992). Ecology of soil microflora and mycorrhizal symbionts in degraded forests at two altitudes. Bio Fertil Soils 12: 272-278.

Kauri T. (1982). Seasonal fluctuations in the number of aerobic bacteria and their spores in four horizons of a beach forest. Soil Biol Biochem 14: 185-190.

Kayang H. (2001). Fungal and bacterial enzyme activities in *Alnus nepalensis* D. Don. European Journals of Soil Biology 37(2): 175-180.

Kayang H., Sharma G.D., Laloo R.C. and Mishra R.R. (2000). Weight loss and concentration of nutrients during the decomposition of leaf litter of *Alnus nepalensis* D. Don in subtropical forests of North East India. Vasundhara 5: 1-10.

Kennedy A.C. and Smith K.L. (1995). Soil microbial diversity and the sustainability of agricultural soils. In: The significance and regulation of soil biodiversity. Kluwer Dordrecht p 75-86.

Kourtev P.S., Ehrenfeld J.G. and Haggblom (2001). Experimental analysis of the effect of exotic and native plant species on the structure and function of soil microbial communities. Soil Biol Biochem 35: 895-905.

Legendre P. and Fortin M.J. (1989). Spatial pattern and ecological analysis. Vegetation 80: 107-138.

Liu X., Lindemann W.C., Whilford W.G. and Steiner R.L. (2000). Microbial diversity and activity of disturbed soil in the Northern Chihuahuan Desert. Biol Fertil Soils 32: 243-249.

Lundgren B. (1982). Bacteria in a pine forest soil as affected by clear cutting. Soil Biol and Biochem 14: 537-542.

Lupiwayi N.Z., Rice W.A. and Clayton G.W. (1998). Soil microbial diversity and community structure under wheat as influenced by tillage and crop rotation. Soil Biol Biochem 30: 1733-1741.

Lynch J.M. and Whipps J.M. (1990). Substrate flows in the rhizosphere. Plant Soil 129: 1-10.

Mishra R.R. (1966). Seasonal variation in fungal flora of grasslands of Varanasi (India). Tropical Ecology 7: 100-113.

Mishra R.R. and Kanaujia R.S. (1972). Studies on certain ecological aspects of soil fungi. Tropical Ecology 13: 5-11.

Moore R., Anderson A.R., Ray D., Walker C., Pyatt D.G., Evans H.F., Carter C.I., Straw N.A., Wairhouse D. and Winter T.G. (1995). Soil biodiversity: A literature review. Scottish Natural Heritage, No.17.

Pankhurst C.E., Ophel-Keller K., Doube B.M. and Gupta V.V.S.R. (1996). Biodiversity of soil microbial communities in agricultural systems. Biodiversity and Conservation 5: 197-209.

Parkin T.P. and Shelton D.R. (1992). Spatial and temporal variability of carbon furan degradation in soil. J Environmental Qualitative 21: 672-678.

Priha O., Graystone S., Pennaner T. and Smolandes A. (1999). Microbial activities related to C and N cycling and microbial community structure in the rhizospheres of *Pinus sylvestris, Picea aber* and *Betula pendula* seedlings in an organic and mineral soil. FEMS Microbiol Ecol 30: 187-199.

Pugh G.J.F. (1962). Studies on fungi in coastal soil. II Fungal ecology in a developing salt marsh. Trans Br Mycol Soc 45: 560-566.

Rehder H. and Schafer A. (1978). Nutrient turnover studies in alpine ecosystems. IV. Communities of the Central Alps and a comparative survey. Oecologia (Berl) 34: 309-327.

Reid W.V. (1994). Setting objectives for conservation evaluation. In: Systematics and Conservation evaluation (Forey PL, Humphries CJ and Vane–Wright RI eds), Systematics Association special Vol 50, Clarenden Press, Oxford p 1-13.

Robertson G.P., Klingensmith K.M., Klug M.J., Paul E.A., Crum J.R. and Ellis B.G. (1997). Soil resources, microbial activity and primary production across an agricultural ecosystem. Ecol Appl 7: 158-170.

Rover M. and Kaiser E.A. (1999). Spatial heterogeneity within the plough layer, low and moderate variability of soil properties. Soil Biol Biochem 31: 175-187.

Saksena S.B. (1955). Ecological factors governing the distribution of soil microfungi in some forest soils of Sagar. J Ind Bot Soc 34: 262-268.

Sami A., Vare H. and Strommer R. (2000). Soil microbial activity and biomass in the primary succession of a dry heath forest. Soil Biol Biochem 32: 1091-1100.

Schinner F. (1978). Streuabbauunter suchungen inder Alainegion oberen subalpinen. Stufe. In: Okologische Analsen von Almflachen in Gasteiner Tal (Cernusca A ed), Universitatsuer lag wags Innsbruck p 251-258.

Shukla A.K. and Mishra R.R. (1992). Influence of soil management systems on the microfungal communities of potato fields. Cryptogamie Mycol 13(2): 135-144.

Singh K.P. (1968). Litter production and nutrient turnover in two deciduous forests of Varanasi. Proc Symp Recent Adv Trop Ecol 655-665.

Singh S.S., Tiwari S.C. and Dkhar M.S. (2001). Evaluation of soil degradation, using physicochemical, biochemical and biological parameters in the humid tropics of Arunachal Pradesh. Ann For 9(2): 287-292.

Sparling G.P. (1985). The soil biomass. In: Soil organic matter and Biological activity (Vaughan and Malcolm RE eds), Martinus, Nihoff/Dr W Junk, Dordrecth, Boston, Lancaster p 223-262.

Stoyan H., De-Polli H., Bohm S., Robertson G.P. and Paul E.A. (2000). Spatial heterogeneity of soil respiration and related properties on the plant-scale. Plant Soil 222: 203-214.

Thakur A.K. and Morris (1928). A biochemical study of some soil fungi with special reference to ammonia production. J Indian Inst Sci All p 141.

Tiwari S.C. (1988). Studies on microbial communities and their activities in soil of pineapple (*Ananas comosus* L.) plantations. Ph.D. Thesis, North Eastern Hill University, Shillong, India.

Tiwari S.C. and Sharma G.D. (1998). Altitudunal variations in dehydrogenase and urease activity and microbial population in soils of the Eastern Himalayas. Journal of Hill Research 11(1): 22-25.

Tiwari S.C., Tiwari B.K. and Mishra R.R. (1986). Temporal and depth-wise variations in carbon dioxide evolution and microbial population in pineapple plantation soils. Journal of Soil Biology and Ecology 6: 67-76.

Tiwari S.C., Tiwari B.K. and Mishra R.R. (1987a). The influence of soil moisture regimes on population and activity of soil microorganisms. Plant and Soil 101: 133-136.

Tiwari S.C., Tiwari B.K. and Mishra R.R. (1987b). Temporal and depth-wise variations in urease and dehydrogenace activities and bacterial population in pineapple plantation soils. Proceeding Indian Academy Sciences 53(B): 173-176.

Tiwari S.C., Tiwari B.K. and Mishra R.R. (1991). Seasonal variations in microfungal population in pineapple (*Ananas comosus.* L) orchard soils. Acta Botanica Indica 19: 55-61.

Waksman S.A. (1927). Principles of Soil Micorobiology. Bailliere Tindall and Co, London.

Warcup J.H. (1950). The soil plate method for the isolation of fungi from soils. Nature (London) 166: 117-118.

Wardle D.A. (1992). A comparative assessment of factors which influence microbial biomass carbon and nitrogen levels in the soils. Biological Reviews 67: 321-358.

Wardle D. and Giller K. (1996). The quest for a contemporary ecological dimension to soil biology discussion. Soil Biochemistry 28: 1549-1554.

Wicklow D.T. and W.F. Whittingham (1974). Soil microfungal changes among the profiles of disturbed conifer-hardwood forests. Ecology 55: 3-16.

Wohlrab G., Tuveson R.W. and Olmstead C.E. (1963). Fungal populations from the early stages of succession in Indian dune sand. Ecology 44: 734-740.

Young I.M. and Ritz K. (1998). Can there be a contemporary ecological dimension to soil biology without a habitat? Soil Biol Biochem 30: 1229-1232.

Zeller V., Bardgett R.D. and Tapeiner U. (2001). Site and management effects on soil microbial properties of sub-alpine meadows, a study of land abandonment along a north-south gradient in the European Alps. Soil Biol Biochem 33: 639-649.

Chapter 14

Biodiversity and Forest Cover Dynamics of Dry Deciduous Forests of Eastern Ghats (Andhra Pradesh)

V. Krishna Prasad, T. Rajagopal,* Yogesh Kant and K.V.S. Badarinath

National Remote Sensing Agency

Department of Space, Government of India, Balanagar, Hyderabad - 500037

*(*Department of Botany - Taxonomy Laboratory, Osmania University, Hyderabad)*

Studies on biodiversity and forest cover dynamics, using remote sensing data, gain importance in the context of increasing deforestation rates and in studies relating to the monitoring, maintenance and conservation of biodiversity. In the present study, an attempt has been made to characterise overstorey diversity from vegetation maps derived from IRS-1C LISS III satellite data. Textural measures, semivariograms and NDVI texture analysis have been used to study the diversity of different forest communities. Biomass estimates have been made for different forest stands, using regression equations relating girth measurements obtained through field studies. To identify the priority areas for the rehabilitation of forest areas that are prone to high rates of deforestation, an index known as, Forest Rehabilitation Need Index (FRNI) has been implemented. The results of the study through the semivariogram approach have suggested decreased semi-variance with increased lag distance for mixed dry deciduous forests, followed by mixed scrub forests and dry deciduous forests. Stands of mixed dry deciduous forests showed a low NDVI texture, dry deciduous forests a medium NDVI texture, and mixed scrub forests a very high texture number, which has been attributed to high anthrapogenic interference in case of scrub forests, when compared to mixed dry deciduous forests and dry deciduous forests. The average mean carbon storage calculated from biomass values suggested 64.34 t $ha^{-1}C$ for dry deciduous forests, 129.0 t $ha^{-1}C$ for mixed dry deciduous forests, and 0.02 t $ha^{-1}C$ for mixed scrub forests. Results of the study through FRNI calculation suggested that one of the mandal Marredumulli attained a higher rank of FRNI, implying that the forests of that area need immediate rehabilitation.

INTRODUCTION

Large-scale deforestation of forests and the rapidly accelerating loss of species, populations and natural habitats are causing unprecedented changes in the diversity, dynamics and stability of ecosystems. The concern over loss of species, ecosystems and natural habitats led to the notion of Biodiversity, about 15 years ago. The ecological effects of large-scale deforestation include decreased biodiversity (Wilson, 1988), disruption of hydrological regimes (Shukla *et al.*, 1990), degradation of soil conditions (Buschbacher *et al.*, 1990) and changes in the balance of greenhouse gases in the biosphere, which may induce climate changes (Post 1990, Dale, 1990). Taxonomy is the art of classifying diversity and is closely linked with the inventory and monitoring of biodiversity. Remote sensing forms a valuable tool in the mapping and monitoring of biodiversity and provides valuable information to quantify spatial patterns of species richness, ecological processes that determine species richness, anthrapogenic factors that cause loss of species richness and to predict the response of species to global climate changes. In the study area, large-scale deforestation, through the removal of fuel wood and small timber in excess of the capacity of the existing forest, in addition to shifting cultivation, had reduced the forest cover considerably. In the present study, an attempt has been made to utilise the spectral and spatial information for studying vegetation structure and anthrapogenic interference in tropical dry deciduous forests of the East Godavari District (Eastern Ghats) - Andhra Pradesh.

STUDY AREA

The study area, popularly known as Rampa Forests, forms the hilly northern part of East Godavari District (Eastern Ghats), Andhra Pradesh, India. With respect to forest resources, the Rampa forests are rich and dense, with plenty of fertile soil, higher rainfall and better conditions of plant growth. Rampa forests (Agency) encompass seven mandals, each of which is mostly inhabited by forests. The forests of the study area correspond to the Southern dry mixed deciduous forests (type 5A/C3; group 5, subgroup 5A) and the Southern tropical forests (group 6, subgroup 6A), as classified by Champion and Seth (1968). The species composition is diverse, mostly dominated by dry deciduous species and at some places, occupied by moist mixed species. The major species include *Xylia xylocarpa, Terminalia alata, Anogeissus latifolia, Dillenia indica, Memycelon edule, Strychnos nuxvomica, Mangifera indica, Feronia elephantum, Cleistanthes collinus, Pterocarpus marsupium, Ficus retusa, Pavetta indica, Canthium dicoccum, Chloroxylon swietenia, Lannea coromandelica, Macaranga peltata, Mitragyna parviflora, Grewia tilaefolia, Madhuca latifolia, etc.* In addition, scrub forests having species of *Carissa spinarum, Zizyphus oenoplia, Phoenix humilis, Dichrostachys cinerea, Prosopis julifera, Acacia nilotica, Randia dumetorum, Cordia myxa*, etc., were also predominant at some places. Under the Rampa Agency, Rolla S. Rao (1982) after intensive floristic studies, identified various forest ranges as hotspot areas of floristic diversity, and suggested that they should be declared as Biosphere Reserve Clusters.

Different mandals have undergone substantial changes with respect to Land use/Land cover (LU/LC) in the recent past, due to intensive shifting cultivation and increasing urbanization.

DATA SETS AND METHODOLOGY

Satellite data pertaining to IRS-1C LISS-III data acquired on 4th February, 1996, path/row 103/60, 61 have been used for delineating the study area and for studying LU/LC statistics. Surveying of India's top sheets at a 1:250,000 scale have been used as secondary data. Visual interpretations in conjunction with Digital analysis of satellite data have been carried out, using field data for deriving information on the spatial extent of different LU/LC classes. Various LU/LC categories, in the study area include agricultural lands, forests, plantations, shifting cultivation areas and water bodies, etc. Registering the image to the top sheets generated mandal masks. The training areas have been defined for different LU/LC categories, based on ground information. Training area statistics, in terms of mean standard deviation, and variance and covariance matrices, have been generated and the data have been classified using the maximum likelihood algorithm.

Reconnaissance surveys of the classified data have been carried out to confirm the obtained results. The area estimates generated mandal-wise, have been compared with the Handbook of Statistics of East Godavari District and State Forest Department working plans with the purpose of comparing the relative deviation.

1. Semivariogram Approach

Semivariogram is a graphical representation of the spatial variability in a given set of data. Semivariance or g (h) is calculated as = $1/(2n - h)$ S [Z (x_i) - Z (x_i+ h)]^2, where h is the lag distance over which semivariance is measured, n is the number of observations used in the estimate of g (h), and Z is the value of the variable of interest at the spatial position (xi). The quantity, Z (xi + h) which is the variable value at distance h from xi. g (h), estimates the variability of radiance, Z as a function of spatial separation (Journel and Huijbregts, 1978, Cohen *et al.*, 1990).

2. NDVI Texture Analysis

NDVI texture analysis has been done for studying the structural changes in the variability caused due to biotic interference in different forest stands of dry mixed deciduous, dry deciduous and scrub forests. It consists of taking the sum of the absolute values of differences between all adjacent pixels within a window and assigning the value to the central pixel (Rubin, 1990). Using this algorithm and a moving 3 × 3 pixel window, a texture analysis has been performed on different forest stands, using the NDVI texture image.

3. Biomass Estimations and Carbon Storage

Based on stratified random sampling, different homogenous vegetation strata have been delineated and at each site, taxonomic surveys have been carried out for studying community characteristics. The major associations obtained through community analysis, have been used for stratification of the vegetation into different communities. The girth values have been used to derive the biomass for bole, branches, roots and leaves for individual species through regression equations (Singh and Misra, 1979; Singh, 1990). At each site, girth measurements for each of the species have been noted (gbh) and those with girth below 30cms are considered as saplings. Individual species in each site were categorised into different girth classes. Mean girth of individual species has been used in regression equations, to get component-wise estimates of the bole, branch, root and leaf of that girth class and finally, the value is multiplied by the density of trees or shrubs in that class. The obtained biomass values have been used for the computation of carbon storage. Carbon content is assumed to be 47% of the dry biomass (Westlake, 1963).

4. Calculation of Forest Rehabilitation Need Index

Difficulties in modelling deforestation lie in the high degree of complications and connections involved in the socio-economic phenomena that control deforestation. The site and magnitude of land-use changes are the result of human dimensions and can be modelled using socio-economic indicators (Apan, 1996). In the present study, factors that are causing deforestation have been studied by linking socio-economic parameters and land use changes, for identifying the priority areas that need immediate rehabilitation.

FRNI is calculated as (W_1 X BPLFI) (W_2 X SEI)

where

FRNI	=	Forest Rehabilitation Need Index
W_1	=	Relative weight of BPLFI
BPLFI	=	Bio-physical and Legal factors index
W_2	=	Relative weight of SEI
SEI	=	Socio-economic indicator
$W_1 + W_2$	=	1

$0 < FRNI < 1$

A. Calculation of biophysical eactors legal factors index (BPLFI)

The ratio of net area to be reforested to the total area which forms the main component of BPLFI, has been calculated by delineating the exclusion areas, derived from satellite data and by applying the prohibitive criteria, based on land use/land cover (LU/LC) statistics.

$$BPLFI = TA - (EA_a + EA_b + EA_c)/ TA$$

where

TA = Total area under cultivation

EA_a = Exclusion Area a - taken as plantation areas

EA_b = Exclusion area b - taken as crop lands

EA_c = Exclusion area c - taken as fallow lands

B. Calculation of agricultural extensification

Since deforestation in the study area is mainly affected by agricultural extensification, it has been incorporated in calculating FRNI. Agricultural extensification, as a function of land availability is calculated as a function of percent tribal population and percent agricultural labourers involved in shifting cultivation.

$$LA = TL - CL$$

where

LA = land availability,

TL = total area of arable land

CL = area under cultivation

Thus, deforestation is linear function of population growth, when LA is large and as a non-linear accelerating function, when LA is small.

C. Socio-economic indicators (SEI)

Since, mostly tribal population (SEI-A) are involved in the shifting cultivation process, percentage tribal population in the study area has been used as one of the socio-economic parameters in quantifying deforestation. Also, since the agricultural extensification process is directly linked with agriculturists and agricultural labourers (SEI-B) in the area, percent population of people practicing agriculture and percent agricultural labourers per mandal, have been taken into consideration as socio-economic indicators, while calculating the FRNI.

Thus, the modified FRNI is calculated as

$$FRNI = (W_1 \times BPLFI) + (W_2 \times LA) + (W_3 \times SEI\ A) + (W_4 \times SEI\ B)$$

The weightage factors assigned for FRNI calculation in the present case have been taken as W_1= 0.5, W_2= 0.3, W_3= 0.1, W_4= 0.1.

RESULTS AND DISCUSSION

The results obtained by classifying the study area from IRS-1C LISS III satellite data are shown in Table 14.1. The different forest stands identified by stratifying the vegetation, through visual interpretation of satellite imagery, in conjunction with field based phytosociological studies, suggested twelve different forest stands in the study area (Table 14.2). The textural analysis from the semivariogram approach suggested decreased semivariance with increased lag distance for mixed dry deciduous forests, followed by mixed scrub forests and dry deciduous forests. Due to high homogenity in plant biomass spatial variability, forest cover types, with respect to mixed dry deciduous forests, showed a low texture number, when compared to dry deciduous forests and scrub forests. Stands of dry deciduous forests showed a medium texture, and mixed scrub forests a very high texture number, which has been attributed to high anthrapogenic interference in case of scrub forests, when compared to mixed dry deciduous forests and dry deciduous forests. The average mean carbon storage calculated from biomass values suggested 64.34 t $ha^{-1}C$ for dry deciduous forests, 129.0 t $ha^{-1}C$ for mixed dry deciduous forests and 0.02 t $ha^{-1}C$ for mixed scrub forests (Table 14.3). Results of the study, through FRNI calculation by incorporating area statistics derived from satellite data in conjunction with satellite data, suggested that one of the mandal Marredumulli attained a higher rank of FRNI (Table 14.4), suggesting that the forests of that area need immediate rehabilitation.

The overall results of the study, through semivariograms and NDVI texture analysis, suggest that high spatial resolution, in conjunction with advanced textural measures, can be used for studying canopy stand structure, with respect to homogenity or heterogenity in general, and inter-stand variability, in particular. Also, the present study suggests that satellite data can be effectively utilised for studies relating to biodiversity and forest related phenomena.

Table 14.1: Area Statistics of Different Mandals Derived from IRS -1C LISS -III Satellite Data. (in sq kms)

Category	*Y.Rama-varam*	*Devi-patnam*	*Rampa-Chodavaram*	*Raja-Vommangi*	*Adda-tigala*	*Ganga-varam*	*Marredu-milli*
Sandy areas	7.77	0.77	1.66	0.44	1.37	0.03	0.19
Crop lands	6.83	20.98	12.52	75.94	28.60	27.70	.6.67
Plantations	152.20	108.01	181.60	115.28	190.78	119.76	6.67
Deep Water	0.00	11.87	4.36	0.00	0.00	23.60	0.00
Shallow Water	2.44	1.09	0.36	3.22	1.98	0.82	15.98
Dry Deciduous forests	99.90	1.09	0.36	3.22	1.98	0.82	15.98
Mixed Dry deciduous forests	137.50	32.96	55.73	60.56	84.30	51.35	273.00
Burnt areas	88.10	16.70	40.17	13.09	52.72	21.45	38.30
Degraded Forests	219.30	35.29	56.52	30.65	124.50	65.20	220.42
Fallow	65.90	61.56	224.30	11.10	91.60	62.70	21.00
Total	779.7	309.9	618.269	458.63	617.1	410.5	900.53

Table 14.2: Species Composition of Forest Stands

Stand	Vegetation Type	Species Associations
1.	Mixed dry deciduous I	Dominant: *Xylia xylocarpa* (Roxb.) Taub. *Dillenia pentagyna* Roxb., association Codominant: *Syzygium cuminii* (L.) Skeels., *Chloroxylon swietenia* DC., *Diospyros montana* Roxb., *Combretum roxburghii* Spreng., etc.
2.	Dry deciduous III	Dominant: *Anogeissus latifolia* (DC.) Wallich ex Beddeome. Community. Codominant: *Holarrhena pubescens* (Buch.Ham) Wallich ex Don., *Cleistanthes collinus* (Roxb) Benth ex Hook f., *Grewia tilaefolia* Vahl.,Vahl., *Woodfordia fruticosa* (L). Kurz..
3.	Dry deciduous I	Dominant: *Terminalia alata* Heyne ex Roth., community Codominant: *Lannea coromandelica* (Houtt.) Merril., *Anogeissus latifolia* (DC.) Wallich ex Beddeome.,*Grewia tiliaefolia* Vahl., *Pterospermum xylocarpum* (Gaerth.) Sant.et Wagh., *Adina cordifolia* (Roxb).Hook.f.ex Brandis., etc.
4.	Mixed dry deciduous III	Dominant: *Chloroxylon swietenia* DC., *Mitragyna parviflora* (Roxb). Korth. Codominant: *Xylia xylocarpa* (Roxb.) Taub., *Syzygium cuminii* (L.) Skeels, *Madhuca longifolia* (Koenig) MaccBride., etc.
5.	Dry deciduous IV	Dominant: *Anogeissus latifolia* (DC.) Wallich ex Beddeome. - *Terminalia alata Heyne* ex Roth., association. Codominant: *Dillenia pentagyna* Roxb., *Holarrhena pubescens* (Buch.Ham.) Wallich ex. Don.
6.	Mixed dry deciduous II	Dominant : *Xylia xylocarpa* (Roxb.) Taub. - *Anogeissus latifolia* (D.C). Wallich ex Beddoeme., community. Codominant: *Mitragyna parviflora* (Roxb). Korth., *Terminalia alata* Heyne ex Roth, *Soymida febrifuga* (Roxb.) Adr. Juss., etc.
7.	Dry deciduous II	Dominant: *Cleistanthes collinus* (Roxb.) Benth ex Hook f., community Codominant: *Strychnos nuxvomica* L., *Holarrhena pubescens* (Buch.Ham.) Wallich ex Don., *Bridelia retusa* (L.) Sprengel., etc.
8.	Dry deciduous V	Dominant: *Lannea coromandelica* (Houtt.) Merriil., - *Anogeissus latifolia* (DC) Wallich ex Beddeome., association Codominant: *Garuga floribunda* Decne., *Grewia hirsuta* Wahl., *Cordia dichotoma* Forst., *Pterospermum xylocarpum* (Gaerth.) Sant. et Wagh., etc.
9.	Mixed Scrub I	Dominant: *Zizyphus oenoplia* (L.) Mill, *Cliestanthes collinus* (Roxb.) Benth ex Hook f., etc. Codominant: *Tarenna asiatica* (L.) Kuntze ex K. Schum., *Dichrostachys cinerea* (L.) Wight et Arn., stunted trees of *Holarrhena pubescens* (Buch.Ham.) Wallich ex Don., etc.
10.	Mixed dry deciduousIV	Dominant: *Xylia xylocarpa* (Roxb.) Taub. Community Codominant: *Soymida febrifuga* (Roxb.) Adr. Juss, *Dillenia pentagyna* Roxb., *Strychnos nuxvomica* L., *Adina cordifolia* (Roxb.) Hook.f. ex Brandis., etc.

Stand Vegetation Type	*Species Associations*
11. Mixed Scrub II	Dominant: Zizyphus oenoplia (L). Mill., *Phoenix humilis* Royle.,etc,. Codominant: *Cliestanthes collinus* (Roxb.) Benth ex Hook f., *Dichrostachys cinerea* (L.) Kuntze ex K. Schum.,*Cordia dichotoma* Forst., *Carrisa spinarum* L., etc.
12. Dry deciduous VI	Dominant: *Holarrhena pubescens* (Buch.Ham.) Wallich ex Don. *Grewia tilaefolia* Vahl. association Codominant: *Grewia hirsuta* Wahl., *Pterospermum xylocarpum* (Gaerth.) Sant. et Wagh., etc., *Canthium dicoccum* (Gaertn) T et B., etc.

Table 14.3: Carbon Dynamics of Vegetation of Study area

	Biomass t/ha	*C t ha^{-1}*
1. **Mixed Dry deciduous Forests**		
Above ground vegetation	239.9	129.0
Below ground vegetation	34.6	
Total Vegetation	274.5	
2. **Dry deciduous Forests**		
Above ground vegetation	112.7	64.34
Below ground vegetation	24.2	
Total vegetation	136.9	
3. **Mixed Scrub Forests**		
Above ground vegetation	1.69	0.972
Below ground vegetation	0.38	
Total vegetation	2.07	

Table 14.4: FRNI Values

Mandal Name	BPLFI	LA	SEI (A)	SEI (B)	FRNI
Y. Ramavaram	0.35	0.25	0.07	0.04	0.71
Marredumilli	0.42	0.26	0.06	0.04	0.77
Addatigala	0.25	0.13	0.08	0.04	0.44
Devipatnam	0.38	0.23	0.07	0.05	0.51
Rampachoda - Varm	0.16	0.06	0.08	0.04	0.23
Rajavommangi	0.28	0.03	0.06	0.03	0.68
Gangavaram	0.24	0.08	0.09	0.04	0.69

REFERENCES

Apan Armando A. (1996). Forest Rehabilitation Need Index (FRNI) for tropical areas : Concepts and Derivation Using Satellite Data and GIS Vol 11(2), Geocarto International p 43-45.

Buschbacher R., Uhl C. and Serrao E.A.S. (1990). Abandoned pastures in Eastern Amazonia. II. Nutrient stocks in the soil and vegetation. J Ecol 76: 682-699.

Champion H.G. and Seth S.K. (1968). A revised survey of the forest types of India, Government of India Publications, New Delhi p 297-299.

Cohen Warren B., Spies T. and Bradshaw G.A. (1990). Semivariograms of digital imagery for the analysis of conifer canopy structure. Remote Sensing Env 34: 167-178.

Dale V.H. (1990). Strategy for monitoring the effects of land use change on atmospheric CO_2 concentrations. Proceedings of Global Natural Resource Monitoring and Assessment. Preparing for 21st century, Venice Italy, 25-30 Sep 1989 - American Society for Photogrammetry and Remote Sensing, Bethesda, Maryland Vol I p 422-431.

Journel A.G. and Huijbregts Ch.J. (1978). Mining Geostatistics. Academic Press, New York p 14-33.

Post W.M., Peng T.H., Emanuel W., King A.W., Dale V.H. and Delnglis (1990). The global carbon cycle. American Scientist 78: 310-326.

Rolla S. Rao (1982). Wild plant resources and the proposed biosphere reserves of the Eastern Ghats of Northern Andhra Pradesh. Proc Eastern Ghats Resources: 165-172.

Shukla J., Nobre C. and Sellers P. (1990). Amazon deforestation and climate changes. Science 247: 1322-1325.

Singh Lalji (1990). Biomass, production and nutrient dynamics in a dry tropical forest. Ph.D thesis. Banaras Hindu University, Varanasi, India p 214.

Singh K.P. and Misra R. (1979). Structure and functioning of natural, modified and silvicultural ecosystems in Eastern Uttar Pradesh, Final technical Report.

Westlake D.F. (1963). Comparisons of plant productivity. Biol Rev 38: 385-425.

Wilson E.O. (ed) (1988). Biodiversity. National Academy Press, Washington DC.

Chapter 15

Animal Biodiversity in North-Eastern Hill Region with Special Reference to Economic Livestock

N. D. Verma and B. P. S. Yadav
ICAR Research Complex for NEH Region
Umroi Road, Umiam - 793 103, Meghalaya

The North Eastern Hill (NEH) Region of India represents unique geographical features, characterised by undulating hilly topography, high rainfall and humidity with a wide range of altitudinal variations. The climate varies from the subtropical plains to the temperate and alpine hills. The grasslands and rangelands, as well as forests favour biodiversity of animals. Animal biodiversity has existed in wild form since times immemorial, and some of them have been domesticated, as demanded by mankind for their food, as draft, animals, packing, pleasure and other usefulness. The pressure to produce enough food for the growing population forced man to develop and evolve suitable breeds of livestock. Mankind is in constant struggle to produce enough food to meet the needs of the burgeoning population and thus, many modern and elite breeds/ species of animals of today (are developed), descended from the wild forms of biodiversity, earlier through constant selection and breeding. Many large grazing animals, being developed especially wild ungulates and predators, have vanished due to hunting pressure and deforestation. Since the dawn of agriculture, animal diversity has been an integral part of farming in hills, so as to support rural stability and for sustaining agriculture (Alderson, 1981). The region supports a rich biodiversity of wild, semi-wilds besides domesticated livestock and poultry. In the present paper, emphasis is mainly focussed on the domestic and semi-domestic animal biodiversity of economic importance.

INTRODUCTION

The North-Eastern region is one of the distinctive regions of the country. The whole region forms a compact geographical unit by itself. The topography, especially the hilly region, is unique of its type with its peaks and valleys covered with dense forests and grasslands, along with many rivers sneaking through these hills. The climate of this region varies from the sub tropical plains to the temperate and alpine hills, which can accommodate animals accustomed to live in snow (Yak) or to hot humid conditions (Buffaloes). Though, the human habitations are scattered and geographically isolated from one another, because of undulating terrain, this region is enriched with versatile and rare flora and fauna.

Forests largely cover the entire North-East region. The total forest cover which is approximately 54 % (NEC, 1995) is distributed in different states, like Arunachal Pradesh (62%), Assam (39%), Manipur (68%), Meghalaya (38%), Mizoram (76%), Nagaland (52%) and Tripura (60%). The type of forest varies from evergreen dense forests to natural grasslands, which are an excellent habitat for a great variety of animal population (wild and domestic).

The NEH region supports a rich biodiversity of wild, semi-wild, besides domesticated livestock and poultry. Although almost all types of wild animals are found in this region, some particular species like red pandas, hog badgers, ferret badgers and crestless porcupines are specific to this area. Among the semi-wild animals, three types of goat antelopes (serow, goral and tabins), musk deer, yak and mithun are distributed in this region. The livestock population (domesticated) of the region includes cattle, buffaloes, sheep, goats, equines, pigs, poultry, yak and mithun (Table 15.1). Though the NEH region is full of vivid and rich animal biodiversity which can be economically utilised by mankind, the present paper mainly focusses on the domestic and semi-domestic animal biodiversity of the region.

Table 15.1: Livestock wealth in North-Eastern states (in thousand)

States	*Cattle*	*Buffaloes*	*Sheep*	*Goats*	*Horses and ponies*	*Pigs*	*Yaks*	*Total Poultry*
1	2	3	4	5	6	7	8	9
Arunachal Pradesh	311	12	28	108	5	243	8	1121
Assam	7278	623	67	2134	14	642	-	11451
Manipur	770	141	16	44	1	383	-	3018
Meghalaya	587	28	15	194	5	280	-	1590
Mizoram	50	5	1	20	2	82	-	838
Nagaland	203	15	1	72	-	314	-	1124
Tripura	828	16	3	441	1	89	-	1865
All India	198567	76769	44837	99405	782	10760	36	258335

CATTLE

Two types of domestic cattles now living are *Bos indicus* (humped cattle of tropical origin) and *Bos taurus* (humpless cattle of temperate origin). Besides these, *Bos gaurua* (The Gaur) or *Bos frontalis* (The Gayal), locally called Mithun, which exists in semi-domestic form, is important for conservation and development. Mithun, a rare bovine species of Arunachal Pradesh, Manipur, Mizoram and Nagaland

(Anon, 1989-91) is only confined to the North-Eastern States of India. Two prominent breeds of Mithun namely, *Arunachalese* and *Nagamese* have been identified (Verma and Yadav, 1995), whereas *Bos sondaicus* (the banteng) and *Bos bison* (wild buffalo) are found in wild conditions. *Bos grunniens* (Yak) is the most important cattle of the high hills of Arunachal Pradesh and Sikkim. This animal has multipurpose utility for food, fur and wool draft. It is the most important species of the highlands and snow-bound hills under the hypoxic environment and is the lifeline of the people. This longhaired oxen-like animal, whose natural habitat is Central Asia, is found in Arunachal Pradesh and Sikkim areas in the NEH Region. It is a beast of burden and a milk producer. It is a sure-footed pack animal useful over the dangerous mountain passes in the Himalayas and an animal of multipurpose uses in the high hills. Swamp/water buffalo (*Bubalus-bubalis*) of Manipur and other valley areas in the region, is another important bovine species which needs attention for its conservation and development. With its long sturdy legs, it is ideally suited for the cultivation of half-submerged rice fields; it can walk through the mud and water, completely at ease.

PIGS

Swine, ungulate of the family Suidae is another important animal used for meat in this region. It is an animal of choice for the tribal population. Pigs are believed to have been domesticated in Eastern Asia, as early as 2900 B.C., and the domesticated pig (*Sus domesticus*) has descended from several wild species like *Sus cristatus, Sus scrofa* (East India swine), *Sus vittatus* and *Sus salvanius*. Among these, the Indian wild pig (*Sus cristatus*) and East India Swine (*Sus vittatus*), existing in the NEH region are very important. Another important variety among the swine species is the Mini pig of Nagaland and the Burma Mizo of Mizoram and the adjoining areas in the region.

SHEEP AND GOAT

Sheep and goat have descended from the wild relatives of two genera *Ovis* and *Capra,* respectively, from its wild relatives. The mountain sheep of non-descript type, known as the local breed of Arunachal Pradesh, is an important animal, which thrives well on hills. The Tibetan sheep, distributed in Northern Sikkim, as well as East and West Kameng districts of Arunachal Pradesh, are shorn twice a year, producing 400-900 g wool per clip. Bonpala sheep an important small ruminant, found in Southern Sikkim, produce is about 400 to 500 g wool per year.

The goat is an important livestock in subsistence agriculture, due to its unique ability to adapt and maintain itself in harsh and difficult environments. Black Bengal and the local Assam goat are commonly available in the valleys and low hills, whereas the Chegu breed of high mountain regions are found in the Tibetan border of the NEH region.

OTHERS

Local breeds of horses and ponies are one of the important drag and pack animals of the hills, whereas dogs are used for hunting and ornamental purposes, besides being the source of food for tribes in Nagaland and other areas of the region. Wild rabbits and hares, from which breeds can be evolved through selection and breeding are also found in jungles and forests. The elephant is another important animal of the region, existing as wild life, as well as in domestic form, for carrying wood load in forests and heavy load in hilly terrains. It is an outstanding beast of burden in India. Two species of elephant, the African *(Loxodonta africana)* and Indian (*Elephas maximus*) exist. However, the Indian elephant is available in the NEH region and it is the best animal for carrying timber in the forest.

POULTRY

Poultry, a common name, includes the domestic bird, chicken, ducks, turkeys, quail, geese, etc. The wild or jungle fowl viz., *Gallus lafayetti*, *G. sonnerati* and *G. varius* are important. The fowl was first domesticated for cock fighting to have pleasure. The domestic fowl (*Gallus gallus murghi*) originated in this region. The desi breed of chicken, Aseel and Chittagong, is said to be more or less of pure characteristics, exist in the region and need attention and conservation for future development.

CONCLUSION AND FUTURE APPROACH

India has a vast diversity of livestock which is reflected in the large number of breeds or strains of cattle, buffalo, sheep, goat, horse, pigs, poultry, dogs, elephant, rabbits and the other rare species of yak and mithun in the region.

In recent years, the earth's most beautiful and dramatic NEH region of India has witnessed a potential threat to its animal biodiversity, because of indiscriminate deforestation and hunting. It is quite alarming to note that some of the finest breeds and species of animals of this region are on the verge of extinction. The protection and preservation of the existing animal biodiversity is an urgent need, so as to meet the requirements of the growing population of the country. Saving the biodiversity (diverse genetic resources) is a pre-requisite for successful livestock improvement programmes. The Policy Makers, Planners, Governmental organisations and Non-Governmental Organisations should give priority towards preserving the precious animal biodiversity of the NEH region.

A breed or population is said to be endangered if the number of animals available for future breeding have decreased below a certain minimum member. Alderson (1981) and Maijala (1982) have given estimates of the endangered population for five livestock animal species (Table 15.2), whereas Smith (1984) estimated the minimum size of a breeding unit and the number of breeding animals that should be replaced annually, to keep the inbreeding levels to about 0.2% a year, through *in situ* conservation (Table 15.3).

Table 15.2: Estimates of endangered population size in livestock species

Species	*No of breeding females (per 1000 animals)*	
	Alderson (1981)	*Maijala (1982)*
Cattle	750	1000
Sheep	1500	500
Pigs	150	200
Horses	1000	Not available
Goats	500	200

Table 15.3: Minimum number of animals required for conservation

	Cattle		**Sheep**		**Pigs**		**Poultry**	
	Male	*Female*	*Male*	*Female*	*Male*	*Female*	*Male*	*Female*
Size of breeding unit	10	26	22	60	44	44	72	72
No. of breeding animals introduced per year	10	05	22	12	44	18	72	72

Comprehensive surveys, from time to time are warranted to evaluate the status of animals to be boosted up from all corners and to conserve and protect the species and breeds, which are on the way to extinction. If this diversity is lost, both nature and mankind will be crippled. Therefore, preservation of animal biodiversity is a necessity for genetic improvement, so as to meet future unforeseen requirements of mankind, for cultural, educational and economic development.

REFERENCES

Alderson L. (1981). The conservation of animal genetic resources in the United Kingdom. FAO Animal Production and Health Paper 24: 53-76.

Anon (1989-91). Annual Report. National Research Centre on Mithun, Jharnapani, Nagaland, India.

Maijala K. (1982). Preliminary report of the working party on animal genetic resources, Session 1. Commission on Animal Genetics, EAAP, G 1.2. Leningrad.

North-Eastern Council (NEC) (1995). Basic Statistics of the North-Eastern Region.

Smith C. (1984). Estimated cost of genetic conservation in farm animals. FAO Animal Production and Health Paper 44(1): 21-30.

Verma N.D. and Yadav B.P.S. (1995). Habitat and breed characteristics of Mithun (*Bos frontalis*) in India. Asian livestock Vol XX (9): 101-102.

Chapter **16**

Sacred Groves of Meghalaya: A Religious Way of *in situ* Conservation

Hiranjit Choudhury*, Sanjiban Goswami* and Suman Kumaria
Department of Botany, North-Eastern Hill University, Shillong - 793 022, Meghalaya
**Department of Botany, St. Edmund's College, Shillong - 793 003, Meghalaya, India*

The conservation of natural resources, which have some religious beliefs attached with them, has been in practice in India for a long time. Conserving the forest and its ecosystem in the form of sacred groves through traditional practices in Meghalaya is worth mentioning. Sacred groves have been traditionally protected and managed by the local communities with a wide range of management practices. The groves are classified into three categories, based on the traditional forest classification system and location viz., *Law Lyngdoh*, *Law Niam* and *Law Kyntang*. These sacred groves serve various important ecological services like the protection of land/soil from erosion, protection of water, *in situ* conservation and are therefore, rich repositories of different flora and faunal species. Conservation of sacred groves in the state is intimately interrelated with the cultural life of the indigenous people. Various religious beliefs, rites and rituals are associated with conservation. As most of these groves are representatives of the relict climax vegetation, they shelter plants and animal species that are not found elsewhere and serve as a refuge, providing safe sites for a variety of species.

INTRODUCTION

The conservation of nature and natural resources with some, religious beliefs and taboos has been in practice in our country since time immemorial. One of the many forms of worship of nature in India is the tradition of the consecration of certain forest(s) or land area(s) to a deity or ancestral spirit. These patches of forests are designated as "sacred groves". These groves are small patches of native vegetation types, traditionally protected and managed by the local communities with a wide range of management practices. They may vary in size, from less than a hectare to up to a few square kilometers, and are therefore, island-like in landscape.

Meghalaya, situated in the North-Eastern part of India, has a good number of sacred groves that are scattered all over the districts of the Khasi and Jaintia Hills (Syngai, 1999). The sacred groves of Meghalaya are public or village or community forestlands, set aside for religious purposes, under the traditional land tenure system (Gurdon, 1975). According to custom, it is an unforgivable offence to cut down trees or even pick flowers and fruits from such sacred groves except for cremation and religious purposes, of course with the due permission from the *Lyngdoh* (Priest/religious head). These groves are classified into three categories, based on the traditional forest classification system and location, viz., *Law Lyndoh, Law Niam* and *Law Kyntang* (Tiwari *et al.*, 1995, 1998a, b; Syngai, 1999). In a place, ruled by the *Lyndoh*, the sacred groves are called *Law Lyndoh*. In places, where traditional religion (*Niam trai*) plays a major role, the groves are called *Law Niam* and in areas where a village headman rules the village, the groves are called *Law Kyntang*. Though variously named, all these sacred groves have the same status. These groves have been closely interwoven with the socio-cultural life of the people, and various forms of rites, rituals and religious ceremonies have remained associated with these forests. Besides, a number of religious beliefs and taboos, which vary from region to region, and even from grove to grove, have also been attached to these sacred groves and are being perpetuated from generation after generation. These religious beliefs and practices have been playing a key role in protecting and preserving the groves in their pristine form, since ages.

There is either minimal or no extraction of forest produce from these groves and as they are representatives of the native relict climax vegetation, therefore, they shelter plant and animal species that are not found elsewhere (Haridasan and Rao, 1985) and as such, they are very rich in biodiversity. Thus, these groves serve as a refuge, and provide safety for a variety of species of flora and fauna for reproduction (Darlong, 1995). They help in the maintenance of viable populations of pollinators and predators, conserve germplasm (Khiewtam, 1986) and serve as potential centres of propagules dispersal, and the restoration (Hughes and Chandran, 1998) of wasteland and fallows. Besides, due to their dense vegetation, often spreading over large areas, they also provide several important ecosystem services.

The sacred groves of Meghalaya, for years, have been playing a vital role in environmental and biodiversity conservation. In spite of their conservational, cultural and aesthetic importance, the sacred groves of Meghalaya are but, little studied. Out of 79 reported sacred groves (Tiwari *et al.*, 1995), only the Mawphlang (Fig. 16.1) and the Cherrapunjee (Mawlongsyiem) sacred groves have drawn the attention of a few researchers. Hazra (1975) published a taxonomic account of the sacred grove at Mawphlang. Various ecological aspects of this grove, such as succession, community characteristics, gap phase regeneration and regeneration ecology of dominant tree species have been studied (Khan *et al.*, 1986, 1987; Khiewtam, 1986; Khiewtam and Ramakrishnan, 1993; Barik *et al.*, 1992, 1996a, b) and Rao *et al.*, (1990, 1997).

Fig. 16.1. Mawphlang sacred grove

CONCEPT OF SACRED GROVES–AN INDIGENOUS KNOWLEDGE SYSTEM

The sacred groves of Meghalaya portray a long tradition of conservation of nature, based on sound ecological fundamentals being practiced by the indigenous tribal people of the state viz., the Khasis and Jaintias since time immemorial. Keeping in mind, the difficulty in making the common people understand the importance of the environmental and conservational value of forests, the forefathers and the heads of these tribal communities probably devised a simple and easy way for their conservation and perpetuation by attaching various religious beliefs and taboos with them. As time passed, sacred groves became a part and parcel of the cultural life of the people in the entire Khasi and Jaintia Hills. Most of these sacred groves are situated near the summit of hills, which were the sources of many important rivers. These hills are mainly composed of oak and *Rhododendron* trees (Gurdon, 1975).

Most of the sacred groves studied, so far in the state revealed that the individual groves have a comparatively bigger size than their counterparts in the rest of the country, where the groves are generally very small. For instance, Mawphlang covers an area of 75 hectares (Tiwari *et al.*, 1998a), Sunapur covers 5.76 sq km, Raliang covers an area of 1.20 sq km (Syngai, 1999) and many others, which cover more than a square kilometer. Considering the vast area of these groves, their location (on the critical sites in catchments areas) and the religious sanctions attached to them, the concept of sacred grove conservation in the state of Meghalaya is strongly believed to have been an indigenous knowledge system, conceived, developed and perpetuated by the indigenous tribal people of the state, mainly the Khasis and the Jaintias (Tiwari *et al.*, 1998a, b; Syngai, 1999).

GENERAL STATUS OF SACRED GROVES IN MEGHALAYA

Sacred groves have always been a part of the cultural life of the people of Meghalaya (Khiewtam and Ramakrishnan, 1989; Tiwari *et al.*, 1995, 1998a, b).

The available information suggests that before the advent of Christianity in Meghalaya, each village had a sacred grove, ranging in area from 0.04 sq km to 12 sq km. A large number of the sacred groves are located in the Cherrapunji region of the East Khasi Hills District. Here, there are two types of groves: (i) those that are considered traditionally sacred, and therefore, ideally not even a fallen twig could be removed from within, and where a whole variety of religious beliefs and practices exist, and (ii) those that are protected reserve forest groves, declared to be so in recent times by the village headman with the concurrence of the village council, where limited extraction of timber is permitted (Khiewtam and Ramakrishnan, 1989). Although there is a legal provision for registering all the sacred groves, private plantations and private forests, etc. with the concerned authorities of the Autonomous District Council(s), yet, there is no complete and comprehensive list of sacred groves available with the State Government or the Autonomous District Council(s). In 1984, the State Forest Department estimated the area under sacred groves to be about 1000 sq km (Anonymous, 1984). The Jaintia hills may have over 200 groves, each no more than 3-4 ha large (Burman, 1992). Ramakrishnan (1995) reported that there are 21 sacred groves in Cherrapunji region with varied degrees of human disturbance. Tiwari *et al.*, (1995) documented 79 sacred forests in the state, irrespective of their status and ownership. These have been located and denoted on a geographical map of Meghalaya. These workers also studied the area of individual groves and their canopy cover. The area varied between 0.01 ha and 900 ha. Out of 79 sacred groves, canopy cover was estimated for 56 sacred groves. Only 7 belonged to the undisturbed category (Tiwari *et al.*, 1995, 1998a, b) and the rest had been exploited by man exhibiting some degree of disturbance (Table 16.1). Of the total area covered by the 79 groves (10,511 ha), 138 ha (1.3%) were undisturbed, about 4420 ha (42.1% of the total area) had dense canopy, 2765 ha (26.3%) sparse canopy and 3,188 ha open canopy (Tiwari *et al.*, 1998a, b).

Table 16.1. Status of sacred groves in Meghalaya (data for 56 groves out of 79)

Canopy cover (%)	*Category*	*Number of groves*	*Percentage area of groves*
100	Undisturbed groves	7	12.5
<100 and >40	Dense canopy	14	25.0
<40 and >10	Sparse sacred grove	11	20.0
<10	Open category	24	42.5

(Source: Tiwari *et al.*, 1998a, b)

LEGAL STATUS OF THE SACRED GROVES IN THE STATE

As compared to the other parts of the country, a sharp contrast is noted as far as the legal status of groves of this state is concerned. The sacred forests of Meghalaya enjoy adequate legal support as these are covered under the United Khasi and Jaintia Hills Autonomous District (Management and Control of Forests) Act, 1958 and the Garo Hills Autonomous District (Management and Control of

Forest) Act, 1961. Both these Acts were passed in pursuance of paragraph 11 of the Sixth Schedule of the Indian Constitution and were extended to all the forest lands of the state, except 722.36 sq km (constituting only ca 8% of the total forests of the state), which are directly under the control of the State Forest Department (Anonymous, 1984).

As per the above Act(s), the sacred groves are to be managed by the *Lyngdoh* (Priest/religious head) or other person(s) to whom the religious ceremonies for the particular locality or village or villages are entrusted and in accordance with the customary practice(s) and the rules framed by the Executive Committee of the concerned Autonomous District Council, from time to time. As per Section 7 of the above Act(s), no tree shall be felled in these forests without the prior sanction of the concerned Chief Forest Officer of the District Council or any officer duly authorised by him, in black and white. Further, Section 9 of the Act states, "No tree(s) shall be felled or removed from *Law Lyngdoh, Law Kyntang* and *Law Niam*, except for purposes connected with religious functions or ceremonies, recognised and sanctioned by the *Lyngdoh* (Priest/religious head) or other persons, in accordance with section 4(b)". There is a provision for the registration of *Law Lyngdoh, Law Kyntang* and *Law Niam* with the District Council in the United Khasi-Jaintia Hills Autonomous District (Management and Control of Forests) Rules, 1960.

SACRED GROVES AND THEIR ROLE IN THE ECOSYSTEM

Protection of Land, Soil and Water

According to one report (Syngai, 1999) the sacred groves in the state cover an estimated area of about 10,000 ha, and about 60 of them, with an area of about 6,500 ha, are located at the source of perennial streams. Tiwari *et al.*, (1995,1998a, b) reported that most of the sacred groves in the state (66 out of 79), covering an estimated area of 10,251 ha, are located on the catchment areas of major rivers and rivulets. At least 58 groves with an area of about 9,621 ha, are located at the origin of perennial streams, while about 38 sacred groves, covering an area of 6,454 ha, are on steep hill slopes. For instance, *Lum Shyllong-Nongkrim** sacred groves are located in an area where, from as many as, eight streams* originate and supply water to a number of downstream human habitations. The other common sites of occurrence of sacred groves are the hill slopes of *Mih Myntdu* (e.g., Trepale Jowai sacred grove). Being located on these critical sites, which are vulnerable to degradation, the sacred groves protect the[1] land and soil from erosion and help in maintaining the quality of water in the downhill streams.

In Situ Conservation of Biodiversity

Literature provides many reports on the general aspects of the sacred groves, the ecological value of the species within them, and the rich biodiversity of the

---

1 * Lum Shyllong Sacred Grove and source of ten streams according to Syngai (1999)

groves, etc. In many landscapes, the sacred groves are representatives of the relict climax vegetation that existed once upon a time and has now been destroyed through a variety of perturbations caused by human activities. In that sense, these groves act as islands of biodiversity and have the potential for biodiversity conservation in a variety of ways (Ramakrishnan, 1998). Most of the sacred groves of Meghalaya, from that perspective, are the repositories of rich biodiversity and most often, represent the climax vegetation of the area, as cited by Ramakrisnan (1998). These groves are the abode of a number of rare and endangered flora and fauna. At least, 50 rare and endangered plant species of the state are now confined only to these sacred groves (Haridasan and Rao, 1985), although some reports suggest that these groves remain the last refuge for 700 rare plant species (Saraswati, 1998). A baseline floristic survey in various sacred groves of the state, by various workers, revealed as high as 514 species representing 340 genera and 131 families in these forests. The highest number of species (39) belongs to Orchidaceae followed by Poaceae, Rubiaceae (28 species each) and Rosaceae (26 species). Among the tree species, the Fagaceae members are dominant over others in most of the sacred groves. Epiphytic flora is quite abundant (Fig. 16.2 and 16.3). Various types of lianas (Fig. 16.4) and ferns are also found in these forests. The vegetation of undisturbed sacred groves is generally very dense and well stratified. It has four major strata viz., canopy layer, sub-canopy layer, shrub layer and ground flora. Emergent trees constitute the top canopy layer (Fig. 16.4), the tree species like *Rhododendron arboreum* and *Myrica esculenta* constitute the sub-canopy layer (Fig. 16.5) in most groves, and the shrub layer is composed of a number of shrubs and saplings of the tree species. Ground flora consist of grasses, herbs, pteridophytes and bryophytic species, along with the seedlings of the trees (Fig. 16.6). Even some saprophytic angiospermic plants, for instance *Monotropa* sp. (Fig. 16.7), an interesting angiospermic saprophyte found abundantly in the Mawphlang sacred grove are present in the grove.

In general, species diversity in the sacred groves was reported to be much greater than in the other forests. For instance, Rao *et al.*, (1990) reported the

Fig. 16.2. An epiphytic angiosperm (*Hedychium* sp.)

Fig. 16.3. Some epiphytic orchids and ferns of the grove

Fig. 16.4. An emergent tree with epiphytic mosses & lianas

Fig. 16.5. Small tree and shrubs

Fig. 16.6. Grass, herbs and pteridophytes (*Lycopodium* sp. with strobili)

Fig. 16.7. An angiospermic saprophyte (*Monotropa* sp.) of the grove

Species Evenness Index (SEI), Shannon Index (SI) of general diversity and Species Richness Index (SRI) of a sacred grove at Mawphlang as 3.60, 1.00 and 0.96, respectively. On the other hand, in two other forests at Upper Shillong and Malki, all the diversity indices were comparatively low (SEI - Upper Shillong 2.20, Malki 1.30; SI - Upper Shillong 0.80, Malki 0.60; SRI - Upper Shillong 0.84, Malki 0.86). Besides flora, these groves are home to a number of faunal elements. Not much literature is available on the faunal biodiversity of the sacred groves. Herpetological Research and Conservation Project in North-East India has initiated the development of an inventory on herpetofauna (amphibia and reptiles) and according to the project report, the diversity of amphibians in the sacred groves of Cherrapunji is quite high. *Calotes maria*, a lizard endemic to the Khasi Hills and Meghalaya *Icthyophis* sp. occurs in the sacred groves in the Khasi Hills (Ahmed, 2001a, b, 2002).

Regeneration Potential and Population Structure

The regeneration potential of the Mawphlang sacred grove was compared with that of the two other forests (Barik *et al.*, 1996a). The density of shade-intolerant species like *Schima khasiana* was more in the other disturbed unprotected forests, while seedlings of shade-tolerants like *Quercus* spp. were more abundant in the sacred groves. In general, high seed predation/disappearance and competition of various resources among the seedling populations, especially for light and space, interfere with the success of regeneration in the sacred forests (Barik *et al.*, 1996 b; Rao *et al.*, 1997). Khan *et al.*, (1987) studied and compared the population structure of tree species with that of a disturbed forest at Upper Shillong, by plotting the proportion of seedlings, saplings, small trees and big trees as a percentage of the total tree density. The structure was an upright pyramid, in case of the disturbed forest and it was an inverted pyramid, in case of the sacred grove.

FRAGILITY OF SACRED GROVES

The fragility of sacred groves can be well understood in the context of Cherrapunji, as large numbers of them are located in this region of Meghalaya. The climax forests in Cherrapunji are now represented only by relic sacred grove forests. A typical example is Mawsmai sacred grove. It is situated in one of the wettest place in the world with an annual average rainfall of about 10370 mm, and in some exceptional years, it could be as high as 24550 mm, as in 1974. Obviously, the humid sub-tropical forests which were extensive in the yesteryears, but developed on highly leached nutrient deficient soils are therefore, very fragile (Khiewtam and Ramakrishnan, 1993). The fragility of these humid forests is compounded by the limestone formations underneath them, having extensive underground tunnels of stalactites and stalagmites (Kratz topography). Despite the formation of limestone, the soil is highly acidic (pH 3.9-5.2) and with poor nutrient supply (Ramakrishnan, 1992). Perturbation of these forests were reported to result in degraded seral grasslands, representing arrested successional stages (Ram and Ramakrishnan, 1992), and in extreme cases, leading to a balded landscape or transformed to arrested seral grass lands (Fig. 16.8) (Ramakrishnan and Ram, 1988). The Cherrapunji ecosystem, now represents desertified conditions which are due to the immense damage inflicted through deforestation in some distant past and the present standing is such, that it now refuses to recover to its original state. The fact that the "jhum" cultivation (shifting agriculture), which is the predominant land use system in the region, has been banned by the village council around Cherrapunji for quite sometime now, indicates the role played by this land use system in the formation of the present landscape.

The sacred grove forest vegetation itself, is somewhat stunted because of unbalanced soil characteristics and extreme climatic conditions, but the species diversity have been reported to be very high, with a variety of trees, shrubs,

Fig. 16.8. Arrested seral grassland

herbs, climbers and epiphytic species, forming a tightly packed system (Ramakrishnan, 1995, 1996). Nutrient release rate in the grove was found to be high due to a very high annual litter fall and rapid litter decomposition rate, as compared to lower montane rainforests elsewhere (Khiewtam and Ramakrishnan, 1993). The soil itself showed less amounts of nutrients to support the large biomass of the sacred grove. The fine root mat developed on the mineral soil was important for supporting the large above, ground biomass and for the tight cycling of the nutrients. The nutrients released from the decomposing litter were short-circuited back into the living biomass through the surface root mat, even before the nutrients could leach out into the mineral soil. Operating under highly stressed environmental conditions, this delicately balanced sacred grove forest ecosystem is highly fragile. The sharp boundary between these sacred groves and the balded landscape in the periphery indicates that the system will never recover through the natural process of regeneration. These sacred groves are island-like or like individual tesserae in a landscape mosaic. Studies of island biogeography have shown that such fragments of forests may suffer extinction of species and may be vulnerable to invasion by aggressively introduced exotic weed-like species or tree species and both natural calamities or man caused perturbations, such as fire, windstorms, diseases, predation by hunters and poachers and deliberate destruction (Hughes and Chandran, 1998). In many sacred groves, for instance, patchy growth of *Pinus kesiya* can be seen around the periphery (Fig. 16.9), which may lead to the invasion of the grove, in case any major disturbance, like, fire etc. occurs, as *P. kesiya* is a quick secondary coloniser of degraded sites.

SOCIAL AND CULTURAL ASPECTS OF SACRED GROVES OF MEGHALAYA

The North-Eastern Hilly Regions of India have a rich human cultural diversity,

Fig. 16.9. A few Pinus kesiya plants at the periphery of the grove

with over 100 different tribes originating from different parts of South and Southeast Asia, each bringing with them their own distinct cultures, evolved and adapted to local situations in isolation from one another due to linguistic barriers. In this context of bio-cultural diversity, the concept of 'sacred grove' is embedded in the agroforest landscape (Ramakrishnan, 1998). Conservation of sacred groves in the state is intimately interrelated with the cultural life of the people (Khiewtam and Ramakrishnan, 1989; Tiwari *et al.*, 1998a, b).

The traditional religions instilled in the people and the specific rites and rituals to be organised are performed inside or by the side of the sacred groves. The priest, together with the headman and the members of the particular village, are made to shoulder the responsibilities to perform such rites and rituals. They vary from place to place and grove-to-grove, depending upon the socio-cultural set up and religion of the local populace. As far as rituals are concerned, they are performed rigidly as per the prescribed traditional religion(s).

The sacred groves of Meghalaya have in fact, been in existence since millennia, which is substantiated by the presence of tall 'monoliths' by the side of sacred groves, for example, the monoliths at Mawsmai on the east of Mawlongsyiem Sacred Grove and the monoliths at Nartiang, which are supposed to be the tallest in the state. These monoliths were erected in memory of the departed souls of the, then most prominent persons. The common belief amongst the Khasis and Jaintias is that the departed souls of the near and dear ones rest in peace in these groves for a few months till their permanent resting places are prepared (Syngai, 1999). Therefore, the local inhabitants believe that the *Sylvan* deity would be offended if any kind of forest produce is extracted from these groves and would punish the offender by twisting his/her head. They also believe that the *Sylvan* deity would protect their village(s) from natural calamities, sickness, and invasion from enemies, etc., if the rites and rituals are performed religiously every year. These sacred groves are normally under the direct control of the Village council, and the priest is responsible for the performance of the rites and rituals in accordance with the traditional religion.

EROSION OF THE VALUES OF TRADITIONAL INSTITUTIONS AND MODERN PERCEPTION ABOUT SACRED GROVES

Tradition recognises forests as one of the main life-supporting systems and as climate regulators; it recognises the direct and indirect benefits they provide to mankind, such as food and materials, but mainly for the role they play in soil conservation and in the water cycle. Not much literature is available about the origin of sacred groves in the state. Perhaps, the forefathers of the indigenous tribes of this state realised the importance of forests and their role as life supporting system and were very wise in visualising the betterment of the future generations, through protection of the sources of water by conserving the forest groves. Even during those days, they probably could anticipate that it was not enough to merely declare those forests as protected groves for the purpose of their conservation. Therefore, they probably started various rituals and ceremonies to proclaim to all the people that these forests were 'sacred'. The rites and rituals have since then been observed and performed religiously and regularly in different groves, and in different parts of the state, especially in the Khasi and the Jaintia Hills. Thereafter, all these groves have been designated as 'Sacred Groves' (Syngai, 1999). Since millennia these traditional beliefs have been instrumental in the conservation of the sacred groves throughout the state. Unfortunately, the sanctity and values of the traditional beliefs attached to these groves are gradually being eroded with, the advent of modernity and education, access to once inaccessible sacred groves owing to the development of road communication, change from traditional religion to Christianity and increasing craving of the people for a better way of life. Further, over the years, the establishment of administration and judiciary bodies by the government has resulted in a considerable diminution of the influence of traditional administrative and judiciary institutions engaged in the management of the groves. All these factors have brought about a change in the attitude of the traditional societies in many parts of the state, leading to large-scale destruction of sacred groves. In addition, economic constraints caused by smaller land holdings have resulted in the encroachment of sacred grove areas for agriculture. Consequently, a plot of shifting cultivation could be found, in certain cases, inside the sacred groves, which were once totally prohibited for all uses other than religious purposes (Tiwari *et al.*, 1998a, b). The decline in the traditional beliefs is more obvious from the fact that, in most of the areas, people responsible for the observance of rituals seem to have forgotten the rites and procedures for performing the rituals (Tiwari *et al.*, 1995, 1998a, b). Traditional rituals are still performed in the undisturbed sacred groves, in accordance with customary beliefs. In moderately disturbed groves with good canopy cover, the traditional rituals are performed, but with less rigidity as compared to the undisturbed groves and no traditional rituals are performed in the sparse and degraded groves.

FUTURE APPROACHES–CONSERVATION STRATEGIES

Rapid and continued erosion of the traditional religious beliefs is accompanied by the brisk and unrelenting decline in the quality and the number of sacred groves. In the present context, the increasing trend of pure materialistic attitudes of the educated and improperly educated indigenous people and the poor socio-economic conditions of the uneducated indigenous rural people have to a large extent compounded the problem of sacred grove degradation in the state. Therefore, there is an urgent need to conserve and manage these groves for the present and the future. A concerted effort in this direction should be well coordinated, involving a variety of relevant governmental agencies, non-government organisations (NGOs), voluntary sectors supported by NGOs, media, grass-root level action groups, a variety of socio-cultural and religious bodies, the scientific community at large and finally, the people's active participation. The following strategies could be adopted:

- The younger generations, who consider religious beliefs central to sacred grove conservation as superstitions, should be made aware of the biodiversity and ecosystem values of the sacred groves and the long-term economic and ecological benefits derived from these groves.
- The socio-economic condition of the rural populace must be considered in detail, so as to achieve successful conservation. For this purpose, site-specific conservation/rehabilitation strategies of groves or degraded lands around the groves could be adopted. For instance, restoration activities of the degraded areas around the groves, based on agroforestry, like the production of cash crop(s) (e.g., cashew nut or mango, etc.) should be carried out. The cash crops would provide an economic income to the local people and would permit the restoration of a vegetation cover in the degraded sites. This may also act as a 'buffer zone' around the sacred groves, which, in turn may reduce the pressure on the sacred grove/site by generating income benefits for the people. This strategy has been found useful in some sacred grove areas in Ghana. CIPSEG Project (Co-operative Integrated Project on Savannah Ecosystem in Ghana) has shown that the rehabilitation of degraded areas around the sacred sites/groves is possible, if based on cultural beliefs, which tie in well with the religious and spiritual views shared by a specific community (Schaff, 1998).
- Site-specific conservation/restoration strategies involving Geller's Scheme (Geller, 1992) which encompass different aspects like the status of the grove and its conservation, and the socio-economic conditions of the people managing the grove, etc. may also be adopted. Following this Scheme, the sacred groves, having dense canopy cover, which are by and large undisturbed, may be given Level-I intervention under multiple tiers intervention programmes, where interventions are least intrusive and involve maximum number of people. The activities under such intervention programmes would include behavioural prompting through awareness programmes and some nominal incentives to the people in the form of

community development programmes. The disturbed sacred groves, which may require higher level and more effective intervention processes, may be brought under Level-II of Geller's scheme. This would require higher costs of restoration, in terms of materials and personnel. The activities under this scheme may involve any one or a combination of the following options- (i) immediate rehabilitation of the degraded groves through artificial/ natural/aided natural regeneration programmes; (ii) regenerate the adjoining village forests and ensure their effective management to meet the firewood, fodder and non-wood requirements of the villagers, so that the pressure on the sacred groves is minimised to the maximum; (iii) convert them into nature reserves and manage them with 'people's participation', on the principles of JFM (Joint Forest Management); (iv) constitute them as outdoor recreation areas and environmental research and education centres with a provision to share the revenue accrued with the local people.

- Sacred groves which are important, as far as wild life is concerned and which have the potential to act as wild life corridors, may be identified and restored/managed by the Wild life Wing of the State/Central government. Further, an institutional linkage, involving the local village-level traditional institutions, state, national and international organisations is required to be established, to launch joint and coordinated sacred grove conservation (Tiwari *et al.,* 1998a, b).
- Joint Forest Management could be an effective device for conserving the sacred forests. Though the system of JFM is originally a tribal practice, yet very few are aware or have heard of the concept in Meghalaya. The tribal communities here are losing their ethos of community living (Mukhim, 2003). This system perhaps says, that tribes should respect community living, which includes participative decision-making, aimed at the best interests of the community. This alone could be an effective survival strategy for the community, and at the same time achieving the conservation of natural resources, including sacred groves. Hence, it is very important to revive this community participation and concept of Common Property Resources in the management of all resources and that would inevitably require mass awareness of the evils of individual decision-making and individual exploitation of resources. Concept of Common Property Resources prevails and is in practice in many villages in the Khasi and the Jaintia Hills, and Mawphlang is one of the best examples (Nongkynrih, 2003).
- Peoples' active participation and involvement of local communities and distict councils have been found effective in the conservation and management of natural resources. For instance, in Zimbabwe, centralised management failed to check poaching of wild life. So local communities and district councils were made partners in a new programme called 'CAMPFIRE'. Hunting quotas were sold to hunters, and the revenue shared with communities and local bodies, which now have a stake in checking poaching, resulted in a wild life boom there (Aiyar, 2002).

REFERENCES

Ahmed M.F. (2001a). Poorly known endemic amphibians of North-East India. Froglog 43.

Ahmed M.F. (2001b). An assessment of ten poorly known endemic amphibians of Northeast India, India. A report prepared for the Smithsonian Institution, Washington DC.

Ahmed M.F. (2002). Inventory and natural history of herpetofauna of the Orang National Park, Assam, India. Technical Report, Aaranyak, Guwahati.

Aiyar S.S.A. (2002). Glocalisation: Globalisation plus localisation. In: Centre for Civil Society; Liberty and Society Seminar, The Economic Times, New Delhi, India p 62-66.

Anonymous (1984). Meghalaya forests. Chief Conservator of Forests. Meghalaya, Shillong.

Barik S.K., Pandey H.N., Tripathi R.S. and Rao P. (1992). Microenvironmental variability and species diversity in tree fall gaps in a sub-tropical broad-leaved forest. Vegetatio 103: 31-40.

Barik S.K., Rao P., Tripathi R.S. and Pandey H.N. (1996a). Dynamics of tree seedling populations in a humid subtropical forest of Northeast India, as related to disturbance. Can J For Res 26: 584-589.

Barik S.K., Tripathi R.S., Pandey H.N. and Rao P. (1996b). Tree regeneration in a subtropical humid forest: Effect of cultural disturbances on seed production, dispersal and germination. J Applied Ecology 33: 1551-1560.

Burman R.J.J. (1992). The institution of sacred grove. J Indian Anthropological society 27: 219-238.

Darlong V.T. (1995). Wildlife preservation and community action. In: Ecorestoration of Degraded Hills (Tiwari BK and Singh S eds) Kushal Publications, Shillong, Meghalaya, India.

Geller E.S. (1992). Applications of behaviour analysis to prevent injury from vehicle crashes. Cambridge Centre for Behavioural Sciences, Cambridge, Massachusetts p 147-159.

Gurdon P.R. (1975). The nature races of India: The Khasis. Cosmo Publication, New Delhi, India.

Haridasan K. and Rao R.R. (1985). Forest Flora of Meghalaya. Vol 1, Bishen Singh, Dehradun, India

Hazra P.K. (1975) Law Lyndoh (sacred grove), Mawphlang. Government of Meghalaya, Shillong, India.

Hughes J.D. and Chandran M.D.S. (1998). Sacred groves around the earth: An overview. In: Conserving the Sacred for Biodiversity Management (Ramakrishnan PS, Saxena KG and Chandrashekara UM eds), Oxford and IBH Publishing Co Pvt Ltd, New Delhi, India p 69-86.

Khiewtam R.S. (1986). Ecosystem function of protected forests of Cherrapunji and adjoining areas. Ph D thesis, North-Eastern Hill University, Shillong, Meghalaya, India.

Khiewtam R.S. and Ramakrishnan P.S. (1989). Socio-cultural studies of the sacred groves at Cherrapunji and adjoining areas in Northeastern India. Man in India 69: 64-71.

Khiewtam R.S. and Ramakrishnan P.S. (1993). Litter and fine root dynamics of relict sacred grove forests of Cherrapunjee in North-Eastern India. Forest Ecology and Management 60: 327-344.

Khan M.L., Rai J.P.N. and Tripathi R.S. (1986). Regeneration and survival of trees, seedlings and sprouts in tropical deciduous and sub-tropical forests of Meghalaya, India. Forest Ecology and Management 14: 293-304.

Khan M.L., Rai J.P.N. and Tripathi R.S. (1987). Population structure of some tree species in disturbed and protected subtropical forests of northeast India. Acta Oecologica Oeol Applic 8: 247-255.

Mukhim P. (2003). Community participation in natural resources management. In: Seminar on 'People's Participation in Natural resources Management' organised by Government of Meghalaya Soil and Water Conservation Department, Shillong, Meghalaya, India p 8-9.

Nongkynrih A.K. (2003). Customary common property resources management. In: Seminar on 'People's Participation in Natural resources Management' organised by the Government of Meghalaya Soil and Water Conservation Department, Shillong, Meghalaya, India p 10.

Ram S.C. and Ramakrishnan P.S. (1992). Fire and nutrient cycling in seral grasslands of Cherrapunji. International J Wildland Fire 2: 131-138.

Ramakrishnan P.S. and Ram S.C. (1988). Vegetation, biomass and productivity of seral grasslands of Cherrapunji in North-East India. Vegetatio 74: 47-53.

Ramakrishnan P.S. (1992). Shifting agriculture and sustainable development of North-Eastern India. UNESCO-MAB Series, Paris, Parthenon Publ, Carnforth, Lancs, UK p 424 (republished by Oxford University Press, New Delhi, 1993).

Ramakrishnan P.S. (1995). Ecology and traditional wisdom. In: UNESCO-sponsored conference 'The Cultural Dimension of Education and Ecology', Indira Gandhi National Centre for the Arts, New Delhi, India.

Ramakrishnan P.S. (1996). Conserving the sacred: From species to landscape. Nature and Resource 32: 11-19.

Ramakrishnan P.S. (1998). Conserving the sacred for biodiversity: The conceptual framework. In: Conserving the Sacred for Biodiversity Management (Ramakrishnan PS, Saxena KG and Chandrashekara UM eds), Oxford and IBH Publishing Co Pvt Ltd, New Delhi, India p 3-15.

Rao P., Barik S.K., Pandey H.N. and Tripathi R.S. (1990). Community composition and tree population structure in a sub-tropical, broad-leaved forest along a disturbance gradient. Vegetatio 88: 151-162.

Rao P., Barik S.K., Pandey H.N. and Tripathi R.S. (1997). Tree seed germination and seedling establishment in tree fall gaps and understorey in a subtropical forest of Northeast India. Australian J Ecology 22: 136-145.

Saraswati B. (1998). The logos and the mythos of the sacred grove. In: Conserving the Sacred for Biodiversity Management (Ramakrishnan PS, Saxena KG and Chandrashekara UM eds), Oxford and IBH Publishing Co Pvt Ltd, New Delhi, India p 31-46.

Schaaf T. (1998). Sacred groves in Ghana: Experiences from an integrated study project. In: Conserving the Sacred for Biodiversity Management (Ramakrishnan PS, Saxena KG and Chandrashekara UM eds), Oxford and IBH Publishing Co Pvt Ltd, New Delhi, India p 145-150.

Syngai D. (1999). Sacred groves of Meghalaya. In: Workshop on 'People's Participation in Biodiversity Conservation', Shillong, Meghalaya, India p 70-76.

Tiwari B.K., Barik S.K. and Tripathi R.S. (1995). Sacred groves of Meghalaya: Status and strategies for their conservation. North-Eastern Hill University, Shillong.

Tiwari B.K., Barik S.K. and Tripathi R.S. (1998a). Biodiversity value, status and strategies for the conservation of sacred groves of Meghalaya, India. Ecosystem Health 4(1): 1-20.

Tiwari B.K., Barik S.K. and Tripathi R.S. (1998b). Sacred groves of Meghalaya. In: Conserving the Sacred for Biodiversity Management (Ramakrishnan PS, Saxena KG and Chandrashekara UM eds), Oxford and IBH Publishing Co Pvt Ltd, New Delhi, India p 253-262.

Chapter 17

Conservation of Medicinal Plants Diversity of Meghalaya in India

H. Kayang

Department of Botany, School of Life Sciences, North-Eastern Hill University Mawlai Umshing, Permanent Campus, Shillong - 793 022, Meghalaya, India

An exploration was undertaken for the identification of medicinal plants frequently used by the Khasi, Jaintia and Garos communities of Meghalaya. A total of 298 medicinal plant species for treating different diseases were identified (124 species were used by the Khasi, 89 species by Jaintia and 85 species by Garos). Of the 298 medicinal plant species mentioned, 5 species were common to all the tribes, 9 species were common to the Khasi and Jaintia tribes, 13 species were common to the Khasi and Garos and 5 species were common to the Jaintia and Garos tribes. Based on the survey and findings, various conservation and management steps have been discussed to protect medicinal plants.

INTRODUCTION

Meghalaya, which lies between 25°47″ - 26°10″ N latitude and 89°45″ – 92°45″ E longitude and covers an area of 22,549 km^2, comprises the South Garo Hills, West Garo Hills, East Garo Hills, West Khasi Hills, East Khasi Hills, Ribhoi and Jaintia Hills districts. The altitude ranges from 50 – 1960 m. It is bounded on the North, East and West by Assam and on the South, by Bangladesh (Fig. 17.1). The State is in an emerging state of development, even though the traditional culture and heritage have thrived through the passage of time and other inclusions of western culture. The people in the State who belong to the 3 major tribes like Khasis, Jaintias and Garos have their own healing practice and health care system, where many plant species are used. The majority of the population are rural folk. People who are employed or under any service, dominate the urban sectors. The rural folk lead a very simple life with their main occupations ranging from agriculture to cattle rearing. Poultry, silkworm rearing and piggery are the other important occupations followed, but more emphasis is being laid on the

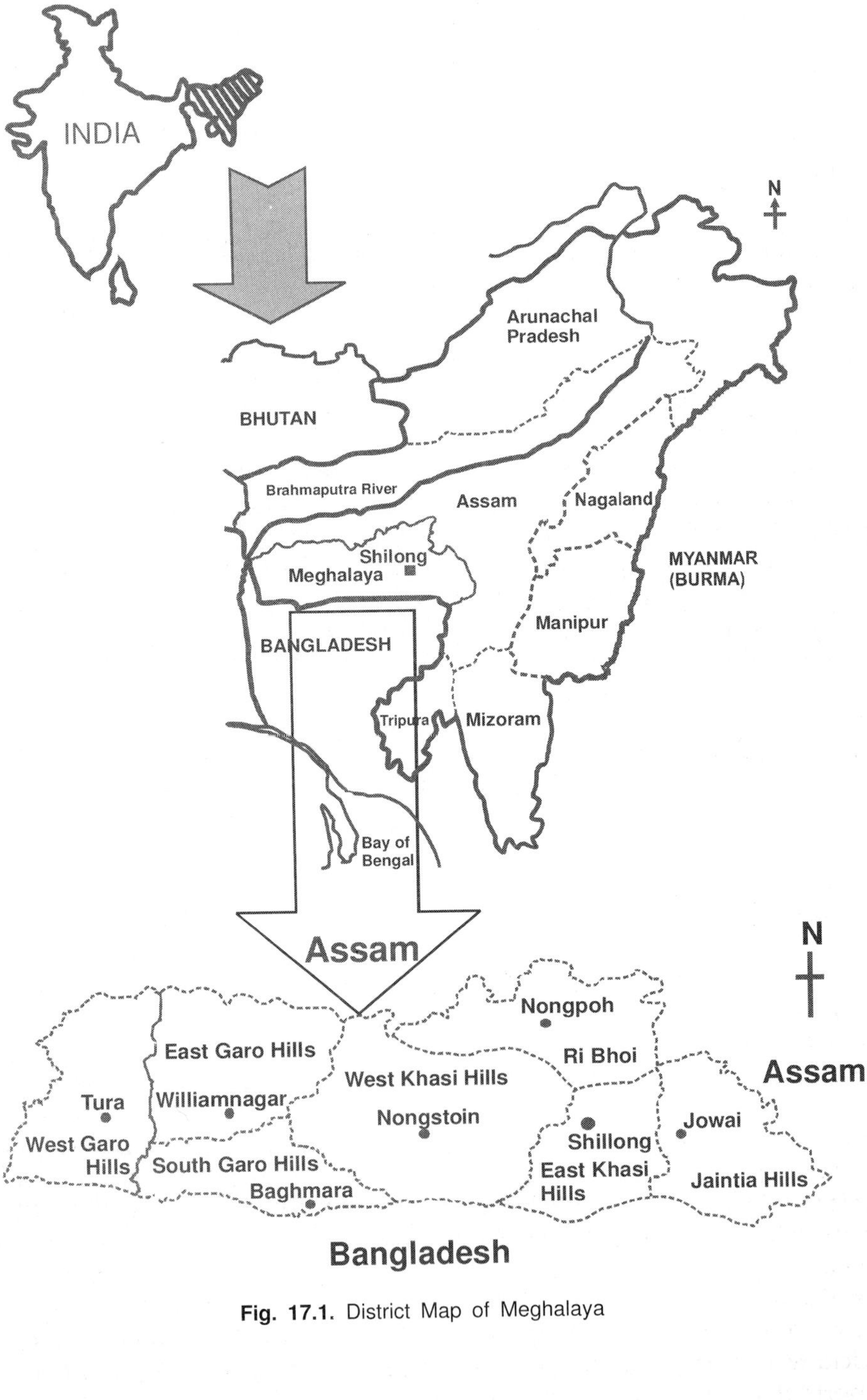

Fig. 17.1. District Map of Meghalaya

agricultural sector. Food crops like paddy, potato, maize, vegetables and others horticultural crops are usually grown commonly.

The people in the State are generally very knowledgeable about the wild medicinal plants around them, many of which have local names and are important to them medically or feature in folklore. This traditional knowledge is the best starting point for effective *in situ* conservation, which requires accurate and up-to-date information on the status of medicinal plants and, the extent and nature of the plants used by local communities and the capacity of the resource base to support different economic activities. This knowledge can be used in evaluating and creating awareness of the importance of medicinal plants, as it is generally easier for the general public to relate to this, than to the results of scientific trials.

The State possesses a variety of plant wealth that is yet to be fully tapped on a commercial scale, which in turn could accrue abundant benefits to the farmers too. In fact, Meghalaya has a great potential for the plantation of medicinal plants because of ideal agroclimatic conditions and suitable soil. Nature, in its generous abundance, has bestowed on Meghalaya a unique array of vegetation, ranging from tropical and sub-tropical to temperate or near temperate (Kanjilal *et al.*, 1982; Elias, 1994). This is due to the diverse topography, varied and abundant rainfall and differential climatic and edaphic conditions in the state, and within small regions.

In a wider context, there is a growing demand for plant based medicines, health products, pharmaceuticals, food supplements, cosmetics, etc. in the national and international market. Conservation and sustainable use of medicinal plants are issues on which immediate focus is required in the context of conserving biodiversity and promoting and maintaining the health of local communities, besides generating productive employment for the poor, with the objective of alleviating poverty in tribal and rural areas.

The international trade in medicinal plants is estimated at US $70 billion and is growing at a rate of seven percent annually. The Planning Commission of the Government of India has plans to increase the trade in medicinal plant extracts to Rs. 3,000 crores by the year 2005, and Rs. 10,000 crores by the year 2010. New opportunities are thus being created that could lead to the generation of employment in the medicinal plant sector (Bhattacharya and Mittra, 2002). Domestic demand of medicinal plants in 1999 – 2000 was estimated at Rs. 1,099 crores, which is expected to rise to Rs. 2,000 crores by the year 2004-2005, as per CERPA estimates (Anon, 2001).

The tradition of collecting, processing and applying plants and plant-based medications, long and carefully maintained by individuals with a profound knowledge, has been handed down from generation to generation among the tribals. The value of this ethnomedicine and traditional pharmacology is now being increasingly recognised in modern human and veterinary medicine (Maydell, 1990; Hamann, 1991; Anon, 1992.). These medicinal plants mostly grow naturally and are collected by the local people and sold to traders who, in turn, sell them to the pharmaceutical and cosmetic industry and to exporters. These important resources have declined significantly in the recent past, which has led to near extermination of some of the most valuable plants. Recently, it has been noted that increasing biotic influences, including socio-economic development, and unrestrained commercial exploitation of forest wealth have threatened the survival of the rich genetic resources of this region, amounting to a great loss of national heritage. However, the most important cause of depletion and danger for medicinal plants comes from man–made pressures. Illegal and unscientific collections, over-exploitation and trade have resulted in the extinction of at least, some species.

At present, the medicinal plants sector is not well organised and needs special attention. Cultivation is clearly a sustainable alternative to the present collection of medicinal plants from the wild. This can be a potential provider of returns to the farmers. No serious efforts have been made in the State to promote the cultivation of medicinal plants of importance. The only efforts made are those of the Department of Environment and Forests and the Non-Government Organisations. Similarly, there is a serious lack of awareness among the people about the potential and scope of medicinal plants.

The present study is focussed on the dual objectives of identification of the concerned species and their analysis in terms of plant parts used and uses for different ailments, etc. with a view to prioritise and select medicinal plants for development in the state. It is not possible here to describe all the medicinal plants of the State. However, a list of very significant medicinal plants is presented here. The aim is to highlight the most frequently used medicinal plants in the tribal areas of Meghalaya, and stress the need for sustainable management of medicinal plants resources.

MATERIALS AND METHODS

Along with a survey of locally available information on the use of traditional herbal medicine collected through personal interview, records in the State's Departments, reports and literatry research, fieldwork was also considered necessary. Short field visits were made to specific areas. During the field visit, interaction with knowledgeable local people and other stakeholders were made to collect information and to solicit their views on medicinal plants and their conservation. The methodology followed during the fieldwork was mainly based on detailed questionnaires and a standard format. Carefully planned fieldwork, spread over four years from 1998 to 2002, was carried out in different tribal pockets of Meghalaya. The purpose of the study was not only the collection of first hand information about the relationship of medicinal plants with the community, but also to verify the already published data, wherever possible.

Studies were carried out among different tribes of the State. The plants were identified, using relevant flora and by matching the specimens in the herbaria of the Botanical Survey of India at Shillong and at North Eastern Hill University, Shillong. Where necessary, interpreters were employed in order to acquire details of uses and other information on plants and their environment. Tribal markets or weekly *haats* were also visited to study the plants and plant products sold there. As the rural folk of the region are largely dependent on wild plants and plant products for their existence, their local markets are full of wild vegetables, fruits and medicinal plants. These markets are both permanent as in Shillong, Sohra, Nongstoin, Jowai, Nongpoh, Tura, Williamnagar, Baghmara and other big towns, or are held on a fixed day each week in small villages. These weekly *haats* are tapped for their rich source of information. The vegetables and fruits collected from the wild and their products are the commonest commodity in these tribal markets, as well as the domesticated variety of crops and animals.

RESULTS AND DISCUSSION

The plants listed were widely used as medicines in Meghalaya, before the advent of chemical medicines. They have important components, which can cure an extensive range of diseases (Table 17.1 – 17.3). Two hundred and ninety eight medicinal plant species (124 species used by the Khasis, 89 species by the Jaintias and 85 species by the Garos) with their uses to treat different diseases were identified. Of the 298 medicinal plant species mentioned, 5 species were common to all the tribes, 9 species were common to the Khasi and Jaintia tribes, 13 species were common to the Khasi and Garo tribes and 5 species were common to the Jaintia and Garo tribes.

Table 17.I: Medicinal Plants used by the Khasi tribe

Sl. No.	*Botanical Name*	*Local Name*	*Family*	*Trend of Occurrence*	*Parts used*	*Used as Medicine*
1.	*Acorus calamus* (L.)	U-bet	Araceae	C	Lf	Influenza and headache
2.	*Adenostemma lavenia* (L.)	Soh-byrthit	Compositae	C	Lf	Cuts, wounds and antidote in insect bite
3.	*Adiantum phillipense* (Linn.)	Tyrkhang khyllai	Adiantaceae	R	Lf	Bone fracture
4.	*Ageratum conyzoides* (Linn.)	Ksang-dngiem	Asteraceae	A	Lf	Cuts and wound
5.	*Ajuga bracteosa* (Wall.ex. Benth)	Tiew khmut tuta	Lamiaceae	F	Lf	Astringent and respiratory congestion
6.	*Albizzia chinensis* (Osb)	Dieng phallut	Mimosaceae	C	Bk	Ringworm and antidote in insect bite
7.	*Allium hookeri* (Thw)	Ja uat	Liliaceae	C	Bu	Burns and boils
8.	*Allium tuberosum* (Roxb)	Jyllang	Liliaceae	A	Wp	Hypertension
9.	*Alnus nepalensis* (D. Don)		Betulaceae	C	Rt, Bk	Diarrhoea
10.	*Alysicarpus monilifer* (DC)		Fabaceae	F	Wp	Snake bite
11.	*Ambrosia artimisifolia* (L.)	'bat japan rit	Asteraceae	C	Lf	Cuts and wounds

Sl. No.	Botanical Name	Local Name	Family	Trend of Occurrence	Parts used	Used as Medicine
12.	*Amomum aromaticum* (Roxb)	Ilashi saw	Zingiberaceae	A	Rh	Nausea and vomiting
13.	*Anaphalis adnata* (Wall.ex.DC)	Skhor blang	Asteraceae	C	Lf	Haematoma
14.	*Arisaema jacquemontil* (Bl)	Hadembsein	Araceae	C	Rt	Ring worm and skin diseases
15.	*Antidesma bunius* (L.)	Soh-syllai	Euphorbiaceae	A	Lf	Rheumatism
16.	*Artemisia nilagirica* (Cl)		Asteraceae	A	Wp	Anthelmintic and bodyache
17.	*Astilbe rivularis* (Buch.-Ham)	Pdah	Saxifragaceae	F	Lf	Toothache, blood purification
18.	*Averrhoa carambola* (L.)	Sohpyrshog	Averrhoaceae	A	Fr	Jaundice and measles
19.	*Azadirachta indica* (A. Juss.)	Dieng nim	Meliaceaeae	F	Lf	Diarrhoea and dysentery
20.	*Bauhinia variegata* (L.)	Dieng tharlong	Caesalpiniaceae	C	Fl	Piles and dysentery
21.	*Begonia josephi* (D. Don)	Jajew	Begoniaceae	C	Wp	Stomach ache and indigestion
22.	*Begonia palmate* (D. Don)	Jajewmaw	Begoniaceae	C	Lf	Food poisoning and vomiting
23.	*Begonia rubrovenia* (Hk)	Jajew	Begoniaceae	C	Lf	Fever, giddiness and headache
24.	*Berberis wallichiana* (DC)	Dieng niangmat	Berberidaceae	F	St	Dysentery and conjunctivitis
25.	*Betula alnoides* (Buch Ham)	Dienglieng	Betulaceae	C	Rt	Indigestion and flatulence
26.	*Biophytum sensitivum* (L.)DC		Oxalidiaceae	F	Lf	Headache, giddiness and fever
27.	*Bonnaya reptans* (Spreng.)	Kra-thang syndat	Scrophulariaceae	F	Lf	Urinary ailments
28.	*Buddleja macrostachya* (Benth)	Jalong krem	Buddlejaceae	F	Lf	Venereal disease
29.	*Cannabis sativa* (L.)	Kynja	Cannabiaceae	A	Lf, Fr	Skin diseases and stomach ache
30.	*Capsicum annuum* (L.)	Soh-mynken syiar	Solanaceae	C	Fr	Antidote in snake bite
31.	*Casearia vareca* (Roxb)		Flacourtiaceae	F	Lf	Amoebic dysentery and intestinal worm
32.	*Cinnamomum glanduliferum* (Meissn)	Diengsing	Lauraceae	C	Lf	Fever, cold, cough and rheumatis
33.	*Cinnamomum pauciflorum* (Nees)	Dieng tarthia	Lauraceae	C	Bk, Sh	Antiseptic
34.	*Cinnamomum tamala* (Nees and Eberm.)	La tyrppad	Lauraceae	C	Lf	Toothache
35.	*Citrus latipes* (Swingle)	Sohkymphor	Rutaceae	C	Fr	Gout, rheumatism and ringworm

Sl. No.	Botanical Name	Local Name	Family	Trend of Occurrence	Parts used	Used as Medicine
36.	*Clerodendrum colebrookianum* (Walp.)	Sla jarem	Verbenaceae	C	Lf	High blood pressure
37.	*Clematis gouriana* (Roxb)	Sladienglum	Ranunculaceae	C	Rt	Cough and cold
38.	*Colocasia esculcenta* (Schott)	La wang	Araceae	F	Lf, Rh	Fever and rickets
39.	*Conyza bonariensis* (Linn.)	'bat symbai pum pum	Asteraceae	F	Lf	Astringent
40.	*Costus speciosus* (J. Konig ex Retz. Smith)	Sla pangmat	Zingiberaceae	F	Rh	Bronchitis, inflammation, anemia and rheumatism
41.	*Crassocephalum crepidioides* (Benth)	Jali	Asteraceae	F	Lf	Constipation and other stomach disorders
42.	*Curcuma angustifolia* (Roxb)	'niang-soh-pet	Zingiberaceaae	F	Lf	Gripe
43.	*Curcuma domestica* (Velaton)	Shynrai stem	Zingiberaceae	A	Rh	Skin diseases and bone fractures
44.	*Cuscuta reflexa* (Roxb)	Jawieh raid	Cuscutaceae	F	Wp	Gastric ulcer and rheumatic pain
45.	*Daphne bholua* (Ham ex.Don)		Thymelaeaceae	F	Rt	Intestinal troubles.
46.	*Datura stramonium* (L)	Tiew shulim	Solanaceae	F	Lf	Paralysis, stroke and rheumatism
47.	*Delphinium altissimum* (Wall)	Bat soh-plihrit	Ranunculaceae	C	Lf, Rt	Plaster to glandular swellings
48.	*Dendrobium moschatum*	Tiew dieng	Orchidaceae	F	Ju	Ear problem
49.	*Desmodium gangeticum*(L.)DC		Papilionaceae	F	Rt	Dysentery
50.	*Dicranopteris linearis* (Burm.f)	Tyrkhang	Gleicheniaceae	C	Lf	Epilepsy and fits
51.	*Dicrocephala bicolor* (Roth)	Liang poh tiew	Asteraceae	C	Lf	Cuts, wounds bleeding
52.	*Drymaria cordata* (L.) Willd.ex Roem and Schult)	Bat-nongrim	Caryophyllaceae	A	Ju	Burns, skin diseases and antidote in snake bites
53.	*Elephantopus scaber* (L.)	'Batshrut sriang	Asteraceae	C	Lf	Abortion, urinary disorders and as a contraceptive
54.	*Eleusine indica* (L.)	Lang krai	Poaceae	C	Ju	Jaundice
55.	*Elsholtzia blanda* (Benth)	Bat-skain	Lamiaceae	C	Ju	Mosquito bites and mosquito repellent
56.	*Engelhardia spicata* (Bl)	Dieng lyba	Juglandaceae	C	Inf	Scabies and skin diseases
57.	*Erigeron karvinskianus* (DC)	'Bat tiew star	Asteraceae	C	Lf	Cuts and wounds as an astringent

Sl. No.	*Botanical Name*	*Local Name*	*Family*	*Trend of Occurrence*	*Parts used*	*Used as Medicine*
58.	*Erythrina arborescens* (Roxb.)	Dieng-song	Papilionaceae	F	Lf	Skin diseases
59.	*Eupatorium adenophorum* (L.)	Bat iong	Asteraceae	A	Lf	Cuts and wounds
60.	*Ficus virens* (Ait)	Diengsoh phohkhlaw	Moraceae	C	Lf	Loss in appetite
61.	*Flemingia ontanana* (Willd)		Fabaveae	F	Rt	Body ache and muscular pains
62.	*Garcinia cowa* (DC)	Soh synrum	Clusiaceae	F	Fr	Headache, cold and dysentery
63.	*Garuga pinnata* (Roxb)	Dieng khiang	Burseraceae	F	Fr, Ju	Indigestion, Conjunctivitis and asthma.
64.	*Gaultheria fragrantissima* (Wall)	'La thynrait	Ericaceae	A	Lf	Bone fractures, rheumatism and sprain
65.	*Glochidion zeylanicum* (Juss.)	Jalwai	Euphorbiaceae	F	Lf	Dysentery and stomachache
66.	*Gmelina arborea* (Roxb)	Dieng laphiang	Verbenaceae	F	Be	Purgative and insect stings
67.	*Gomphostemma parviflorum* (Wall.ex Benth)		Labiatae	F	Lf	Headache, giddiness and fever
68.	*Gynocardia odorata* (R.Br)	Sohliang	Flacourtiaceae	F	Sd	Leprosy, nausea and rheumatism
69.	*Hedere nepalensis* (K.Koch)	Soh poramshre	Araliaceae	F	Be	Purgative and mumps
70.	*Hedyotis scandens* (Roxb)	Mo-shoh shu	Rubiaceae	C	Lf	Gastric troubles, cough and cold
71.	*Hedyotis verticillata* (L.) Lam	Jyrmi skei	Rubiaceae	F	Lf	Reduces body temperature
72.	*Hibiscus rosa sinensis* (L)	Tiew china	Malvaceae	C	Fl	Urinary disorders and skin diseases
73.	*Hodgsonia heteroclite* (Roxb.)	Soh risa	Curcubitaceae	F	Rt	Fever
74.	*Holmskioldia sanquinea* (Retz.)		Verbenaceae	F	Rt	Fever
75.	*Houttuynia cordata* (Thunb)	Jamyrdoh	Saururacae	A	Lf	Blood purification, sores and boils
76.	*Hedychium coronarium* (Koen)	Kymbat khawiang	Zingiberaceae	A	Rh	Respiratory ailment and bodyache
77.	*Hedychium gracile* (Roxb)	Shynrai khlaw	Zingiberaceae	A	Rh	Chest pain
78.	*Hedychium spicatum* (Buch-Ham ex Sm)	Sying khlaw	Zingiberaceae	A	Rh	Sore throat and stomachache
79.	*Hypocharis radicata* (L.)	Bat jhur kthang	Asteraceae	C	Lf	Stomachache
80.	*Indigofera tinctoria* (L.)		Papilionaceae	C	Rt	Wounds
81.	*Ipomea uniflora* (Roem and Schult)	Tiew turoi	Convolvulaceae	C	Lf	Cholera and vomiting

Sl. No.	Botanical Name	Local Name	Family	Trend of Occurrence	Parts used	Used as Medicine
82.	*Itea chinensis* (HK and Arn)	Dieng myllong	Saxifragaceae	C	Lf	Skin diseases
83.	*Kaempferia rotunda* (L.)	Ingsmoh	Zingiberaceae	C	Rh	Stomach trouble and as general tonic
84.	*Lindera pulcherrima* (Benth)	Sia-sia	Lauraceae	F	Bk	Rheumatic pains
85.	*Litsea khasiana* (Meissn)	Dieng mosu	Lauraceae	C	Rt	Chronic bronchitis.
86.	*Lycopodium clavatum* (L)	Tmain khla	Lycopodiaceae	C	Wp	Skin diseases and rheumatism
87.	*Lygodium scandens* (L)	Tyrkhang kiow dieng	Schizaeaceae	C	Wp	Swelling, skin diseases and Bone fractures
88.	*Maesa indica* (Roxb) Wall	Dieng soh jala	Myrsinaceae	C	Be	Vermifuge
89.	*Mahonia nepalensis* (D.Don)	Dieng-tiang-mat	Berberidaceae	F	Ju	Eye diseases
90.	*Mahonia pycnophylla* (Fedde)	Ningmat	Berberidaceae	F	Ju	Eye diseases
91.	*Mallotus philippensis* (Lamk) Muell	Dieng chandan	Euphorbiaceae	F	Fr	Tapeworm
92.	*Melia azedarach* (L.)	Dieng ja rasang	Meliaceae	F	Bk	Anthelmintic and febrifuge
93.	*Mimosa pudica* (L)	Bat iambait	Momosaceae	C	Lf	Boils, sores and diabetes
94.	*Nepenthes khasiana* (Hk.f.)	Ksetphare	Nepenthaceae	R	Ju	Stomachache, eye sores or urinary troubles
95.	*Osbeckia stellata* (Buch.Ham ex D. Don)	Soh-lyngkthut	Melastomaceae	A	Lf	Wounds of various types and snake bites
96.	*Oxalis corniculata* (L.)	Jabuit	Oxalidaceae	A	Wp	Diarrhoea
97	*Pandanus fasciculus* (Lamk.)	Sla shylliah	Pandanaceae	F	Ju	Skin diseases and leprosy
98.	*Parochetus communis* (Buch. Ham ex D. Don)	Khia-knoi	Papilionaceae	F	Ju	Stomachache
99.	*Phenera khasiana* (Baker)	Jarmi bin khlaw	Caesalpiniaceae	F	Sd	Demulcent in dried and cracked skin
100.	*Piper griffthii* (DC)	Mrit khlaw	Piperaceae	F	Sd	Cough and cold
101.	*Pithecellobium bigeminum* (I) Mart	Dieng yap yar	Mimosaceae	C	Sd	Blood purification
102.	*Plantago erosa* (Wall.)	Shkor blang	Plantaginaceae	C	Lf	Bandaging of wounds, boils and removal of pus
103.	*Pouzolzia hirta* (Bl.)	Memsleh	Urticaceae	C	Rt	Hair tonic and hair growth
104.	*Potentilla fulgens* (Hk.)	Lynniang	Rosaceae	C	Rt	Diarrhoea and diabeties.

Sl. No.	Botanical Name	Local Name	Family	Trend of Occurrence	Parts used	Used as Medicine
105.	*Pseudognaphalium luteoalbum* (L.) Hall et Burtt	Tiew kubi	Asteraceae	C	Lf	Gout and rheumatism
106.	*Psidium guajava* (L.)	Soh priam	Myrtaceae	C	Lf	Chronic dysentery
107.	*Rhus semialata* (Murr)	Sohma	Anacardiaceae	C	Fr	Stomachache and diarrhoea
108.	*Ricinus communis* (L.)	Ryndia khlaw	Euphorbiaceae	F	Lf	Rashes and skin diseases
109.	*Rorippa indica* (L.)	Tyrso khlaw	Brassicaceae	F	Sd	Bleeding gums in scurvy
110.	*Rorippa nasturtium-aguaticum* (L.) Hayak	Tyrso-um	Brassicaceae	C	Wp	Pneumonia or other Pulmonary ailments
111.	*Rubia cordifolia* (L.)	Rhoi	Rubiaceae	C	Lf	Ulcers and insect stings
112.	*Rubus ellipticus* (Sm)	Soh-shiah	Rosaceae	F	Fr	Dysentery
113.	*Schefflera hypoleuca* (Kurz) Ham	Sla tymphu	Araliaceae	F	Rt	As tonic after child birth
114.	*Schima wallichii* (Choisy)	Diengngan	Theaceae	C	Lf	Flatulence
115.	*Schisandra neglecta* (Sm.)		Schisandraceae	F	Lf	High fever
116.	*Smilax glabra* (Roxb.)	Khong	Smilacaceae	C	Ju	Skin diseases
117.	*Taxus baccata* (L.)	Dieng she Blei	Taxaceae	R	Lf	Tumours and ulcer
118.	*Terminalia chebula* (Retz.)	Soh salukah	Combretaceae	C	Fr	Diuretic and conjunctivitis.
119.	*Thallictrum foliosum* (DC)	Jatira khlaw	Ranunculaceae	F	Rt	High blood pressure and malarial fever
120.	*Toddalia asiatica* (Retz)	Soh sat khlaw	Rutaceae	C	Rt	Malaria and other periodic fevers
121.	*Vitex negundo* (L.)	Tyllai skip	Verbenaceae	F	Lf	Muscular pain and swelling
122.	*Zanthoxylum acanthopodium* (DC)	Ja-iur	Rutaceae	F	Sd	Stomach disorders, fish poison and vermicide
123.	*Zingiber zerumbet* (Sm.)	Ing-Blei	Zingiberaceae	A	Rh	Cough, cold and relieves stress
124.	*Zizyphus mauritiana* (Lam.)	Sohbroi	Rhamnaceae	A	Fr, bk	Fever and diarrhea

Abbreviations Used: A- Abundant; C- Common; F- Frequent; R – Rare; Lf-leaf; Bu-bulb; Bk-bark; Wp-whole plant; Rt-root; Rh- rhizome; Fr-fruit; Fl- flowerl; St-stem; Sh- shoot; Ju-juice; Inf-infloresecence; Be – berries; Sd- Seed; La-latex; Tu-tuber; Pit-pitcher; Pu-pulp.

Table 17.2: Medicinal Plants used by the Jaintia tribe

Sl. No.	*Botanical Name*	*Local Name*	*Family*	*Trend of Occurrence*	*Parts used*	*Used as Medicine*
1.	*Acasia pennata* (L.) Willd	Chiah poswa	Mimosaceae	F	Lf, La	Skin diseases
2.	*Acorus calamus* (L.)	Bat-ksuid	Araceae	C	Rt	Chest congestion, Malaria and headache
3.	*Adiantum phillippense* (L.)	Krah keriah	Adiantaceae	C	Wp	Bone fracture and sprain
4.	*Aegle marmelos* (L.)	Sohbel	Rutaceae	F	Fr	Diarrhoea and dysentery
5.	*Ageratum conyzoides* (L.)	Sla iewtung	Alliaceae	A	Lf	Cuts and wounds
6.	*Alchornea tillaeafolia* (Muell.-Arg.)	Ksiang lenti	Euphorbiaceae	F	St	Sprains
7.	*Allium tuberosum* (Roxb.)	Jyllang	Asteraceae	R	Wp	Cough and cold
8.	*Alpinia bracteata* (Rosc.)	Latara	Zingiberaceae	F	Rh	Toothache
9.	*Alpinia galangal* (L.)	Phlang sow	Zingiberaceae	F	Rh	Skin diseases
10.	*Amorphophallus bulbifer* (Roxb.)	Thynrewpsen	Araceae	F	Rt, St	Skin diseases
11.	*Aristolochia catcartii* (Hk.f and T)	Patiksang	Aristolochiaceae	R	Rt	Food poisoning
12.	*Aristolochia saccata (Wall.)*	Krahlahit	Aristolochiaceae	R	Tu	Haemorrhage and stomachache
13.	*Aristolochia tagala* (Cham.)	Khurthlong	Aristolochiaceae	F	Rt, Fr	Rheumatism and bone fracture
14.	*Asparagus racemosus* (Willd.)	Phlang chokriawsea	Aspagaraceae	A	Rh	Urinary trouble
15.	*Borreria articularis* (L.) Willd.	Phlang bhoi	Rubiaceae	C	Lf	Stomach disorder
16.	*Botrychium ternatum* (Thund.)	Tyrkhang	Botrychiaceae	C	Rt	Dysentery
17.	*Buddleja macrostachya* (Benth.)	Jalong krem	Buddlejaceae	F	Lf	Skin diseases
18.	*Camellia kissi* (wall.)	Sli sohlwit	Theaceae	F	Rt	Skin diseases
19.	*Careya urens* (L.)	Styngkrain	Barringtoniaceae	F	Bk	Dysentery
20.	*Cassia tora* (L.)	Tawblei	Caesalpiniaceae	F	Lf	Ringworm and skin diseases
21.	*Centella asiatica* (L.) Urban	Khlein syiar	Apiaceae	A	Wp	Boil, tumours and dysentery
22.	*Centranthera grandiflora* (Benth.)	Phlang stem	Scrophulariaceae	F	Rt	Insect stings
23.	*Cinnamomum camphora* (L.) Nees and Eberm	Dieng-pingwait	Lauraceae	F	Lf	Cough, cold and fever
24.	*Cissus repens* (Lam.)	Jiaherew	Vitaceae	F	Rt	Boils and muscular pain
25.	*Citrus medica* (L.)	Soh-kwit	Rutaceae	C	Fr	Fever, headache and body ache

Sl. No.	Botanical Name	Local Name	Family	Trend of Occurrence	Parts used	Used as Medicine
26.	*Clerodendrum viscosum* (Vent.)	Slijerem	Verbenaceae	C	Lf	Fever, diarrhoea and dysentery
27.	*Clerodendrum wallichii* (Merr.)	Lekynrum chia blang	Verbenaceae	C	Lf	Skin disease and germicide
28.	*Curcuma domestica* (Velaton)	Shynrai	Zingiberaceae	A	Rh	Cough, cold, allergy and boils
29.	*Curcuma ontana* (Rosc.)	Chyrmit khlow	Zingiberaceae	F	Rh	High temperature and headache
30.	*Curcuma zeodaria* (Rosc.)	Chyrmit loom	Zingiberaceae	C	Rh	Bone fractures
31.	*Cuscuta reflexa* (Roxb.)	Jawieh raid	Cuscutaceae	F	Wp	Balm on body and rheumatic pain
32.	*Daphne cannabina* (Hook.f)	Murit	Thymeliaceae	C	Bk	Toothache
33.	*Davallia trichomanoides* (Bl)	Krah beuh	Davalliaceae	F	Lf	Food poisoning
34.	*Drymaria cordata* (L.) Willd. Ex Roem and Schult.	Slia-slia	Caryophyllaceae	C	Wp	Dysentery, burn and skin diseases
35.	*Emblica officinalis* (Gaertn.)	Sohmylleng	Euphorbiaceae	A	Fr	Conjunctivitis
36.	*Eryngium foetidum* (L.)	Dhonia khlow	Apiaceae	F	Lf	Headache
37.	*Eupatorium adenophorum* (L.)	Phlang Burma	Asteraceae	C	Lf	Cuts and wounds
38.	*Eupatorium odoratum* (L.)	Phlang Dkhae	Asteraceae	C	Lf	Cuts and wounds
39.	*Fagopyrum cymosum* (Trev.)	Jarain	Fagaceae	A	Lf, Rt	Intestinal worms and dandruff
40.	*Flemingia vestita* (Backer.)	Sohphlang	Leguminasae	F	Tu	Anthelminthic
41.	*Garcinia lancifolia* (Roxb.)	Soh-suit	Clusiaceae	C	Fr	Stomach trouble
42.	*Gaultheria fragrantissima* (Wall.)	Lathynrait	Ericaceae	F	Lf	Arthritis, rheumatism and other ailments of the joints
43.	*Gongronema nepalense (Wall.)*	Krah skun	Asclepiadaceae	C	Lf	Boils and ringworms.
44.	*Habenaria crniifera* (Lindl.)	Phlang stem	Orchidaceae	C	Rh	Stomach disorders
45.	*Hedyotis uncinella* (Hk.and Arn.)	La jam	Rubiaceae	C	Lf	Insect stings
46.	*Houttuynia cordata* (Thunb.)	Jamyrdoh	Saururaceae	A	Wp	Lowering blood sugar
47.	*Hydrocotyle javanica* (Thunb.)	Kynbat-syiar	Apiaceae	C	Wp	Fever
48.	*Ipomea pileata* (Roxb.)	Chermew skun	Convolvulaceae	C	Lf	Skin disease
49.	*Ixora acuminata* (Roxb.)	Krah kshiang	Rubiaceae	F	Lf	Diarrhoea, dysentery and influenza

Sl. No.	Botanical Name	Local Name	Family	Trend of Occurrence	Parts used	Used as Medicine
50.	*Kaempferia rotunda* (L.)	Phlang san	Zingiberacae	C	Lf, Rh	Stomach disorders, swellings and boils
51.	*Lindernia ruelloides* (Colsm.)	Krah thang shyndat	Scrophulariaceae	F	Lf	Urinary complaints
52.	*Macropanax dispermus* (Bl.)	Sli soh resa	Araliaceae	F	Lf	Insect sting
53.	*Mahonia acanthifolia* (G. Don)	Krah niang mat	Berberidaceae	C	St, Bk	Eye diseases
54.	*Mimosa pudica* (L.)	Krah meriu	Mimosaceae	C	Rt	Snake bite, boils and sores
55.	*Mussaenda frondosa* (L.)	Chermew kynthah blang	Rubiaceae	F	Lf	Body swelling and jaundice
56.	*Nepenthes khasiana* (Hk.f.)	Tiew rakot	Nepenthaceae	R	Pit	Indigestion and kidney trouble
57.	*Osbeckia nepalensis* (Hk.f.)	Sli lekum	Melastomataceae	C	Rt	Bodyache and swelling
58.	*Paedaria foetida* (L.)	Nangra puhung, Batiewtung	Rubiaceae	F	Lf, Rt	Skin diseases and stomach disorders
59.	*Panax pseudoginseng* (Wall.)	Jynseng	Araliaceae	R	Rh	General tonic to relieve stress
60.	*Passiflora edulis* (Sims.)	Soh brab	Passifloraceae	C	Lf	Liver tonic, heart problems and blood pressure
61.	*Phlogacanthus thyrsiformis* (Hardw.)	Jia merembut	Acanthaceae	F	Lf	Indigestion and stomach trouble
62.	*Polygonum capitatum* Buch.-Ham.ex D. Don.)	Samtympei	Polygonaceae	C	Lf	Stomachache
63.	*Polygonum nepalensis* (Meissn.)	Ja-ut	Polygonaceae	A	Lf	Blood pressure
64	*Polygonum perfoliatum* (L.)	Shrat	Polygonaceae	C	Lf	Dysentery and diarrhoea
65.	*Portulaca oleracea* (L.)	Jehu-sia	Portulacaceae	F	Wp	Gonorrhoea and blood purification
66.	*Pouzolzia hirta* (Bl.)	Taknor	Urticaceae	C	Lf	Hair tonic
67.	*Pueraria phaseoloides* (Benth.)	Sliniang terhu	Fabaceae	C	Lf	Toothache
68.	*Rhododendron arboreum* (Sm.)	Tiewsaw	Ericaceae	R	Lf	High blood pressure and diarrhoea
69.	*Rhus semialata* (Murr.)	Sohma	Anacardiaceae	R	Fr	Diarrhoea and dysentery
70.	*Sarcandra glabra* (Thunb.)	Soh-krismas	Chloranthaceae	F	Lf	Fever and blood deficiency
71.	*Scoparia dulcis* (L.)	Krah lebekor	Scrophulariaceae	F	Lf, Inf	Kidney and urinary trouble
72.	*Smilax perfoliata* (Lour.)	Lekynrum doh	Smilacaceae	F	Fl	Tonic and blood purification
73.	*Solanum barbisetum* (Nees.Ed.)	Sohshiah	Solanaceae	F	Lf	Low blood pressure and blood sugar

Sl. No.	Botanical Name	Local Name	Family	Trend of Occurrence	Parts used	Used as Medicine
74.	*Solanum khasianum* (Clarke.)	Sohpdok	Solanaceae	F	Fr	Toothache
75.	*Solanum xanthocarpus* (Schrad and Wendl.)	Sohngang	Solanaceae	F	Be	Low blood pressure
76.	*Sonchus asper* (L.)	Jalynniar	Asteraceae	A	Lf	Anti-diabetes
77.	*Spilanthes acmella* (L.)	Hooin	Asteraceae	C	Fl	Toothache
78.	*Swertia chirata* (Roxb.)	Batwieh	Gentianaceae	A	Lf	Vermicide
79.	*Taxus baccata* (L.)	'She Blai	Taxaceae	R	Bk	Anti-carcinogenic properties and tonic
80.	*Thunbergia grandifolia* (Roxb.)	Chermew lemah	Acanthaceae	F	St	Eye diseases
81.	*Thysanolaena maxima* (Roxb.)	Synsar	Poaceae	C	St	Eye diseases
82.	*Tinospora cordifolia* (Wall.) Miers	Jyrmibteng	Menispermaceae	F	Oil	Fractures and dislocated bones
83.	*Triumfetta rhomboidea* (Jacq.)	Sli sko	Tiliaceae	F	St	Stomachache
84.	*Urena lobata* (L.)	Sli skha	Malvaceae	F	Lf	Diarrhoea
85.	*Viburnum foetidum* (Wall.)	Sohlang	Caprifoliaceae	F	Fr	Skin diseases
86.	*Zanthoxylum acanthopodium* (DC.)	Jaiur	Rutaceae	F	Fr	Fever, cough and cold
87.	*Zingiber officinale* (Roscoe)	Ingbah	Zingiberaceae	A	Rh	Fever, cough and cold
88.	*Zingiber rubens* (Roxb.)	Ingmakhir	Zingiberaceae	A	Rhi	Fever, cough and cold
89.	*Zingiber zerumbet* (L.) Sm	Ing-Blei	Zingiberaceae	F	Rh	Fever, cough and cold

Abbreviations Used: A- Abundant; C- Common; F- Frequent; R – Rare; Lf-leaf; Bu-bulb; Bk-bark; Wp-whole plant; Rt-root; Rh- rhizome; Fr-fruit; Fl- flowerl; St-stem; Sh- shoot; Ju-juice; Inf-infloresecence; Be – berries; Sd- Seed; La-latex; Tu-tuber; Pit-pitcher; Pu-pulp.

Table 17.3: Medicinal Plants used by the Garo tribe

Sl. No.	Botanical Name	Local Name	Family	Trend of Occurrence	Parts used	Used as Medicine
1.	*Abroma angusta* (L.)	Bon khopai	Sterculiaceae	F	Oil	Fever, ring worm and scabies
2.	*Acanthus leucostachysis* (Wall.)	Sam sikal	Acanthaceae	F	Lf	Swelling, fever and toothache
3.	*Achyranthus aspera* (L.)	Minamkachi	Amaranthaceae	F	Rt	Leprosy
4.	*Acorus calamus* (L.)	Betse	Araceae	C	Rt	Cough and cold

Sl. No.	Botanical Name	Local Name	Family	Trend of Occurrence	Parts used	Used as Medicine
5.	*Aegle marmelos* (L.) Corr.	Belethi	Rutaceae	C	Lf, Rt, Bk	Melancholia, fevers, palpitation of the heart, ophthalmic and ulcers
6.	*Anthocephalus chinensis* (Lam.)	Mi-bol	Rubiaceae	F	Lf, Bk	Astringent, febrifugal and anti-diuretic properties
7.	*Antidesma burnius* (L.)	Bol-aborak	Euphorbiaceae	F	Lf	Syphilitic ulcers
8.	*Arisaema jacquemontii* (Bl.)	Jinjok	Araceae	C	Tu	Ringworm and skin diseases
9.	*Aristolochia cathcartii* (Hk.f and T)	Baro-warkhut	Aristolochiaceae	R	Rt	Stomach ailments
10.	*Artemisia vulgaris* (L.)	Sak-sak	Asteraceae	A	Lf	Headache and severe bleeding
11.	*Asparagus filicinus* (L)	Riching	Asparagaceae	A	Tu	Gripe in infants
12.	*Asparagus racemosus* (Willd.)	Som riching	Asparagaceae	A	Rt	Fever
13.	*Boerhavia diffusa* (L.)	Samdelma	Nyctaginaceae	C	Lf	Rheumatic
14.	*Bombax ceiba* (L.) SW.	Bolchu	Bombaceae	C	Fl, Bk	Astringent and coetaneous troubles Demulcent, emetic, tonic and blood dysentery
15.	*Bonnaya reptans* (Spreng.)	Sam-reng chick	Scrophulariaceae	F	Lf	Snakebite
16.	*Butea monosperma* (Lam.)		Papilionaceae	F	Sd	Delirium
17.	*Calotropis ontanan* (Br.)	Akom-aring	Asclepiadaceae	F	Lf	Malaria
18.	*Canscora andrographioides* (Griff.)	Sak sre	Gentianaceae	F	Rt	Cuts, wounds and skin diseases
19.	*Capparis assamica* (Hk.f and T.)	Mantori	Capparaceae	F	Lf	Headache and bodyache
20.	*Cassia fistula* (L.)	Soneru	Caesalpiniaceae	A	Rt, Bk, Pu	Purgative, tonic and febrifuge
21.	*Centella asiatica* (L.) Urb.		Apiaceae	A	Wp	Dysentery and diarrhoea.
22.	*Chonemorpha fragrans* (Moon.)	Kotchibeta	Apocynaceae	F	Rt, St	Stomach disorders, chest pain, and rheumatism
23.	*Citrus latipes* (Swingle.)	Tanaka	Rutaceae	C	Fr	Gouty and rheumatic joints
24.	*Clematis ontana* (Buch Ham.ex DC)		Ranunculaceae	F	Rt	Cough and cold

Sl. No.	Botanical Name	Local Name	Family	Trend of Occurrence	Parts used	Used as Medicine
25.	*Costus speciosus* (J.Konig ex Retz) Smith	Karami	Zingiberaceae	F	Rt	Bronchitis, inflammation, anemia and rheumatism
26.	*Crepis fuscipappa* (Benth)	Pon bihar	Asteraceae	C	Lf	Earache
27.	*Curcuma aromatica* (Salisb)	Tikegopl	Zingiberaceae	C	Lf	Gastric troubles, asthma, tuberculosis and blood impurity
28.	*Deeringia amaranthoides* (Lamk)	Sanum	Amaranthaceae	F	Lf	Fever, headache, nose bleeding and dysentery
29.	*Desmodium laxiflorum* (DC)	Bhutu hom	Papilionaceae	C	Rt, Lf	Stomach disorder
30.	*Disporum calcaratum* (D.Don)	Tike jakriting	Liliaceae	F	Tu	Eye diseases
31.	*Dracaena ensifolia* (Wall.)	Milam	Liliaceae	F	Lf	Cold, malaria and rheumatism
32.	*Drymaria cordata* (L.) willd.ex Roem and Schult.)		Caryophyllaceae	A	Lf	Snakebites
33.	*Elephantopus scaber* (L.)	Achaksn	Asteraceae	F	Rt	Heart and liver problems
34.	*Emblica officinalis* (Gaertn.)	Amalaki	Euphorbiaceae	A	Fr	Skin diseases and blood pressure
35.	*Fagopyrum cymosum* (Trev.)		Polygonaceae	A	Lf	Tonic
36.	*Garcinia cowa* (Roxb.ex DC)	Tekra rengron	Clusiaceae	C	Bk	Mosquito repellent
37.	*Geodorum purpureum* (Br.)	Matea bas	Zingiberaceae	F	Lf, Tu	Malaria and whooping cough
38.	*Globba clarkei* (Baker.)	Dike holdiram	Zingiberaceae	F	Lf, Rt	Dysentery
39.	*Hedyotis scandens* (Roxb.)	Sam rating	Rubiaceae	F	Lf	Cough and cold.
40.	*Holarrhena antidysenterica* (L.) Wall	Bol-matra	Apocynaceae	A	Bk	Amoebic dysentery
41.	*Homalomena aromatica* (Schott.)	Roathi	Araceae	A	Rh	Swelling, pimples and skin sores
42.	*Houttuynia cordata* (Thunb.)		Saururaceae	A	Wp	Anti-diabetes
43.	*Hydrocotyle javanica* (Thunb.)	Mana-muni	Apiaceae	C	Lf	Cough, cold and fever
44.	*Itea chinensis* (Hk and Arn.)	Myllone	Saxifragaceae	F	Lf	Skin diseases
45.	*Ixora acuminate* (Roxb.)	Saoltua	Rubiaceae	F	Lf, Fl	Blood purification
46.	*Jasminum lanceolaria* (Roxb)	Pipli	Oleaceae	F	Lf, Rt	Ringworm
47.	*Justicia gendarussa* (L.f.)	Dochenpok	Acanthaceae	C	Lf	Bone fractures and dislocation of bones

Sl. No.	*Botanical Name*	*Local Name*	*Family*	*Trend of Occurrence*	*Parts used*	*Used as Medicine*
48.	*Lasia spinosa* (Thw.)	Timulona	Araceae	F	Rh	Intestinal worm
49.	*Litsea khasiana* (Meissn)		Lauraceae	A	Oil	Deodorants, blisters and boils.
50.	*Melia composite* (Willd.)	Sural	Meliaceae	C	Lf	Gastric ulcers.
51.	*Milletia pachycarpa* (Benth)	Khariu	Papilionaceae	C	Lf	Vermifuge
52.	*Molineria crassifolia* (Baker.)	Rekosi	Poaceae	F	Lf	Diarrhoea
53.	*Mycetia longifolia* (Wall) O. Ktz	Janthro	Rubiaceae	F	Lf	High fever or blood pressure
54.	*Natsiatum herpeticum* (Buch Ham ex Arm)		Icacinaceae	F	Wp	Influenza
55.	*Nepenthes khasiana* (Hk.f.)	Mimankuchi	Nepenthaceae	R	Lf	Indigestion and kidney problem
56.	*Oenanthe stolonifera* (Wall.)	Bopo goli ting	Apiaceae	F	Lf	Stomachache and constipation
57.	*Oldenlendia diffusa* (Roxb.)	Chenong	Rubiaceae	F	Lf	Sore eyes and other eye diseases
58.	*Oldenlendia nudicaulis* (Roxb.)	Chenogn-ri	Rubiaceae	F	Lf	General debility
59.	*Ophiopogon intermedius* (D. Don)	Ticea ohik	Liliaceae	A	Lf	Cuts and wounds
60.	*Ophiorrhiza subcapitata* (Wall.)	Samachik	Rubiaceae	A	Rt, Lf	Fever, sore throat, tonsils and facial blemishes
61.	*Paedaria foetida* (L.)	Gandharadal	Rubiaceae	C	Lf	Dysentery or indigestion
62.	*Parabaena sagittaria* (Miers)	Chiongbombuelu	Menispermaceae	F	Rt	Skin diseases
63.	*Phlogacanthus tubiflorus* (Nees)	Som rongtek	Acanthaceae	F	Lf	High fever
64.	*Pilea lancifolia* (HK. F)	Sam-rongtek	Urticaceae	F	Lf	Fever and antidandruff
65.	*Plumbago zeylanica* (L.)	Agea	Plumbaginaceae	C	Rt	Piles
66.	*Pogostemon parviflorus* (Benth.)	Sam-sanum	Lamiaceae	C	Lf	Headache
67.	*Polygonum chinensis* (L.)	Samichang	Polygonaceae	C	Rt, Lf	Urinary disorders
68.	*Polygonum nepalensis* (Meissn.)	Samichang	Polygonaceae	A	Wp	Blood pressure
69.	*Pothos kunstleri* (Hk.f)	Garore	Araceae	F	Lf, St	Toxicity
70.	*Pouzolzia indica* (Gaud.)	Fakruom	Urticaceae	C	Lf, Rt	Urinary and spleen disorders
71.	*Rhaphidophora hookeri* (DC.)	Dhukentri	Araceae	C	Lf	Snake and dog bite
72.	*Rhus semialata* (Murr.)	Khitma	Anacardiaceae	F	Fr	Diarrhoea and dysentery

Sl. No.	*Botanical Name*	*Local Name*	*Family*	*Trend of Occurrence*	*Parts used*	*Used as Medicine*
73.	*Rubus moluccanus* (L.)	Thekhi-sembok	Rosaceae	F	Rt	Blood clotting and to prevent swelling.
74.	*Smilax prolifera* (Roxb.)	Marangwa	Smilacaceae	F	Lf, Rt	Ulcer and skin diseases
75.	*Spatholobus roxburghii* (Benth.)	Maribata	Papilionaceae	F	Bk	Toothache and gum troubles
76.	*Spilanthus acmella* (L.)	Sam atching	Asteraceae	F	Lf	Fever
77.	*Strobilanthes scaber* (Nees.)	Sam siphra, bimchat	Acanthaceae	C	Lf	Itching and skin diseases
78.	*Swertia chirata* (Roxb.) Karst	Chirata	Gentianaceae	A	Lf, St	Anthelminthes and low blood pressure
79.	*Symplocos racemosa* (Roxb.)	Boligpok	Symplocaceae	C	Bk	Indigestion and impaired blood circulation
80.	*Tacca laevis* (Roxb.)	Colbere	Taccaceae	F	Tu	Diarrhoea and blood dysentery
80.	*Terminalia chebula* (Retz.)	Artak, salukal	Combretaceae	C	Fr	Diarrhoea, stomach pain and spleen disorders
81.	*Thunbergia coccinea* (Wall.)	Kakku budu	Acanthaceae	F	Lf, Rt	Bone fracture
82.	*Valeriana hardwickii* (Wall.)		Valerianaceae	F	Wp	Insect stings
83.	*Zanthoxylum rhetsa* (Roxb.) DC	Sumitchory	Rutaceae	A	Lf, Sd	Fever, cough and cold
84.	*Zingiber officinale* (Rosc.)	Ada	Zingiberaceae	A	Rh	Fever, cough and cold
85.	*Zingiber rubens* (Roxb.)		Zingiberaceae	A	Rh	Fever, cough and cold

Abbreviations Used: A- Abundant; C- Common; F- Frequent; R – Rare; Lf-leaf; Bu-bulb; Bk-bark; Wp-whole plant; Rt-root; Rh- rhizome; Fr-fruit; Fl- flowerl; St-stem; Sh- shoot; Ju-juice; Inf- infloresecence; Be – berries; Sd- Seed; La-latex; Tu-tuber; Pit-pitcher; Pu-pulp.

It has been shown through numerous examples that the tribals widely use a large variety of medicinal plants available to them. On one hand such activities have contributed to our knowledge of various uses of biodiversity and on the other hand, they have resulted in the rapid depletion of natural resources. Their demand in the local market has increased, causing a threat to these wild species. Such plants too, may become the medicines of the future. Although these medicinal plants are presently under-utilised, however, to calci to future needs, this invaluable treasure of native diversity needs care, more research and focus on its collection, conservation and usage.

A large number of these species can be cultivated in different districts of the state, as the agroclimatic conditions are suitable for their growth. According to the Planning Commission and National Medicinal Plant Board, among the prioritised species of medicinal plants, more than 10 species are found either wild or in a cultivated condition in Meghalaya (Table 17.4). A few more species, which could be considered for cultivation and development in the states on a commercial basis through Community Forest Management, Self Help Groups and other suitable modes where benefits directly go to the cultivators, have been identified as priority species for conservation and have a high demand in the local and regional market and also for export outside the state. Examples of these plants are *Acorus calamus, Aristolochia* spp., *Centella asiatica, Cinnamomum* spp., *Clerodendrum colebrookianum, Costus speciosus, Curcuma* spp., *Fagopyrum* spp., *Flemingia vestita, Gaultheria fragrantissima, Gynocardia odorata, Houttuynia cordata, Kaempferia* spp. *Litsea* spp., *Lycopodium* spp., *Melia composita, Nepenthes khasiana, Piper longum, Potentilla fulgens, Sarcandra glabra, Smilax glabra, Solanum* spp., *Spilanthes* spp., *Taxus bacata, Terminalia* spp., *Zanthoxylum* spp., *Zingiber* spp. These prioritised species can be grown in a variety of land conditions available, like field margins, existing tree plantations, community forests, protected areas, sacred groves, wastelands, jhum fallows, etc. The productivity of Jhum fallows can thus be increased and can aid in jhum rehabilitation. Further, the rural folk can generate subsistence income through the cultivation of medicinal plants and conserve the genetic resources, while improving the local health status and economy.

Table 17.4: List of medicinal plants prioritised by the Medicinal Plant Board and Planning Commission, Government of India

Aconitum heterophylum Wall.	*Glycirrhiza glabra* L.
Aconitum ferox Wall.	*Gymnema sylvestre* R. Br.
**Aegle marmelos* L. Corr.	*Nardostachys jatamansi* DC.
Andrographis paniculata L.	**Ocimum sanctum* L.
**Asparagus racemosus* Willd.	*Picrorrhiza kurroa* Benth.
Bacopa monnieri (L.), Pennell.	**Piper longum* L.
Berberis aristata DC.	*Phyllanthus amarus* Schum and Thonn.
Cassia angustifolia Vahl.	*Plantago ovata* Forsk.
Chlorophytum arundinaceum Baker	**Rauwolfia serpentina* (L.) Benth.Ex Kurz.
Commiphora wightii (Arn.) Bhandari.	*Santalum album* L.
Coleus barbatus Benth.	*Saraca asoca* (Roxb.) de Wilde.
Convolvulus pluricaulis Choisy.	*Saussurea costus* C. B. Clark.
**Emblica officinalis* Gaertn.	**Solanum nigrum L.*
Embelia ribes Burm. F.	**Swertia chirata* Buch Ham.
Garcinia indica L.	**Tinospora cordifolia* Miers.
**Gloriosa* superba L.	*Withania somnifera* (L.) Dunal.

* The species of Medicinal Plants found in Meghalaya.

CONSERVATION MEASURES

In order to protect the diversity of medicinal plants of the state, the following have been suggested:

The data on medicinal plants will serve as a useful tool to prepare development and action plans for the herbal drug industry, so as to improve and uplift the life and economy of the state. While planning for the development of the local community, along with conservation and protection of all kinds of plants traditionally used by these tribals, what is urgently needed is the integration of all aspects of the diversity of medicinal plants.

There is need to develop a strategy to explore the vast wealth of there medicinal plants, find effective methods of propagation, encourage sustainable harvesting of plants from forests, involve small farmers and communities in their cultivation and conduct studies on the commercially exploitable species. An effective way of giving protection to ethnomedicinal plants is to provide a legal cover to the habitat or the species, so as to enable enforcement agencies to have the authority to control/regulate their enactment.

The demands for medicinal plants are increasing day by day, within and outside the country, and while the bulk of the trade is still from wild harvested sources on forest land, only a very small number of species are cultivated. The expanding trade in medicinal plants has serious implications on the survival of several plant species, with many under serious threating of becoming extinct. A holistic management action plan is necessary to be formulated for the assessment and management of resource base, best harvesting and processing practices, trade issues and aspects dealing with the intellectual property rights on the traditional medicines used by the tribal people. In the medicinal plants industry, the various actions needed include: setting up of a national level authority, responsive marketing cooperatives, organisation of a formal market, disseminating of market information, and creation of international market opportunities, etc.

The active participation and co-operation of the local people should be taken into account while implementating legislation measures, as well as other conservation practices. Ideally, the inclusion of a rare species in the endangered plant programme collection will only be the first step towards its recovery as naturally reproducing wild populations on protected land. At present the emphasis is less on the long-term maintenance of the species in cultivation and more on developing the technology to enhance wild populations and to introduce new populations in protected sites.

To avoid pressure from wild populations, mass propagation of potential species needs to be popularised among the inhabitants, to ensure that adequate quantities are available for future generations. Research priority should be given to develop appropriate technology for propagation, cultivation, processing, chemical characterisation and marketing of medicinal plants, and useful and endangered species. Local people should be trained to propagate, preserve and collect the medicinal plants as a part of extension. They should be educated and provided with the proper guidelines, so that there is a continuous regeneration of wild flora. It is again important that we should not disturb the local forest flora, which is generally susceptible to environmental changes, which may lead to the extinction of natural species.

Protection of medicinal plants based on statistical data on the user list from the ethnomedical practitioners.

In situ and *ex situ* conservation of endangered or likely to be endangered (Rare) species should be immediately started in the appropriate districts of the state. It is extremely important to take urgent steps for their *in situ* conservation, by banning their collection from the wild. Use of spontaneous and wild collected plants must be limited as far as possible and they should be replaced, step by step, by being cultivated. Rural folk should be encouraged to raise their own ethno-medicinal gardens or herb gardens in their vicinity, to ensure conservation of the depleting biodiversity of medicinal plants.

With the erosion of tribal culture, the traditional healers have become a threatened category. Also, the genetic diversity in medicinal plants has diminished due to shifting cultivation and large-scale destruction of their natural location. The over-exploitation of medicinal resources in an unscientific manner, by unskilled labour, and poor natural or artificial regeneration has resulted in the virtual extinction of certain vital species. An understanding of the many aspects of human influences on biodiversity and their underlying driving forces is of crucial importance for setting priorities and directing efforts towards conservation and sustainable use. Therefore, there is an urgent need for a local inventory of medicinal plants, to identify all the species that merit priority and to formulate strategies for the *in situ* conservation and cultivation of these species.

REFERENCES

Anon (1992). Global Biodiversity Strategy. Published by the World Resources Institute (WRI). The World Conservation Union (IUCN) and United Nations Environment Programme (UNEP).

Anon (2001). Demand study for selected medicinal plants, Vol-1. Centre for Research Planning and Action, Hailey Road (CERPA), New Delhi.

Bhattacharya P. and Mittra B. (2002). The Healing Touch. M.P case study of sustainable medicinal plants conservation in JFM areas. Wasteland News, XVII(2): 32-38.

Elias H. (1994). Ki Khanatang u Barim, Don Bosco Publication, Shillong, Meghalaya, India.

Hamann O. (1991). The Joint IUCN – WWF plants conservation programme and its interest in medicinal plants. In: The Conservation of Medicinal Plants (Akerele O, Heywood V and Synge H eds), Cambridge University Press, Cambridge p 13-22.

Kanjilal U.N., Kanjilal P.C., De R.N. and Das A. (1982). In: Flora of Assam Vol I – V, Avon Book Company, Delhi, India.

Maydell Hans J. (1990). Arbres et Arbustles du Sahel: Leurs caracteristiques et leurs utilizations, GTZ.

Chapter 18

Botanic Gardens and Conservation of Plant Diversity

S.C. Sharma and A.K. Goel

Botanic Garden, National Botanical Research Institute, Lucknow - 226 001

Botanic Gardens have a leading role to play in the conservation of plant diversity, which is facing the threat of extinction, especially during the 21st century. We are presently in an era of significant biological changes and losses, due to loss of plant habitats, arising primarily through human interference. It has been estimated that we are losing 60 hectares of forests every minute and one plant species per day throughout the globe. There are about 2.5 lacs plant species, known so far and may be, even more are yet to be discovered and described.

The significant reasons for the loss of plant diversity are attributed to uncontrolled urbanisation, population explosion, industrialisation and tremendous pollution, causing very heavy losses to plant diversity, resulting in environmental degradation at an alarming rate. During the next few decades, as many as one quarter of the plant species could be lost forever, if timely action is not initiated. Such taxa may remain unexplored for their medicinal, economic and ornamental potentialities for human welfare. During the next 30–40 years, 60,000 or more plant species, the world over, will face severe threats to their survival or even become extinct.

On global basis, the International Union for Conservation of Nature and Natural Resources (IUCN) has estimated that about 10% of the world's vascular plant species, total of nearly 20,000–25,000, are presently under varying degrees of threat.

It is now realised by mankind that the preservation of natural heritage is the most vital and essential factor for its survival. Botanic Gardens must be set up to protect wild plants and natural vegetation. There are about 1500 Botanic Gardens and Arboreta throughout the world and many countries are establishing new Botanic Gardens for this noble cause. Today, Botanic Gardens play a very important and increased role in the *ex situ* conservation of plants, for their acclimatisation, multiplication, judicious exploitation and rehabilitation.

In 1987, the Botanic Garden Conservation International was established at the Royal Botanic Gardens, Kew, United Kingdom, to coordinate the Botanic

Gardens in their various R and D activities for *ex situ* conservation programmes. These Botanic Gardens are conserving thousands of endemic, rare and endangered plant species.

Nowadays, Botanic Gardens are considered as the best centres for the *ex situ* conservation of plant genetic resources, categorised under various degrees of threat, either of native or exotic origin. They are the major practitioners of *ex situ* conservation through the enrichment and maintenance of germplasm collections of living plant species, seeds and tissue culture banks.

The Botanic Gardens are treasure of information for the vast conservation resources of scientifically stored and managed plant diversity for the present, as well as for posterity. In modern times, Botanic Gardens should take up database networking of living plant resources that will serve as a feedback for accelerating the conservation programmes for saving plant diversity.

The *ex situ* conservation of some important taxa viz. *Welwitschia mirabilis, Psilotum nudum, Lodoicea maldivica, Nepenthes* and Cycads in various Botanic Gardens, can be cited as significant examples. The present communication deals with various aspects involved in *ex situ* conservation studies on plant diversity in Botanic Gardens.

INTRODUCTION

There are about 2.5 lacs species of plants on the earth, the lone biosphere known so far (Heywood, 1978). On an average, the plant species amazingly possess twice the genetic diversity, as compared to animals. The diversity of plants has been at a greater risk during the last century. The maximum number of species is found in the sub-tropical and tropical regions of the world. Developing or under-developed countries, possessing insufficient financial resources, mainly cover the regions rich in biological diversity. As a result of various development activities, several plant species are facing threats of extinction to an alarming extent in these countries. The mass-extinction episode underway, is largely centred in the tropical and sub-tropical forests, because they comprise at least half of the plant genetic diversity of all species on this globe (Myers, 1988).

Attention is now rapidly turning towards the need for the conservation of a wider range of living plant resources utilised by the mankind, as well as those of wild origin. Extinction rate has accelerated to such an extent that even before new taxa are discovered, they get lost forever, without being tested for their economic or medicinal efficacies.

During the last two million years, the world flora has undergone phenomenal changes, largely due to climatic and environmental factors. This has resulted in the migration of plant species from one region to another and also in local extinction and speciation, finally showing the distribution of plant diversity, as we see in modern times. The richness of the species varies within particular habitat types. Among the tropical forests, there are certain phytogeographical regions, which are well known for their high species diversity. These areas are now considered as hotspots and a refuge where the species are able to survive periods of harsher climatic changes. The species richness of some tropical forests is astounding. The tropical forests are currently estimated to consist approximately 90,000 of plant species, discovered and identified so far, as compared with about 50,000 in the entire northern temperate zones of North America and Eurasia (Walters and Hamilton, 1993).

Tropical forests represent, in terms of biodiversity richness, the most complex and highly dynamic and fragile ecosystem on this planet. They harbour a rich repository of potential raw material and have been meeting the ever-increasing societal demands and aspirations of mankind. The International Union for Conservation of Nature and Natural Resources (IUCN) has estimated that about 10% of the world's vascular plant species, totalling to about 20,000 – 25,000 taxa globally, are under varying degrees of threat.

PLANT DIVERSITY IN INDIA

The Indian subcontinent harbours one of the richest tropical ecosystems in the world with a rich biological diversity. It has been rated tenth among the countries rich in plant diversity, and ranks fourth in Asia. It ranks sixth as far as the centres of agrodiversity are concerned. The relevance of plant diversity lies in the information that is encoded in the genes. This constitutes genetic diversity. The forests in India are a veritable emporium for a large number of economic and medicinal plant species.

The floristic wealth of this country is extremely rich in diversity and endemism. It is represented by over 15,000 species of flowering plants distributed in nine phytogeographical regions of the country (Jain, 1983). Climatically and geographically, the richness of its plant wealth exhibits a higher degree of diversity ranging from arctic, alpine, temperate, sub-temperate, sub-tropical, tropical evergreen and desert vegetations. Out of 15,000 plant species, nearly 4,000 are endemic and 2,500 under various degrees of threat, of which approximately 1,000 species are native to sub-tropical and tropical regions of India. These species inhabit fragile ecosystems and variable habitats. Due to varied topography and significant altitudinal variations, from the sea level to the highest mountain ranges and a long coastal line in peninsular India, desert in the west, cold desert in the north and humid tropical conditions in the eastern regions, plant diversity is highly versatile in the Indian sub-continent (Sharma and Goel, 1990). Further, we may have to say goodbye to 33% of our flowering plant species by the time we achieve zero growth of human population.

CONSERVATION OF PLANT DIVERSITY

Primarily, the Botanical Survey of India (BSI) and various other agencies in India have undertaken intensive exploration for the search of rare, endangered and endemic plant species. Their documentation and screening of published accounts during the last 30 years in India has facilitated the publication of a preliminary list in the form of *Materials for a catalogue of threatened plants in India* by Jain and Sastry (1983), in BSI which paved the way towards the publication of detailed accounts on endangered and endemic plant species (Jain and Sastry, 1984; Nayar and Sastry, 1987, 1988, 1990) in the form of, *Red Data Book of Indian plants* (3 volumes). These accounts have enumerated information on 617 plant species, out of an estimated 1500 taxa falling under various categories of threat. The compilation of the 4th volume of the Red Data Book of Indian plants is in process. Conservation of plant diversity may be achieved by three specific approaches.

- To maintain essential ecological processes, ecosystems and life support systems.
- To preserve the genetic diversity, on which breeding programme for the protection and improvement of the cultivated plant species depend.
- To ensure the sustainable utilisation of species and ecosystems, which support millions of rural communities as well as major industries (World Conservation Strategy).

EX SITU CONSERVATION STUDIES

Ex situ conservation keeps the components of biodiversity alive outside their original habitat or natural environment. Increasing number of taxa, facing challenges in their wild habitat, need managerial intervention to ensure their survival. This may constitute habitat management and effective *ex situ* conservation programmes. *Ex situ* conservation should be initiated for a number of reasons, some of which are enumerated below.

- To ensure and support the goal of species survival, allowing intensive genetic management.
- To maintain germplasm collection in a readily available way for R and D studies and utilisation.
- To develop propagation/cultivation techniques.

But there are a number of problems and limitations associated with *ex situ* conservation:

- Technological failure and conditions not congenial for the survival of some species, out of their natural habitat.
- Vulnerability of managerial changes.
- Risk of genetic contamination.
- Pathogen contamination.
- Proper identification of the species requiring immediate attention and more significantly which are at a higher risk of extinction in nature.

ROLE OF BOTANIC GARDENS IN *EX SITU* CONSERVATION

Botanic Gardens are the best centres for initiating *ex situ* conservation programmes on rare, endangered and endemic plant species of indigenous or exotic origin. IUCN considers that the Botanic Gardens of the world can be developed into a major global force for the conservation of plant genetic resources. This can be accomplished through an integrated approach and programmes on research and education about *ex situ* conservation and plant exploration. The success of such progammes lies in proper coordination, cooperation, exchange of planting material and appropriate planning among the Botanic Gardens (Mohan Ram, 1983).

Plant collections and storage of seed samples of endangered and endemic plant species in the seed banks, as a form of insurance policy, is a valuable function of Botanic Gardens. *Ex situ* conservation and multiplication of such species is only the first step towards saving them from extinction. The second step is ensuring their reintroduction and survival in the appropriate protected sites in natural habitats and ecosystems. Duplicate material should be made available to other Botanic Gardens willing to maintain the genetic purity, specially those in the geographical or climatic range of the species concerned.

Efforts to coordinate the activities of Botanic Gardens at the international level have been undertaken by the Botanic Gardens Conservation International (BGCI) at RBG, Kew, U. K. This secretariat is meant to disseminate information for the promotion and coordination of the *ex situ* conservation programmes of threatened and endemic plant species. BGCI has a worldwide membership of 320 Botanic Gardens.

In India, the *ex situ* conservation activities are in progress at various departments of the Universities; Botanical Survey of India; M S Swaminathan Research Foundation, Chennai; Tropical Botanic Garden Research Institute, Thiruvananthapuram; Orchid Research and Development Station, Tipi, Arunachal Pradesh; National Bureau of Plants Genetic Resources, New Delhi and National Botanical Research institute, Lucknow.

TECHNIQUES FOR *EX SITU* PLANT CONSERVATION

The maintenance of *ex situ* population of plants, carried out by various institutions and forest departments, involves three major steps, such as field gene banks, seed banks and *in vitro* storage methods.

Field Gene Banks

Botanic Gardens, having germplasm collections of diversified groups of plant species, are effectively maintained in field gene banks, and they contain a significant number of individuals of the same species, representing a considerable proportion of the known wild diversity. IPGRI, Rome is coordinating field gene banks of major tropical food crops. In India, this activity is being looked after by NBPGR, New Delhi.

Seed Banks

Seed Banking is the most efficient and effective method for the *ex situ* conservation of sexually reproducing plant species, as the seeds are viable for long time (termed as orthodox seeds). Moreover, the seeds are small and occupy less space. Every seed has a different genetic constitution, and varied samples incorporate a wide range of genetic variability. But many tropical species have seeds, which possess no natural dormancy and die quickly if not allowed to germinate immediately. Species with recalcitrant seeds and those that do not readily produce seeds must be maintained under *ex situ* conditions as growing plants in field gene banks or in the form of living collections. Seeds of orthodox types can be conserved for very long periods at sub-zero temperature, if they are dried to about 5-8% of their moisture contents. Seed viability in medium-term storage 0 to 5°C can be 5–25 years, whereas in long-term storage -10 to -20°C, the viability of seeds may be even much longer. Longevity varies from taxon to taxon. NBPGR, New Delhi is maintaining a seed-bank of crop plants and their wild relatives in India. In India, many Botanic Gardens and research organisations possess seed-banking facilities. But this is an expensive facility and needs utmost care and precision for the maintenance of seeds.

In vitro Storage

Another important method for the preservation of wild plant resources is *in vitro* storage. *In vitro* storage of germplasm refers to the conservation of plants and propagules under laboratory conditions. This technique is specifically suited for the long-term storage of propagules, and for species having recalcitrant seeds, which cannot otherwise be maintained in a seed bank.

In this country, the R & D organisations, like TBGRI, Thiruvananthapuram; Orchid Research and Development Station, Tipi, Arunachal Pradesh; Indian Institute of Horticultural Research, Hassargatta; Central Institute of Medicinal and Aromatic Plants and National Botanical Research Institute, Lucknow are maintaining the *in vitro* germplasm of endemic, rare and medicinally important taxa.

But there is a need for the enhancement of Data Base Information and to generate know-how in a more cooperative and coordinated manner. The involvement of various governmental and non-governmental agencies at various levels is necessary to achieve success for the conservation of plant diversity in the Indian subcontinent (Sharma and Goel, 1994). The networking of Botanic Gardens in India can play a significant role in the *ex situ* conservation of plant genetic resources (Nayar, 1987).

EX SITU CONSERVATION AT NBRI BOTANIC GARDEN

For the conservation of endangered and endemic plant species, NBRI Botanic Garden had taken-up a project on the '*Ex situ*' conservation of endemic, rare, threatened and endangered plant species of sub-tropical and tropical regions of India', sponsored by the Ministry of Environment and Forests, New Delhi. The germplasm collection of the following species has been built-up by organising plant collection tours, for their documentation and studies on acclimatisation and multiplication: *Cycas pectinata* Griff., *Cycas beddomei* Dyer, *Frerea indica* Dalz., *Phoenix rupicola* T. Anders., *Peucedanum dehradunensis* Babu, *Sophora mollis* Royle, *Erythrina resupinata* Roxb., *Rauvolfia serpentina* Benth. ex Kurz, *Commiphora wightii* Bhandari, *Isonandra villosa* Wight, *Adhatoda beddomei* Cl., *Vanilla walkeriae* Wight and *Trachycarpus takail* Bec., etc. These plant species are being acclimatised for their multiplication. The germplasm material of the species, which has been successfully multiplied, is exchanged with other Botanic Gardens and Institutions, within and outside the country. Further, the germplasm collection of academically important taxa viz., *Psilotum nudum*, *Welwitschia mirabilis* and 25 species of Cycads, have also been enriched in this Botanic Garden.

FUTURE APPROACH

The extinction of an indigenous plant species is an irreparable loss for mankind. Growing human population and environmental degradation are constantly adding to the severe pressure on the fragile ecosystem and biodiversity. Efforts should be made to document the tangible values of plant diversity in India. The maintenance of species and ecosystems is the basic necessity and keystone for the sustainable development and the judicious utilisation of plant diversity. Though greater investments towards conservation programmes are needed, increased funding alone cannot achieve the desired results, unless other issues like population control, economic and social/political policies are given due attention and priority, while framing conservational strategies at the regional/national levels. For achieving success in conservation programmes for plant diversity, proper interaction and coordination of various government and non-government agencies is of great significance, along with the proper education and awareness among the masses, about wild plant resources and their diversity and utilitarian aspects. Rapid technological upgradation is very necessary to keep pace with advances in the developed countries in modern times. For the effective conservation of plant diversity, database networking programmes among various Botanic Gardens, at the national and regional levels and know-how towards harnessing the potential for the present and posterity, is most vital and essential. A rationalised and balanced approach to both *in situ and ex situ* methods is the key for the success of the conservation programmes in Botanic Gardens.

REFERENCES

Heywood V.H. (1978). Flowering plants of the world. Oxford University Press, UK p 335.

Jain S.K. (1983). Flora and Vegetation of India—An outline. BSI, Howrah p 25.

Jain S.K. and Sastry A.R.K. (1983). Materials for a catalogue of threathened plants of India. BSI, Howrah p 69.

Jain S.K. and Sastry A.R.K. (1984). The Indian plant Red Data Book—I. BSI, Howrah p 162.

Mohan Ram H.Y. (1983). Role of Botanic Gardens in the conservation of tropical plants. In: Conservation of Tropical Plant Resources (Jain SK and Mehra KL eds), BSI, Howrah p 154-158.

Myers N. (1988). Threatened biotas: Hotspots in tropical forests. The Environmentalist 8: 1-20.

Nayar M.P. (ed) (1987). Network of Botanic Gardens. BSI, Howrah p 272.

Nayar M.P. and Sastry A.R.K. (eds) (1987). Red Data Book of Indian Plants Vol 1. BSI, Howrah p 367.

Nayar M.P. and Sastry A.R.K. (eds) (1988). Red Data Book of Indian Plants Vol 2. BSI, Howrah p 268.

Nayar M.P. and Sastry A.R.K. (eds) (1990). Red Data Book of Indian Plants Vol 3. BSI, Howrah p 271.

Sharma S.C. and Goel A.K. (1990). The need for *ex situ* conservation of endangered plants from tropical India. Botanic Gardens Conservation News 1(6) : 16-19.

Sharma S.C. and Goel A.K. (1994). Role of Botanic Gardens in modern times. Indian J Forest 17(3): 230-238.

Walters M. and Hamilton A. (1993). The vital wealth of plants – WWF and the conservation of plants. WWF, Switzerland p 1-48.

Chapter 19

Shifting Cultivation: Issues and Strategies in the NEH Region

B.P. Bhatt[1], A. Pattanayak[2] and P. Tandon[3]

[1]*Agroforestry Division, ICAR Research Complex for NEH Region, Umroi Road, Umiam - 793 103, Meghalaya, India*

[2]*Plant Breeding Division, ICAR Research Complex for NEH Region, Umroi Road, Umiam - 793 103, Meghalaya, India*

[3]*Botany Department, North-Eastern Hill University (NEHU), Shillong - 793 002, Meghalaya, India*

Considering the high cost, labour and energy input involved in terrace cultivation, and in the absence of other viable alternatives to shifting cultivation, the majority of the population of the North-Eastern hill states continues to depend on shifting cultivation for their subsistence livelihood. Due to limited arable land and increasing population growth, farming on the ecological fragile and marginal mountain lands, including those situated on more than 30% slope, will continue. If shifting cultivation in its present form is allowed to continue, land degradation and the impoverished living conditions of resource-poor upland farmers are bound to worsen with time. Considering the adverse impacts of shifting cultivation, such as loss of precious top soil, nutrients and forest biodiversity, destabilisation of slopes and low productivity, sustainable farming alternatives need to be developed and implemented immediately.

INTRODUCTION

Shifting cultivation is a part of the life of the hilly tribal areas of the country. Shifting cultivation in its traditional form was an ecologically and economically viable system of agriculture, as long as population density was low and the *Jhum* cycle was long enough to maintain soil fertility. However, because of shortened cycles and the systems slow response to the changing requirements of high population pressure, *Jhum* cultivation has lead to catastrophic degradation of biodiversity and the environment. Substantial increase in human and livestock population, lack of land ownership rights, shortening of *Jhum* cycle, bun method of cultivation on hill slopes and hill tops, mining, industrial demand for timber and pulpwood, clearance of forests for agriculture and urban development, along with firewood extraction have all caused land degradation in areas where shifting cultivation is a part of tradition. Due to the depletion of natural resources, the very existence of the tribal populace is being threatened. The state as well as central governments have undertaken various *Jhum* control schemes, such as land development comprising terracing, construction of water harvesting structures, forestry and horticultural development including plantation crops, agricultural inputs supply, demonstration of modern techniques and development of piggery and poultry, etc. In spite of these ambitious programmes, the country is still witnessing large areas practicing *Jhum* cultivation that is dangerous to the sustainability and viability of hill agriculture. The present paper deals with the status of shifting cultivation, its implications on resource degradation along with the prospects of some potential farming systems for sustainable land use. The authors have focussed on the North Eastern Hills (NEH) of India, where shifting cultivation is widespread and traditional.

ISSUES INVOLVED AND THEIR PRESENT STATUS

The land use system in shifting cultivation involves slashing the vegetation, burning the dried slash (Fig. 19.1), raising a mixture of crops on temporarily nutrient enriched soil for a year or two, fallowing the plot for the re-growth of natural vegetation, and eventual return to the same plot for another cropping phase after a few years (Ramakrishnan, 1992). It is used to be an appropriate and sustainable land use practice in diverse socio-economic set-ups, where the dependent human population was within the carrying capacity of a 10-15 year *Jhum* cycle. However, today, scientists view shifting cultivation as environmentally destructive and a faulty land use practice having very low input/output ratio. Shifting cultivation became unsustainable primarily, due to the increase in population that led to an increase in the demand for food. In order to meet the growing food demand, the *Jhum* cycle (the intervening period between two successive slashes) got shortened, which resulted in an overall decrease of crop yield (Anonymous, 1997). This necessitated in bringing more virgin forest areas under shifting cultivation. Thus, the vicious cycle continued

(a)

(b)

(c)

(d)

Fig. 19.1. Stages of slash and burn shifting cultivation (jhum). **(a)** Clearing of forest by felling trees. **(b)** Burning of leaves and other biomass in the field. **(c)** Heat generated due to high intensity fire causes damage to flora and fauna. **(d)** Farmers construct small temporary huts in jhum area for watch and ward of their crop.

and more forest areas were converted to wastelands, as a result of repeated *Jhum* having very short (often 2-3 years) cycles.

MAJOR TYPES OF SHIFTING CULTIVATION

1. Jhum Cultivation

Even today, shifting cultivation is the mainstay of the subsistence of tribal people (Table 19.1). Nagaland has the highest area under *Jhum* cultivation, followed by Manipur, Meghalaya and Arunachal Pradesh. *Jhum* is practiced in a highly complex system with wide variations in cropping pattern. The periodic work *Jhum* carried out in fields has been mentioned in Table 19.2. The cycle of agricultural operations in *Jhum* field includes (i) selection of the forest area before December by the entire village on the basis of the rotation of fields, (ii) cleaning the forest tract by cutting down the jungle during December –January and leaving chipped leaves and twigs to dry till February, (iii) burning the dried debris into ashes around February to mid March, before the onset of the monsoons, (iv) planting seeds of various crops in an intimate mixture with the help of a dibbling stick or sharp knives, (v) weeding, watching and protecting of crops, and (vi) harvesting, threshing and storing.

Table 19.1: Area involved (lac ha) and number of families engaged in shifting cultivation

State	*Area involved*	*No. of Families*
Arunachal Pradesh	2.61	54,000
Manipur	3.60	70,000
Meghalaya	2.65	52,290
Mizoram	0.45	50,000
Nagaland	6.33	1,16,046
Tripura	1.08	43,000

(Source: Anonymous, 2000)

Table 19.2: Periodic work in *Jhum* fields

Month	*Produce harvested*
January/February	Harvesting of chillies, brinjals, pumpkins, turmeric, ginger, other edible roots, shoots, leaves and ferns from old *Jhum* plot
March	Harvesting of bamboo shoots, from the new *Jhum* plot, which come up after burning. Harvesting of chillies, brinjals, pumpkins, turmeric, ginger, other edible roots, shoots and leaves from old *Jhum* plots.
April	Harvesting of chilies, brinjals, pumpkins, turmeric, ginger, other edible roots, shoots and leaves from old *Jhum* plots.
May	Harvesting of watermelons, cucumbers and maize from new *Jhum* fields and continuing harvesting from old fields.
June	Harvesting of sorrel leaves and continuation of harvesting cucumbers and water melons. No further harvesting is done from old *Jhum* fields.
July	Harvesting of maize, beans, millets, pumpkins (harvest begins and continues for 10 months). Also early rice is harvested.
August	Main harvest of rice in some areas. Continue to harvest vegetables, pumpkins, etc.
September	Main harvest of rice in some areas. Continue to harvest vegetables, pumpkins, etc.
October/December	Harvest vegetables, pumpkins, beans, soybeans and turmeric.

The most remarkable feature of shifting cultivation is that all essential crops like rice, maize, tapioca, colocasia, cucurbits, sweet potato, ginger, finger millet, cotton, tobacco and many others are grown in the same field in a mixed land use system, resembling the latest cafeteria system of cultivation. The large number of crop species are effectively managed over both space and time due to sequential harvesting throughout the year. The cultivators erect watchhouses, either on a raised platform or on tree branches, in their respective fields, for watching and protecting the crops. The field is used for a year for mixed copping and subsequently, it is either abandoned or cultivated with one or two selected crops; occasionally some residual crops are collected from the abandoned fields. Owing to leaching, erosion and loss of fertility after one or two years (some time three years), the cultivated areas is abandoned, a new piece of land is then selected to repeat the process, leaving the old one fallow for a period of 3-5 years. The average size of a *Jhum* plot varies from 1.0 to 2.0 ha, and the average family consists of 2 adults and 3 to 4 children. Collection of timber, cane and bamboo for house-building and making household tools and implements are important sources of income. Although practices under shifting cultivation vary widely in different parts of the North East, and the variability in practices are largely tribe-specific, shifting cultivation in any form, invariably involves clearing the vegetation, and then, slashing and burning the plant parts, including debris. The *Jhum* cycle plays a critical role in the natural regeneration and consequently, the *Jhum* economy. About 3-4 decades ago, shifting agriculture was quite sustainable as the forest was allowed to regenerate successfully. Due to shortening of the *Jhum* cycle, productivity has decreased drastically. The *Jhum* cycle has reached the critical limit in parts of Arunachal Pradesh, Assam and Mizoram, where it is as low as 2 to 3 years, even though the maximum range is from 4 to 10 years (Verma *et al.*, 2001).

2. Bun Cultivation

This system of *Jhuming* involves cutting shrubs and grasses, putting dried vegetation in the form of raised beds along the slope and covering the same with soil collected from the surroundings, burning the covered vegetation and planting it in the soil afterwards (Fig. 19.2). Although good yields of crops have been obtained through this system, yet it leads to large amounts of soil erosion. The *bun* size varies according to the hilly terrain and land availability. *Bun* is always prepared along the slope. A large variety of crops are cultivated, and as much as 14 crops are grown sequentially, in this system.

SHIFTING CULTIVATION IN THE NORTH-EAST – A CRITICAL ANALYSIS

1. *Some positive aspects*

Shifting cultivation helps conserve the rich cultural diversity, as *Jhum* is

Fig. 19.2 : Traditional cultivation systems of NEH Region.
Bun a system of cultivation in jhum lands.

interwoven into the culture and tradition of more than 200 tribes which inhabit the North-East region. Shifting cultivation, being a labour intensive but low input based farming system, it provides an assured source of food security to the poor farmers of the hill region. Shifting cultivation, in its traditional form, might also have contributed towards the conservation of agrobiodiversity, particularly the germplasm of native crop varieties. The practice of shifting cultivation in the North-East also represents an effective form of land use, by way of optimum utilisation of space, where as many as 60 different types of crops are cultivated at a time, in the same plot. In the process, a small piece of shifting cultivation plot fulfills almost all the needs of the farmer and minimises his dependency on external inputs. Besides burning, various cultural practices of shifting cultivation help in controlling weeds, soil-borne pathogens and other diseases of crops. Given the difficult terrain of the North-East, application of modern technology has several limitations. *Jhum*, therefore, serves as a low external input agricultural technology.

2. The negative aspects

One of the most important negative environmental impacts of shifting cultivation is the damage that it causes to the soil system. Besides causing air pollution due to burning, shifting cultivation is also responsible for loss of soil nutrients and useful soil fauna and microbes. Burning of slash lowers soil acidity, organic matter and total nitrogen, but enhances phosphorus and cations. Slashing, burning and subsequent cropping result in a significant reduction in the available carbon, nitrogen and magnesium, compared to the pre-cropped stage. Most shifting cultivation practices are a put of subsistence level farming system, having very low output/input ratio compared to other farming systems/methods. The clearing of forest areas for shifting cultivation, at regular and frequent intervals, results in the loss of primary forests and the formation of secondary forests. This causes substantial loss to tree diversity and the associated vegetation which are adapted to the primary forests. Due to the shortening of the Jhum cycle, quite often, the secondary forests also do not get adequate time to regenerate. The repeated use of land with short Jhum cycles, finally converts the Jhum fallows into degraded wastelands. Therefore, shifting cultivation is considered to be the single most important factor causing deforestation and forest degradation in the North-East. Incidentally, the largest wasteland area in the North-East comes under the category of shifting cultivation (Tripathi and Barik, 2003).

3. Status of Shifting Cultivation

There are varying estimates of areas affected by shifting cultivation, by different agencies. The area affected by shifting cultivation, in the North-Eastern region, was reported to be 3.81 million ha, by a task force on shifting cultivation (1983). Other such estimates are 2.80 million ha, by the North-Eastern Council (1975) and 7.40 million ha, by the FAO (1975). However, as per the State of Forest report (1999), the cumulative area affected by shifting cultivation, during the above period, was found to be 1.60 million ha. The extent of area under shifting cultivation is maximum (0.39 m ha) in Nagaland, followed by Mizoram (0.38 m ha) and Manipur (0.36 m ha). These states together, account for about 65% of the total area under shifting cultivation (Anonymous, 1999 - Table 19.3). According to these estimates, the area under shifting cultivation is between 2.80 million ha and 7.40 million ha.

Table 19.3: Area affected by shifting cultivation

State	*Geographical area (sq. km.)*	*Cumulative area of shifting cultivation (1987-1997) (million ha)*
Arunachal Pradesh	83,743	0.23
Manipur	22,327	0.36
Meghalaya	22,429	0.18
Mizoram	21,081	0.38
Nagaland	16,579	0.39
Tripura	10,486	0.06
Total		1.60

(Source: Anonymous, 1999)

Shifting cultivation has been the main source of livelihood for most of the tribes in North-East NEH Region, Shillong, and Eastern hills; and a substantial portion of the total hill population exclusively depends on it for survival. On an average, 3,869 sq km area is put under shifting cultivation every year and an estimated 443,336 households earn their livelihood from shifting cultivation.

4. Land use and soil erosion

High rainfall and steep hill topography is always associated with problems of severe soil erosion, particularly when the land use systems have biotic interference. Quantitative facts on soil erosion hazards from soil erosion studies in the region, indicate that except for forestlands none of the other land uses are safe as they lead to land degradation (Table 19.4). The study also indicated that erosion and runoff increase with up to 45% slope, thereafter, it decreases linearly with further increase in the slope. The soil erosion from hill slopes (60-70%) under first, second year and abandoned *Jhum* was estimated to be 147,170 and 30 t/ha/yr (Singh and Singh, 1981). It is estimated that 181million tons of soil is lost every year from this region, due to this defective system of farming (Satapathy, 2003).

Table 19.4: Soil erosion hazards associated with various land use practices in NEH Region

Land use system/practices	*Experimental plot size*	*Soil loss (t/ha/yr)*	*Average annual rainfall (mm)*
Shifting cultivation	Small	30.2.170.2	1600
Shifting cultivation	Field	5.1-83.3	1600
Tuber crops on raised bed (bun)	Medium	40.0-50.0	1800
Pineapple cultivation along the slope (first 2 years)	Small	24.0-62.6	1600
Homestead areas	Field	6.8	1600
Mixed crop of maize and rice	Small	19.7-21.0	1600
Rice crop on slope	Small	32.9-45.0	1600
Bare fallow	Small	83.8	1600
Cropping systems	Medium	51.0-83.8	1600
Grass cover (planted)	Medium	10.83	1600
Natural bamboo forest	Field	0.04-0.52	1600

Note: Area of small, medium and field size plots were in ranges of 2-5, 16.40 and 69,000 m^2, respectively.

5. Some traditional environment friendly farming systems

Some potential indigenous farming systems of the region are the *Zabo System, Rice Cultivation of Apatani Plateau* (Fig. 19.3), *Agriculture with Alder Trees,* and *Terraced Rice Cultivation* (*Panikheti* – Fig. 19.4), (Prasad and Sharma, 1994).

Zabo, which integrates forests, agriculture, livestock and fisheries with an efficient soil and water conservation base, is practiced in the Phek district of Nagaland. Organic base of farming is the characteristic feature of the *Zabo* system.

The *Apatani Plateau* in Subansiri district of Arunachal Pradesh is known as the rice bowl of the Apatani tribe, who cultivate wet rice with indigenous

Fig. 19.3 : Traditional cultivation systems of NEH Region. Wetland rice cultivation system practiced in Subansiri district of Arunachal Pradesh. Water collected from upper plots is channelized through lower plots and a specific water level is always maintained. The system represents a unique example of water conservation.

19.4: Traditional cultivation system of NEH Region. Panikheti, a system of wetland rice cultivation practiced in Nagaland and Sikkim. In this system also water is collected from top and channelized to the lower plots. Organic farming with long duration traditional rice varieties is practiced in both systems.

knowledge of land and water management. Another potential system is the farming of Alder (*Alnus nepalensis*), in and around agriculture fields, for the restoration of soil fertility and sustainable crop yield. It is even cultivated in *Jhum* fields. The practice is most popular among Angami, Chakhesang, Chang, Yungchinger and Konyak tribes of Nagaland. Khonoma village of Nagaland is proud of its plantation and Alder-based agriculture. Apart from this, Angami and Chakhesang tribes of Nagaland have developed a system of terrace rice cultivation (*Panikheti*). In this system, some forest area is retained at the hilltop to divert the runoff water to rice fields, through channels and sub-channels, while making terraces. Other major indigenous farming systems includes bamboo drip irrigation, which is found in the Jaintia and Khasi Hills of Meghalaya. The system is highly efficient for water management. Water is carried with the help of different forms of bamboo pipes and further distributed into different bamboo sub-channels for use at the required site.

EFFORTS MADE TO CONTROL SHIFTING CULTIVATION IN THE NORTH-EAST REGION

Several *Jhum* and *Jhum*ia rehabilitation schemes have been implemented in the North-Eastern region since independence, by the state and central governments, to control shifting cultivation in the region. The examples of such schemes are Watershed Development Projects in Shifting Cultivation Areas (implemented by the Ministry of Agriculture, Government of India), Soil conservation schemes of the Government of India, Tripura *Jhum*ia Rehabilitation Scheme of the Government of Tripura and New Land-use Policy Scheme of the Government of Mizoram. However, most of these schemes have not been able to achieve the desired level of success.

The concern of the Government of India for controlling shifting cultivation is reflected in its agricultural and forestry policies from time to time. For instance, both the Forest Policies notified during the post-independent period, viz., National Forest Policy, 1952 (NFP 1952) and National Forest Policy, 1988 (NFP 1988) have emphasised the need to control shifting cultivation and rehabilitate the affected areas. The NFP, 1952 discuss "tree lands in agricultural areas and their importance in rural economy" (section 17) and the "concern for damage caused to forests by shifting cultivation" (section 23), and the NFP 1988, states the "concern for adverse effects of shifting cultivation on environment and land productivity" (section 4.7) and "to contain the areas already affected and rehabilitation through energy plantation and social forestry".

In order to deal with the problems of shifting cultivation, the Government of India has also constituted two task forces, one in the Ministry of Agriculture in the year 1983, and the other in the Ministry of Environment and Forests in 2001. Besides estimating the area under shifting cultivation in the country, the task forces have recommended a sustainable approach to manage shifting cultivation by adopting a holistic and integrated approach.

1. Successful experiments having potential to manage shifting cultivation effectively

Since independence several research and extension organisations in the North- East have been trying to develop alternatives to shifting cultivation. Such efforts have resulted in developing ecologically and economically viable land use models, having the potential to work as alternatives to shifting cultivation. These successful experiments need to be adapted widely, depending upon the local socio-economic and biophysical conditions. Some of these successful experiments tried in the North-East have been listed in Table 19.5. The list also includes those successful practices which have been tried and standardised by the people themselves and have been successfully practiced for some time now, as alternatives to traditional *Jhum*.

Table 19.5: Some successful experiments to supplement shifting cultivation

- ICAR 3-tier model (experiment by ICAR)
- NEPED (Nagaland Environment Protection and Economic Development) (experiment by Government of Nagaland)
- SALT (Sloping Agriculture Land Technology) (experiment by GBPIHED)
- SWEET (Sloping Watershed and Environmental Engineering Technology) (experiment by SFRI)
- Intensive watershed based livestock production system (experiment by ICAR)
- *Modified shifting cultivation practices undertaken by* Jhumias *with the introduction of cashcrops like large cardamom, medicinal plants, broom grass, beetle leaf and beetle nut, cinamomum, fruit orchards and orchid cultivation (documented by RCNAEB and SFRI)*

(Source: Tripathi and Barik, 2003)

FUTURE STRATEGIES FOR MANAGING SHIFTING CULTIVATION

There can be two approaches to successfully manage shifting cultivation in North-East India. Considering the socio-cultural importance of *Jhum* in the life of the people of the North-East, the best strategy would be to modify and improve the existing practices of *Jhum* cultivation. Alternatively, new alternative programmes can replace the existing forms of *Jhum*. While modifying the existing *Jhum*, the following aspects must be taken into account:

- soil and water conservation
- maintenance of soil fertility
- crop diversity
- food security
- high-yielding varieties of crops
- low volume and high value cash crops
- market links
- yield optimisation
- conformity with ecological principles

(Source: Tripathi and Barik, 2003)

In hill agroecosystems, agroforestry plays an important role in sustainability, resource conservation and food security. It combines elements of agriculture, whether crops or animals with elements of forestry in production system in its land piece, either simultaneously or sequentially. This land use system is particularly suitable for resource-poor marginal lands and even, for wastelands (Bhatt *et al.*, 2001). Foresters in the past have seen agroforestry, mainly in terms of improving the supply of forest products; agriculturists have seen it as a logical extension of traditional intercropping practices, as an aid to soil conservation measures and as plantation agriculture, such as tea plantations. Soil conservation specialists see agroforestry as a major tool in the maintenance of soil productivity, and ecologists seek the increase in diversity that moving away from monocultures can bring. In rural development, the improvement of land use systems and thus, of rural income and well-being are the primary objectives. In a broader sense, the major function of agroforestry is associated with sustainability for the farmer, stability of resources, production and income, and minimisation of risk. The objective of sustainability is probably the most important function of such kind of land use systems (Bhatt and Verma, 2001).

REFERENCES

Anonymous (1997). Agenda notes for the meeting on WDPCA Project. Ministry of Agriculture (SWC Wing), New Delhi.

Anonymous (1999). State of Forest Report. Forest Survey of India. Ministry of Environment and Forests, Government of India, p 113.

Anonymous (2000). State of the Forest Report 1999. Forest Survey of India, Ministry of Environment and Forests, Government of India, Dehradun, India.

Bhatt B.P. and Verma N.D. (2001). Some multipurpose tree species for agroforestry systems. ICAR Research Complex for NEH Region, Umiam, Meghalaya, p 170.

Bhatt B.P., Singh R., Misra L.K., Tomar J.M.S., Singh M., Chauhan D.S., Dhyani S.K., Singh K.A., Dhiman K.R. and Datta M. (2001). Agroforestry research and practices: An overview. In: Steps towards Modernisation of Agriculture in the NEH Region (Verma ND and Bhatt BP eds), ICAR Publication, Meghalaya, India p 365-392.

Prasad R.N. and Sharma U.C. (1994). Potential indigenous farming systems of the North-Eastern hill region. Technical Bulletin, ICAR Research Complex for the NEH Region, Umiam, Meghalaya, India p 146.

Ramakrishnan P.S. (1992). Ecology of shifting cultivation and ecosystem restoration. In: Ecosystem Rehabilitation, Vol 2: Ecosystem Analysis and Synthesis (Wali, MK ed), SPB Academic Publishing Bv, The Netherlands, p 19-35.

Satapathy K.K. (2003). Soil and water conservation measures for sustainable hill agriculture. In: Approaches for Increasing Agricultural Productivity in Hill and Mountain Ecosystems (Bhatt BP, Bujarbaruah KM, Sharma YP and Patiram eds), ICAR Research Complex for the NEH Region, Umiam, Meghalaya, India p 3-13.

Singh A. and Singh M.D. (1981). Soil erosion in the Northeastern hill region. Research Bulletin No 10, ICAR Research Complex for the NEH Region, Umiam, Meghalaya, India.

Tripathi R.S. and Barik S.K. (2003). Shifting cultivation in North East India. In: Approaches for Increasing Agricultural Productivity in Hill and Mountain Ecosystems (Bhatt BP, Bujarbaruah KM, Sharma YP and Patiram eds), ICAR Research Complex for NEH Region, Umiam, Meghalaya, India p 317-322.

Verma N.D., Satapathy K.K., Singh R.K., Singh J.L. and Dutta K.K. (2001). Shifting cultivation and alternative farming systems. In: Steps towards Modernisation of Agriculture in the NEH Region (Verma ND and Bhatt BP eds), ICAR Publication, Meghalaya, India p 345-364.

Chapter **20**

Wild Edible Fruits and Seeds of North-East India

P.B. Gurung

Herbarium, Department of Botany, Norht-Eastern Hill University, Shillong - 793 022, Meghalaya

In the present era of bioprospecting, the study of the traditional uses of plants, which were hitherto unknown to the world at large, gains more significance. The Northeastern region of India, inhabited by more than a hundred different tribes of various ethnic groups, is a treasure-house of traditional plants. In the backdrop of fast modernisation and the considerable headway achieved in bringing the region in the national mainstream, there is an urgent need to record the treasure of indigenous knowledge in the region.

An account of sixty-three wild fruits and seeds eaten by the indigenous tribes is given; details of local name, scientific name, diagnostic description of plants, method of consumption are also given. This paper includes several hitherto, unrecorded wild edible fruits of the region.

INTRODUCTION

The North-Eastern region of India, comprising of the states of Arunachal Pradesh, Assam, Manipur, Meghalaya, Mizoram, Nagalnad, Sikkim and Tripura, is inhabited by tribals of various ethnic groups viz., Khasi, Garo, Jaintia, Adi, Monpa, Miri, Mishmi, Kachari, Chakma, Mech, Rabha, Chiru, Gangte, Kom, Zou, Mizo, Paite, Pnar, Naga, Bhil, Halam, to name a few. For centuries, the tribals have lived in virtual isolation from the rest of the world. The majority of the ethnic groups reside in villages in remote areas, far away from urban centres. The forests around their rural habitat have provided them with varied and abundant resources. The surrounding plants comprise an integral part of their material and spiritual culture. They have household remedies for common ailments, while the traditional healer takes care of complex ailments and rural health. They also cultivate various indigenous vegetables and fruit crops that represent a distinct plant genetic stock suited to their local conditions. It is estimated that in India, about 800 species are consumed as food-plants, chiefly by the tribal inhabitants (Singh and Arora, 1978). Of these, about 300 species occur in the North-Eastern region (Singh and Arora, 1978; Watt 1889-1899; Kanjilal *et al.*, 1934-1940). A few examples are *Piper peeploides*, used as a condinment in Khasi Hills, *Parkia roxburghii*, grown in country yards for their fruits in the region; *Moghania vestita*, mainly grown as a tuber crop in the Khasi and Jaintia Hills; *Hodgsonia heteroclita*, a cucurbit, cultivated by the Mizo tribe for its oilseeds and edible kernel; *Digitaria cruciata* var. *esculenta*, a minor millet, of Khasi Hills and several species of *Alocasia*, *Colocasia*, *Amorphophallus*, *Dioscorea* and many other species are cultivated by various tribes in the region. 170 ripe fruits, which are pulpy and sweet/sub-sweet are eaten raw, and many of these are used for pickles/vegetables, when unripe; about 15 have edible seeds and are eaten raw or roasted (Arora, 1981).

Abroma augusta Linn.f.

Vernacular names: Dieng-tyrkhum in Khasi; Bon-kopahi in Assamese

Shrub or small tree up to 4.5m high; leaves 10-27.5cm long, obliquely lanceolate; flowers 5cm across, dark red or purple; fruit 3.8-5cm long, 5-winged capsule.

Seeds are edible. Available from August to January.

Actinidia callosa Lindl.

Vernacular name: Mei-soh-khan in Khasi.

Climbing shrubs; leaves elliptic with serrated margins, 10-15 cm long; flowers borne in clusters in leaf-axils, whitish; fruits ovoid, ± 3.5cm long.

Ripe fruits are acidic and edible. This shrubby climber bears fruits in the month of September.

Baccaurea sapida (Roxb.) Muell.-Arg.

Vernacular names: Sohramdieng in Khasi; Gosampe in Garo.

Evergreen tree up to 15m high; leaves 10-22cm long; fruit 1.5-3cm long, spherical, borne on the trunk or branches, yellow when ripe; seeds 3-4, red in colour.

The juicy pulp(aril) around the seeds is delicious, available from June to July, when the trees are in fruit.

Berberis aristata DC.

Thorny shrub up to 3m high; leaves whorled, 3.7-7.5cm × 1.5-3cm, oblanceolate with spiny margins; flowers in drooping racemes, bright yellow; fruits 7-10mm long, ovoid, blue.

Ripe fruits are edible and the plant bears fruits from June to November. After sun drying the ripe fruit is eaten as a blood purifier.

Canarium benghalense Roxb.

Vernacular name: Borsamphol in Garo; Bisjang or Neribi in Assamese.

Tall evergreen trees; leaves compound, 30-90cm long, leaflets, 5-10 pairs, 7.5-17.5cm long, lanceolate, entire; flowers up to 45cm long; fruit is drupe, ± 4cm long, elliptical, dark purple with an aromatic mesocarp and three-angled bony seed.

Fruits are edible and are available from November-January.

Cyathocalyx martabanicus Hk.f. and Thoms.

Vernacular name: Bolong in Garo and Hrieroth in Mizo.

Tall evergreen tree up to 42m high; leaves 10-22.5cm × 3-7.5cm, elliptical; flowers 2-3.7cm long, densely hairy, in groups of 2-3; fruit is an elongated berry, up to 7.5cm long, ripe fruits are red in colour.

Pulp of ripe fruits is fairly sweet, aromatic and delicious to eat. Trees bear fruits from October to November.

Calamus floribundus Griff.

Vernacular name: Soh Thri in Khasi; Sohmil in Garo

Found in the warmer parts of the Khasi, Jaintia and Garo Hills districts of Meghalaya. The pulp inside the outer woody skin is juicy and is much relished by the indigenous population. This thorny shrub bears fruits from February to April.

Castanopsis hystrix DC.

Vernacular name: Chaukhu in Garo; Hingori in Assamese; Thangji in Manipuri.

Evergreen trees up to 36m tall with young parts and inflorescence pubescent; leaves 7.5-17cm long, lanceolate, margins serrated; fruits 1.3cm long, conical to 3-angled, enclosed in a scaly cup with needle-like appendages.

Fruits have a good flavour and are edible. The tree bears fruits from September to October.

Castanopsis indica (Roxb.) DC.

Vernacular name: Dieng-soh-ut in Khasi; Chakkum-chhokran in Garo; Hinguri in Assamese.

A middle-sized or large evergreen tree up to 18m high with a silvery-grey bark; branchlets, petioles, underside of leaves and inflorescence rusty-tomentose; leaves 7.5-19cm long, coriaceous, elliptic-oblong, sharply serrate; fruit is an egg-shaped acorn, 7.5mm long enclosed in a prickly-scaly cup.

Fruits resemble hazelnuts in shape and flavour, and are eaten by the tribals; fruits are available from September to December.

Celastrus paniculata Wild.

Vernacular name: mei-lah-lih in Khasi; Tylle-soehlain in Jaintia.

Large deciduous climbing shrub with the sap turning blue with the touch of a knife blade; leaves serrated, acuminate, obovoid, shortly beaked, 5-10cm long; flowers pale-green, 3.8mm across, in 5-20cm long terminal drooping panicles; capsule 3-valved, 3-6 seeds, completely enclosed in a red arillus.

Unripe fruits are boiled and eaten as a vegetable. Fruits are available from July to March.

Dillenia indica Linn.

Vernacular name: Ao-tenga in Assamese; Panpui in Garo; Dieng-soh-karbam in Khasi; Aitrang in Kuki; Aithlong in Mizo; Hiegri in Manipur.

An evergreen tree, up to 18m high with the spreading branches forming a dense crown; leaves, 20 – 30 × 8-10cm, elliptic with sharply toothed margins; flowers >15cm in diameter, white, fragrant; fruits up to 10cm in diameter, green, round.

Ripe fruits are eaten raw or cooked, commonly eaten mixed with other vegetables and fish curries, also pickled and made into a jelly; the acid juice extract is mixed with sugar and water to give a cooling drink.

Trees bear fruits from February to April.

Dillenia pentagyna Roxb.

Vernacular name: Okshi in Assamese; Achki in Garo; Dieng-sah-bar in Khasi.

Trees up to 18m high, deciduous, with silky young parts; leaves 30-60cm × 10-30cm with parallel venation and sharply-toothed margins; flowers 2.5-3.8cm across, greenish yellow, trees leafless when in bloom; fruits up to 2cm in diameter, round, enclosed in thick and fleshy sepals.

Ripe fruits are pleasantly acidic and edible; young fruits are also eaten raw or cooked.

Fruits are available from March to July.

Dillenia scabrella Roxb.

Vernacular name: Banji-oi in Assamese; Agatchi-badura in Garo.

Trees up to 15m high; leaves 20-37cm x 10-17cm with sharply serrated margins; flowers 3.5-5.5cm across, golden yellow, leafless when in bloom; fruits ± 2cm in diameter, round, enclosed in fleshy sepals.

Fruits are edible. Trees bear fruit from May to July.

Docynia indica (Wall.) Decaisne

Vernacular name: Sohphoh Khasi

Found in the Khasi and the Jaintia Hills, the ripe fruits are greenish-yellow and sour in taste. Pickled fruits are a delicacy in the region. Trees are in fruits from July to October.

Eleagnus conferta Roxb.

Vernacular name: Sohkua in Garo.

This shrubby plant bears fruits copiously from March to May. Ripe fruits are red in colour and are eaten with a pinch of salt. Pickled fruits are great and have ample scope for commercial exploitation.

Fruits are available from March to May.

Elaeocarpus floribundus Blume

Vernacular name: Charphal in Manipuri

Evergreen trees up to 18m high; leaves 7.5-15cm long, elliptical, leathery, margins toothed; flowers 8mm across, white, borne on long stalks in the axils of fallen leaves; fruit 4.5cm long, light green, oblong, with a pleasantly acid mesocarp; seeds wrinkled, 3-grooved.

The pleasantly acid fruit is eaten raw. Cooked or pickled fruits are a delicacy. Fruits are available from August to December.

Elaeocarpus prunifolius Wall.

Vernacular name: Soh-khyllam-ai-blang in Khasi; Dieng-la-khmar in Jaintia.

Evergreen trees up to 12m tall; leaves 5-10cm long, lanceolate, margins entire or shallow toothed; flowers 10-13mm across with white silky petals; fruit ovoid, 1.5-2cm long, stone shallowly wrinkled, 3-grooved.

Fruits are available from June to October and are edible.

Elaeocarpus sikkimensis Mast.

Vernacular name: Seleng in Assamese

Evergreen tree up to 15m high; leaves 10-20cm long, elliptical, margins serrated; flowers 8-13mm across, white; racemes 6-10cm long; fruit, 5cm × 2.5cm; stone wrinkeled.

Fruits are edible and available in the month of July.

Elaeocarpus varuna Buch.-Ham. ex Mast.

Vernacular name: Dieng-ri-lam in Khasi; Gangma-chiring in Garo; Pong-o-test in Naga; Bhadraik in Assamese.

Evergreen tree up to 18m high; leaves 7.5-20cm long, elliptic to lanceolate, margins toothed; flowers 1.3cm across, white with silky sepals and petals; racemes 6-12.5cm long; fruit 1.3-1.8cm long, oblong, drupe, 3-grooved.

Fruits are eaten. Fruiting takes place from July to September.

Euryale ferox Salisb.

Vernacular name: Shriew-u-pubon in Khasi.

Fresh-water aquatic plant; leaves floating, bristly with deep bluish purple undersurface, 15-45cm in diameter, elliptical; flowers 2.5-6.7cm across, violet or variegated; fruit 5-10cm in diameter, spongy, bristly, crowned with persistent sepals, ripen under water; seeds with pulpy arils.

The seeds are eaten raw or after thorough drying and are available from June to October.

Fissistigma verrucosum (Hk.f. & Th.) Merr.

Vernacular name: Jyrmi-soh-ram-khlow.

Large woody climber, branchlets rusty, hairy; leaves, 7.5-15cm × 3.5cm, oblong-lanceolate; flowers, 1.2cm long, pale yellow, fragrant; berries are borne in clusters, each 3.5-7cm in diameter, thickly warty and tomentose.

The aromatic pulp of ripe fruits is much savoured. This woody climbing plant is in fruit from October to January.

Flacourtia jangomas (Lour.) Raeusch.

Vernacular name: Darichik in Garo; Goch-poniol in Assamese; Duk-duki in Tripuri.

Shrub or small tree up to 9m high, thorny; leaves 5-12.5cm × 2.5-5cm, ovoid to oblong, margins blunt, serrated; flowers 2.5-3.8mm across, unisexual; fruit, 1.2-3cm long, top-shaped or spherical, purplish when ripe, 10-14 seeded.

Ripe fruits are eaten, sometimes sold in local markets. Fruiting from August–January.

Garuga pinnata Roxb.

Vernacular name: Dieng-khiang in Khasi; Agri in Garo; Bonkung-asing in Miri; Gendeli poma in Assamese.

Trees up to 12m high; leaves compound, 15-45cm long; leaflets 11-19 in number each 5-15cm long, lanceolate to ovoid, oblique with round toothed margins; flowers 3.8-6.3mm long in clusters at branch tips; fruit 1.8-2.5cm across, greenish yellow, unevenly roundish with a pointed tip; stones wrinkled.

Fruits are eaten raw, cooked or pickled.

Garcinia atroviridis Griff. ex T. Anders

Found in Assam and Arunachal

Evergreen trees up to 15m tall; leaves 15-22.5cm long, oblong to lanceolate, thick, shining, dark green with reddish petioles; male flowers 1.5cm across, deep red; female flowers 3.7cm across with a deep red stigma; fruit, 7.5cm across, round, yellowish green to yellow.

Fruits are edible and extensively used as a sour relish for curries, both fresh and sun-dried.

Garcinia cowa Roxb.

Vernacular name: Kau-gach in Assamese; Rengran in Garo.

Evergreen trees up to 12m high; leaves 5-17.5cm × 2.5-7.5cm, dark green, somewhat leathery, young leaves reddish-brown; flowers yellow, unisexual; fruit, 5cm across, faintly 6-8 lobed, dull red, orange inside, 4-8 seeded.

Fruit is acidic and is eaten. Sun-dried sliced and preserved fruits are consumed to relieve dysentery.

Garcinia lanceaefolia Roxb.

Vernacular name: Dieng-soh-jadu in Khasi; Thisuru in Garo.

Shrubby tree up to 3.6m high; leaves, 5-12.5cm long, lanceolate, dark green, somewhat fleshy; flowers, 7.6mm across, unisexual, solitary or 2-3 together; fruit 2.5cm long, ovoid, orange-red or orange-yellow, 6-8 seeded.

Ripe fruits are acidic and are relished.

Garcinia morella Desv.

Vernacular name: Kuji-thekera in Assamese; Korbomba in Kuki.

Evergreen tree up to 15m high; leaves 7.5-15cm long, lanceolate; flowers greenish-white, in the axils of fallen leaves, unisexual; fruit 2cm across, spherical, yellow when ripe, 4-seeded.

The semi-solid fat obtained from the seeds is used as an alternative to ghee and butter.

Garcinia paniculata Roxb.

Vernacular name: Dieng-soh-longkor in Khasi; Thisru in Garo; Sochopa-tenga in Assamese.

Evergreen tree up to 20m high; leaves 10-22.5cm × 5-10cm, shining; flowers white, unisexual; male flowers 1.3cm across, fragrant, borne in panicles; female flowers larger, but fewer, borne in racemes; fruit 2.5cm across, spherical, fleshy, 4-seeded, yellow when ripe.

Ripe fruits taste like mangosteen and is relished. Fruits available from May to July.

Garcinia pedunculata Roxb.

Vernacular name: Soh-lyntraw in Khasi; Heibung in Manipuri; Bor-thekera in Assamese.

Tree about 18m high with fluted trunk; leaves 15-30cm × 7.5-13.7cm, narrowed into the petiole; male flowers pale green, in 8-12 flowered panicles; female flowers solitary, larger than male; fruit 7.5-11.2cm in diameter; seeds 8-10, immersed in juicy flesh.

The agreeably acidic fruit is consumed raw or cooked. Sliced and dried fruits are used as an alternative to lemons. Fruiting from December to June.

Garcinia spicata Hk.f.

Vernacular name: Dieng-soh-kwang in Khasi.

A middle-sized tree with wide spreading branches and angular branchlets; leaves 7.5-20cm long, elliptic, obtuse; flowers unisexual, 5-8mm across, borne on short pedicels, fasciculate on axillary protuberances (undeveloped branchlets), which frequently lengthen out and form spiciform racemes up to 15cm long; fruit green, spherical, beaked, about the size of a walnut, 1-3 seeded.

Fruits are available during winter and are edible.

Garcinia xanthochymus Hk.f.

Vernacular name: Dieng-soh-khyllung in Khasi; Aruak in Garo; Demphal in Tripuri; Tepor in Assamese.

A middle-sized tree with a dense dark green crown; branchlets quadrangular, dilated below the nodes; leaves 20-45cm × 3.8-10cm, with curved margins, thickly coriaceous, oblong or elliptic-oblong, acute; flowers 1.3cm long, greenish-white, in axillary clusters; fruit 3.7-7.5cm in diameter, crowned by a persistent stigmatic lobe, golden yellow when ripe, 1-4 seeded.

Fruit is pleasantly acidic and is eaten raw or cooked; also used in the preparation of refreshing drinks.

Gnetum gnemon Linn.

Vernacular name: Annel in Kuki; Mejherguti in Assamese.

Shrub or shrubby tree up to 4.5m high with swollen joints; leaves 6-24cm long, elliptical, leathery, glossy green on the dorsal surface; petiole, 5-15mm long; flowers in slender clusters, 2.5-3.8cm long; fruit 1.3-1.8cm long, oblong, apiculate, ripe fruit reddish.

Ripe fruits are eaten raw, cooked or roasted. Young leaves, flowers and fruits are also eaten cooked as a vegetable.

Gnetum montanum Mark.f.

Vernacular name: Jermai-saprah in Jaintia; Thaniping-rhui in Mizo; Mameilet in Assamese.

Large climber with forked branches; leaves 7.5-20cm long, opposite, elliptic, glossy; flowers unisexual, aggregated in circles in the axils of cup-shaped scales; fruit 1.9-3.8cm long, elliptic, fleshy, reddish-orange when ripe, 1-seeded.

Fruits are available from September to October and are eaten.

Grewia elastica Royle

Vernacular name: Dieng-soh-langhri-thaat in Khasi; Bor-khentri in Garo; Bijol-goch in Assamese.

Deciduous tree up to 18m high; leaves, 5-17.5cm long, ovoid to elliptic, dorsal surface green and rough, margins blunt toothed; flowers golden yellow; sepals larger than petals, 5-12mm long; fruit 6mm across, spherical or obscurely lobed, ripe fruits blackish.

Fruits are available from August to November and are eaten.

Grewia hirsuta Vahl.

Shrub up to 2.7m high; leaves 5-12.5cm long, lanceolate or oblong, 3-nerved, margins toothed; flowers white, turn yellow with maturity; sepals much longer than petals, 5-10mm long; fruit 7.5-12mm in diameter with a ring of hair at the base, obscurely 1-4 lobed, reddish-brown when ripe.

The plant is in fruit from November-January and are eaten.

Gynocardia odorata R.Br.

Vernacular name: Sohliang in Khasi; Thithopha-bipha in Garo.

Evergreen tree up to 15m high, easily noticeable with the fruits growing on the trunk and the bigger branches; leaves 15-20cm × 3.7-8.7cm; flowers up to 3.8cm across, pale yellow, fragrant, unisexual; fruits 7.8-12.5cm across, hard, round, ash-coloured; seeds many, 2.5cm long, immersed in a fragrant orange pulp.

Seeds and pulp are eaten after boiling. Boiled seeds are chopped into thin round slices and eaten, sometimes sold in the market. Trees bear fruit from January to March.

Holboellia latifolia Wall.

Vernacular name: Mi-rang-Ksa or Soh-lygn-kait in Khasi.

Evergreen scandent climber; leaves palmately compound, 3-9 leaflets, 7.5-15cm × 1.2-5cm, shape variable; flowers, 1.2-2cm across, greenish purple, fragrant; fruit of 1-3 banana like fruitlets, each 5-10cm long.

The yellow pulp of the fruit is edible. This evergreen climbing plant bear fruits from September to November.

Kadsura roxburghiana Kaempf. ex Juss.

Vernacular name: Mi-ja-ngew in Khasi; Kang-mari in Manipuri.

Evergreen climbing shrub with cracked bark; leaves 7.5-15cm × 3-7.5cm, elliptic-lanceolate; flowers 1.2cm across, whitish; fruit is drupelets, each 7.5mm long, oblong with red juice.

Fruits are eaten. The evergreen climbing shrubs bear fruits from August to November.

Mammea suriga (Buch.-Ham. ex Roxb.) Kosterm

Vernacular name: Suklong in Mizo.

Tree with dense foliage and milky sap; leaves 15-22.5cm long, deep green, leathery, oblong, margins slightly curved; flowers, 1.5cm across, white, speckled with red; fruit 2.5cm long, obliquely ovoid, 1-seeded.

Fruits are edible and available during monsoon.

Mesua ferrea Linn.

Vernacular name: Dieng-ngai in Khasi; Ngai-ching in Naga, Ulhau in Manipuri; Nahor in Assamese.

Evergreen tree up to 24m high, exuding an aromatic oleoresin; leaves 6-17.5cm long, lanceolate, shining dark green above, undersurface with a white waxy coating; young leaves reddish; flowers 3.5-10cm across, white with golden yellow stamens; fruit 2.5-3.5cm long, reddish, ovoid, wrinkled, enveloped by the persistent calyx, 1-4 seeded.

Trees bear fruits from August to October. Fruits taste like chestnut and are eaten.

Microcos paniculata Linn.

Vernacular name: Dieng-soh-lieng-hadem in Khasi; Bolchibins in Garo; Pisoli in Assamese.

Tree or shrub up to 15m high; leaves 7.5-18cm × 2.5-8.7cm, base oblique, 3-nerved; flowers white or yellowish-white, borne in panicles at branch tips;

sepals much longer than petals; fruit, 7.5-10mm in diameter, spherical, wrinkled, purplish.

Fruits are edible and available during the winter.

Miliusa velutina Hk.f. & Thoms

Vernacular name: Bor-samphol in Garo.

Medium-sized deciduous tree up to 12m high with silky hairy shoots and flowers; leaves 7.5-25cm × 5-15cm, fragrant, hairy; flowers up to 1.8cm in diameter, borne singly or 2-6; fruits up to 2cm in diameter, bluish-purple to black, oval, minutely hairy.

Fruits resemble black cherry and are edible. Trees are in fruit from June to October.

Myrica nagi Hk.f.

Vernacular name: Sophie Nam in Khasi.

Ripe fruits are deep red, juicy, sweet and sour and are good for the preparation of both fresh juice and pickles . The trees are in fruit during April-May. Fruits are slightly smaller than those of *Myrica esculenta.*

Myrica esculenta Buch.-Ham ex D. Don

Vernacular name: Sophie Bah in Khasi; Bol miseng in Garo.

Evergreen tree up to 12m high with aromatic bark; leaves 5-17cm long, lanceolate; flowers unisexual; fruit 2-3cm long, greenish yellow when ripe; seed-1, wrinkled.

Ripe fruits are greenish-yellow, juicy and sour in taste and are good for the preparation of both fresh juice and pickles . The tree bears fruits profusely from April to May. Fruits are bigger than those of *Myrica nagi.*

Mahonia napaulensis DC.

Varnacular name: Ding-niang-mat in Khasi

Evergreen shrubby tree up to 6m high; leaves compound, 18-45cm long, 7-17 leaflets with spiny margins; flowers 5-6.5mm in long racemes at branch tips, fragrant; fruits 5-10 mm long, ovoid, purple or blue-black.

Trees are in fruit from March-June and are edible.

Nelumbo nucifera Gaertn.

Vernacular name: Podum in Assamese; Soh-lapudong in Khasi.

Perennial aquatic herb with spongy rootstocks; leaves 30-60mm in diameter, circular, held above water long petioles; flowers 12.5-25cm across, white and rose-pink coloured, fragrant, lifted much above water by long stalks; fruit bell or funnel-shaped, 5-10cm across; seeds, 6-12mm long, bony, ovoid or oblong peeping out through the pits at the flat top of the fruit.

The nut-like seeds are eaten raw, roasted or cooked. The delicious kernels, after removing the green embryos, are eaten raw; also dried and used in bakery. The plant bears fruits from October to December.

Protium serratum (Wall. ex Colebr.) Engl.

Vernacular name: Dieng-soh-mir in Khasi; Thikring in Garo; Murtenga in Assamese; Bil in Mizo.

Evergreen tree; leaves compound, imparipinnate, 15-35cm long; 7-11 leaflets, each 7.5-14cm long, opposite, elliptic-oblong, margins serrated or nearly entire, base oblique; flowers 2.5-3mm across, green, in lax clusters; fruit, 1.2-1.9cm in diameter, globose, red when ripe, containing 1-3 one seeded nuts.

The fruit has a pleasantly acid mesocarp and is eaten. Fruiting in May.

Pterigota alata R.Br.

Vernacular name: Dieng-soh-lakor in Khasi; Mi-bol in Garo; Phunber-pui in Mizo; Pahari in Assamese.

Trees up to 45m high; leaves 10-25cm long, cordate, 3-7 nerved; flowers, 1.5 to 2.5cm across, reddish brown; fruits are long stalked pendant woody pods, 7.5-12.5cm across, seeds ± 40, each 2.5cm long, flattened, winged.

Seeds are eaten after roasting. However, raw seeds are deleterious.

Pyrus pashia Buch.-Ham. ex D.Don

Vernacular name: Soh-jhur in Khasi.

Deciduous tree up to 15m high, somewhat spiny; leaves 5-10cm long, ovoid or lanceolate, margins serrated; flowers 2-2.5cm across, white blotched with pink, fragrant; fruit 1.3-3.8cm across, globose, yellowish brown, speckled with pale spots, fleshy, 4-5 seeded.

Ripe fruits are sour-sweet and are relished; sometimes sold in the market. Fruiting from September to May.

Prunus nepalensis (Ser.) Steud.

Vernacular name: Gakiel in Garo.

Ripe fruits are reddish-black and have a great potential for economic exploitation. The Fruit Preservation Centre of the Government of Meghalaya has successfully produced jams, squashes and ready to serve drinks, using the fruit. Trees are in fruit from late August to late October.

Quercus glauca Thunb.

Vernacular name: Dieng-rih in Khasi; Dieng-sari-war in Jaintia.

Evergreen tree up to 25m high, bark warty; leaves 5-17cm long, lanceolate, upper half of margins toothed; flowers unisexual; fruit an acorn, 1.5-1.8cm long, ovoid.

Acorns (nuts) are eaten. Fruiting takes place from August to January.

Roydsia suaveolens Roxb.

Vernacular name: Tylle-saphlit in Jaintia; Maonl in Tripuri.

Rambling woody climber, white dots on branches and leaves; leaves 12.5-30cm × 5-12cm, oblong, thick, green above, pale underneath; flowers 1-1.6cm across, pale yellow or dull white, fragrant, borne in solitary racemes; fruit 3.2cm across, ellipsoid or ovoid to globular, orange when ripe with yellow aromatic sweet pulp, 1-seeded.

Ripe fruits eaten. Fruiting takes place from August to January.

Randia longispina (Roxb.) DC.

Vernacular name: Dieng-makasing-khlaw in Khasi; Thiskeng in Garo; Monraling in Kuki; Guroi in Assamese.

Shrub or small tree with opposite axillary spines; leaves 2.5-8.8cm long, ovoid, wrinkled; flowers 2cm long, white turning yellow, fragrant; fruit 2.5-3.8cm, ovoid, yellow when ripe, aromatic, many seeded.

Ripe fruits eaten roasted or cooked; raw ones used as fish poison.

Rubus ellipticus Smith

Vernacular names: Sohshiah in Khasi; Sohchi-u-sow in Jaintia.

Prickly shrub up to 2.5m high; leaves compound; 3 leaflets, each 3-10cm long, margins serrated; flowers white, 7.5-15mm across, in clusters; fruit ± 1.3cm across, orange yellow when ripe, made up of a mass of fleshy 1-seeded fruitlets on the persistent calyx.

Fruits are sweet and juicy and offer scope for commercial exploitation.

Saurauia napaulensis DC.

Shrubby tree up to 9m high; leaves, 17-35cm × 5-15cm, silvery underneath, margins toothed; flowers 1.2cm across, borne on long drooping panicles; fruit a berry, 1cm long, oval, 4-5 lobed.

The sweet pulp of ripe fruits is eaten. The plant bears fruits from September to January.

Saurauia punduana Wall.

Vernacular names: Dieng-ja-la-ngap and Dieng-soh-khijut in Khasi.

Evergreen shrubby tree up to 7.5cm high, branchlets minutely scaly; leaves 15-30cm x 6-10cm with toothed margins; flowers 1.8cm across, white, changing with maturity to pink, axillary.

Ripe fruit is eaten. Fruits are available from September to November.

Saurauia roxburghii Wall.

Vernacular names: Bon-posola in Assamese; Bor-jir-sinning in Garo; Dieng-soh-lapied in Khasi; Diaching in Naga.

Evergreen shrubby tree up to 9m high; young parts hairy with minute scales; leaves 12-35cm long, elliptic, margins shallow-toothed; flowers 6mm across, arising from the axils of fallen leaves, white, turning pink with maturity; fruit is a berry, 5mm in diameter.

Ripe fruits eaten. Fruits are available from May to August.

Sterculia guttata Roxb.

Vernacular name: Hirikh in Assamese

Evergreen tree; leaves 12.5-25cm long, ovoid or nearly cordate, rusty tomentose beneath; flowers 1.2-1.9cm long, white or pale yellow, spotted with purple with a repulsive smell, borne in panicles; pods 7.5-12.5cm long, hairy, ovoid, red or deep orange, 3-4 seeded.

Seeds are eaten roasted. Fruiting takes place from August to September.

Sterculia indica Merr.

Vernacular name: Nak-chepta in Assamese; Konkelu, Saglepapio in Arunachal.

Evergreen shrub or small tree, bark warty; leaves 10-40cm long, nearly lanceolate; flowers 1.2-2.5cm across, pale greenish-yellow, on pendant panicles; pods, 7.5-12.5cm long, lanceolate, beaked, crimson.

Tender fruits eaten cooked; mature seeds eaten fried or roasted.

Sterculia lanceaefolia Roxb. ex G. Don

Vernacular name: Mimong-omak in Garo; Nag-phena in Assamese.

Middle-sized tree; leaves 7.5-20cm × 2.5-8.7cm, variable in shape and size; flowers 1.2cm across, brick red, in erect racemes; pods, 5-10cm long, lanceolate, beaked; 4-8 seeded.

Seeds are eaten roasted. Fruiting takes place from May to July.

Sterculia versicolor Wall.

Vernacular name: Star-um in Khasi; Khaipang-thing in Manipuri.

Large tree; leaves compound, 7-9 palmately arranged leaflets; leaflets, 12.5-27.5cm long; flowers 5-6mm long, pale yellow, borne in panicles at branch ends; pods, 7.5-8.7cm long, 5-6 together, hairy; seeds 1.5cm long, oblong, black.

Seeds are eaten roasted.

Sterculia villosa Roxb.

Vernacular name: Dieng-star in Khasi; Ubak in Garo; Odela in Assamese; Heppuk-bang in Naga.

Deciduous tree up to 18m high; leaves 25-45cm across, crowded at branch ends, distinctly 5-7 lobed, ventral surface velvety, leafless during flowering; flowers 6-18mm across, yellowish, borne in panicles at branch ends; 2-5 pods together, bright red, 3.8-8.7cm long; seeds 8-10mm long, black.

Seeds are eaten roasted or baked.

Taxus baccata Linn.

Vernacular name: Dieng-soh-bleh (Diengsableh) in Khasi.

Middle-sized evergreen coniferous tree up to 12m high; leaves 2.5-3.8cm long, strap-shaped, aromatic; flowers unisexual, male and female flowers borne separately; fruit is drupe, 8-10mm long, bright red when ripe, 1-seeded.

The fleshy part of the fruit around the seed is sweetish and is eaten; seed is considered poisonous.

REFERENCES

Arora R.K. (1981). Native Food Plants of the Northeastern Tribals. In: Jain SK (ed), Glimpses of Indian Ethnobotany, New Delhi.

Arora R.K. and Hardas M.W. (1977). Jobs-tears (*Coix lacryma - jobi*) – a minor food and fodder crop of Northeastern India. Econ Bot 31(3): 358-366.

Arora R.K. and Hardas M.W. (1977). *Hodgsonia heteroclita* – an oil-rich cucurbit. J Bom nat Hist Soc 74(3): 559-561.

Kanjilal U.N., Kanjilal P.C., Das A., De R.N. and Bor N.L. (1934 - 40). Flora of Assam. Vols 1-5, Govt Press, Shillong.

Hooker J.D. (1872-79). The Flora of British India. Vols. 1-7, London.

Maheshwari P. and Singh U. (1966). Dictionary of Economic Plants of India. New Delhi.

Singh H.B. and Arora R.K. (1972). Raishan (*Digitaria* sp) — a minor millet of the Khasi Hills, India. Econ. Bot. 26: 376-380.

Singh H.B. and Arora R.K. (1973). Soph-long, Moghania vestita, a leguminous root crop of India. Econ Bot 27(3): 332-338.

Singh H.B. and Arora R.K. (1978). Wild edible plants of India. ICAR, New Delhi.

Watt G. (1889-1899). The Dictionary of Economic Products of India. Vols. 1-6. Supdt, Govt Printing, Calcutta.

Chapter 21

Crop Responses to Elevated CO_2

D.C. Uprety[1], Y.P. Abrol[2] and V.R. Reddy[3]

[1] *Division of Plant Physiology, Indian Agricultural Research Institute, New Delhi-110012, India,*

[2] *Department of Botany, Hamdard University, New Delhi-110062, India*

[3] *Crop Systems and Global Change Lab (USDA), Beltsville, MD 20705, (USA)*

INTRODUCTION

Global climate change, considered to be the most serious threat to the environment, has been the center of scientific debate in the recent years. Climate changes, which affect agriculture, are: (1) Changes in land use and land cover (2) World wide decline in biodiversity (3) Changes in atmospheric composition of gases and (4) Changes in climate.

Human activities have led to the build up of greenhouse gases (GHG) including water vapor, carbon dioxide, methane, nitrous oxide, chloroflourocarbon (CFC), hydrofluoro carbon, perflurocarbon and sulphur hexafluoride. The major anthropogenic activities that have contributed to the increased concentrations of GHGs are the burning of fossil fuels, cutting of forests and grasslands for land clearing and use of CFCs as refrigerants in industrial processes and aerosol sprays (Fig. 21.1). These GHGs are very important for human survival on this planet. Their presence in the atmosphere at the right levels ensures that the

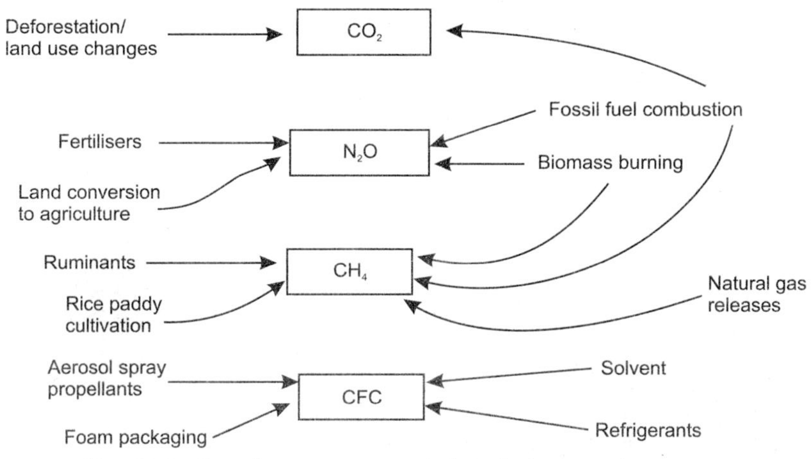

Fig. 21.1. Greenhouse gases and their Anthropogenic sources.

planet is maintained at a temperature of +14°C, warm enough for life to exist. Without the GHG, the mean temperature of the planet will be only –17°C.

The dramatic increase in human population occurred, since pre-industrial revolution (1850) increased the concentration of these gases due to large-scale use of fossil fuels, deforestation, industrialization and urbanization. Prior to the inception of industrial revolution the CO_2 content of the earth's atmosphere had been unchanged for several centuries, at an average of approximately 280μmol mol^{-1}. With the beginning of industrialization the equilibrium was perturbed and CO_2 concentration in the air began to rise subsequent to unprecedented economic growth. Today's crops and natural vegetations are growing at an elevated CO_2 level that has not been experienced by terrestrial vegetation for 26 million years (Pearson and Palmor, 2000). Understanding of plants responses and adaptation strategies to a future increase in CO_2 will help in agroecosystem management and development of plant type for changed climatic condition.

Anthropogenic activities exponentially increased the concentration of atmospheric CO_2 at the rate of 1.8μmol mol^{-1} per year and its level is expected to be 550 μmol mol^{-1} by 2050. According to Long *et al.* (2004) the CO_2 is now higher than it was at any time in the past 36 million years. The fossil fuel adds 5.7Gt whereas deforestation adds 2.3 Gt of CO_2 in the atmosphere per year, accumulating about 8.0 Gt carbon per year with its atmospheric level increasing by 0.5 % per year (Fig. 21.2).

Agriculture provides both sources and sinks of greenhouse gases (GHG). It has been estimated that 20% of the greenhouse effect (radiative forcing) is related to agricultural activities (Cole *et al.*, 1996). The atmospheric concentration of these greenhouse gases has increased since 19th century by 30% for CO_2, 145% for methane and 15% for nitrous oxide (Houghton *et al.* 1996). Per capita CO_2 emission from South Asia was 0.47Tg (1995); whereas its projected value is 0.90Tg for 2050.The total South Asia's regional emission of GHG (in CO_2 equivalent) is 3% of the global total. Agricultural emission is 0.2 tC (tons of carbon) per person from India, 1.2 tC global average per person and 5.0 tC per person for USA (IPCC 2001) (Table 21.1). Lal *et al.* (1998) showed that US agriculture has contributed

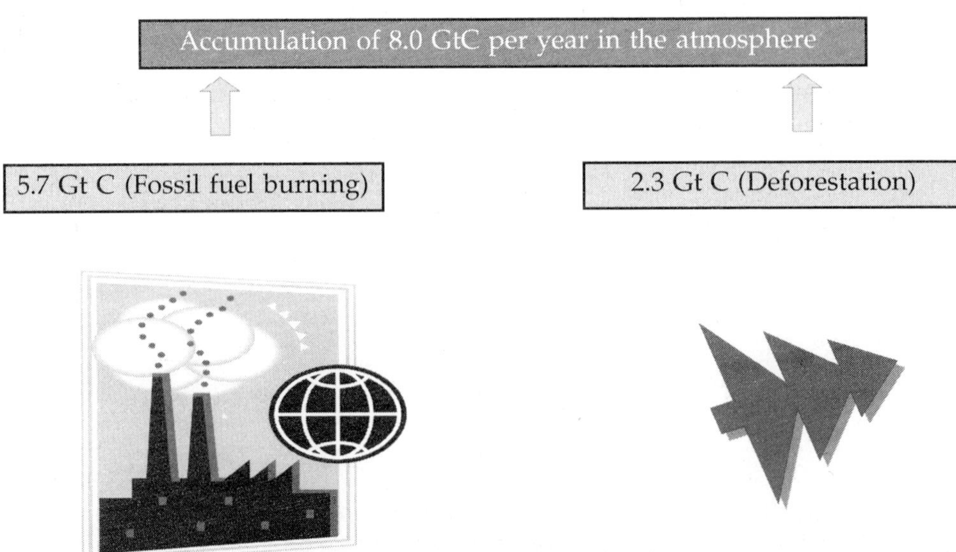

Fig. 21.2. Sources and sinks of carbon.

116 of the total 1596 million metric tones of carbon equivalents (MMTCE) (i.e., 7.3% of US emissions). The global carbon cycle is made up of large carbon reservoirs and its flow is important for agriculture. Estimates of carbon pools indicate the total carbon content in the atmosphere was 750,000 (MMTCE); biomass 550,000 (MMTCE), soil organic matter (1,500,000) and oceans 38,000,000 (MMTCE) (Ocean being the largest pool for carbon) (Eswaran *et al.*, 1993).

Table 21.1. Carbon dioxide emissions in the atmosphere (tons of carbon/person)

1. Agriculture Emission (India)	0.2 TC
2. Global Average of Carbon Emission	2.1 TC
3. Emission of Carbon (USA)	5.0 TC

Agriculture plays an important role in mitigating the greenhouse gas emissions. It brings 32% reduction in CO_2 emissions, 42% of the carbon offsets from biofuel production 16% from reduced CH_4 emissions and 10% from reduced emissions of N_2O. Cole *et al.* (1996) indicated that agriculture may contribute to mitigation of climate change by combining appropriate land with best management practices to increase crop production.

HISTORICAL MEASUREMENT OF CO_2

The measurements of atmospheric CO_2 concentration began at South Pole Antarctica in early 1957 and at Maunaloa (Hawaii) in March 1958 by using the analysis of the air trapped in bubbles in polar ice cores. The ice core studies showed that CO_2 concentration was about 205 μmol mol^{-1} 20,000 years ago. Pre-industrial value was 280μmol mol^{-1} which remained stagnant for 10,000 years. Prior to 1900 the CO_2 concentration in the atmosphere was 290μmol mol^{-1}. Maunaloa studies also observed an increase of 1.0 μmol mol^{-1} CO_2 per year from 1958 to 1982. In 1958, value of atmospheric CO_2 was 316μmol mol^{-1} (Keeling *et al.*, 1995) and in 1995-2000 it was recorded as high as 360μmol mol^{-1} (Collete, 1995) and 372μmol mol^{-1} (Uprety *et al.*, 2004) about 30% more since the beginning of the industrial revolution (Table 21.2 and Fig. 21.3). It is currently increasing at the rate of 1.5 to 1.8 μmol mol^{-1} per annum (Keeling *et al.*, 1995) and would likely to be doubled by the end of 21st century (Houghton, 1996).

Table. 21.2. Atmospheric concentration of carbon dioxide (chronological order).

20,000 years ago	205
1860	274
1900	290
1958	316
1982	340
1990	350
1995	365
2003	371

History of research on plant responses to CO_2 started when deSassure(1804) noted the faster growth in pea plants with an elevated atmospheric CO_2 (8% CO_2). The first description of the CO_2 greenhouse effect was reported by John Tyndall (1861). Swedish Nobel prize winner Swante Arrhinous (1896) and American Geologist T.C. Chamberlin first published an account on greenhouse effect. However, CO_2 fertilization was practised by greenhouse grower over the

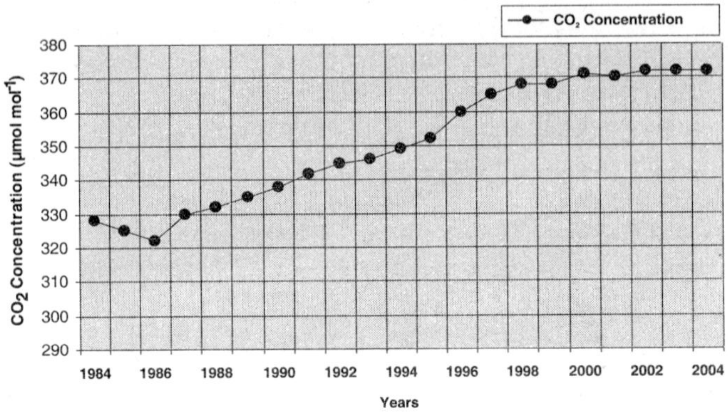

Fig: 21.3. The rise in atmospheric CO_2 at IARI, New Delhi, India, 1984-2004,Winter crop seasons (Uprety *et al*, 2004)

last 200 years using the decomposition of organic matter as a source of CO_2. About 4000 acres of greenhouse crops were being grown with CO_2 enrichment in the Netherlands by 1861. The first CO_2 enrichment experiment in the field was done by Lundegardh (1920-23) on sugarbeet and oats. He recorded 16% increase in sugarbeet root production with 15% increase in CO_2 (from 284 to 328 μmol mol^{-1}), whereas oat production was 30% more for doubling of CO_2 from 282 to 564μmol mol^{-1} (quoted from Wittwer, 1986). Kimball (1983) reviewed 430 observations and reported 33± 6% increase in average yield of various crops due to the doubling of CO_2 above ambient level. Current studies have shown various adaptation strategies in plants to meet the challenge of rising atmospheric CO_2 (Uprety *et al.*, 2004) (Table 21.3).

Table 21.3. Historical observations on CO_2 enrichment in plants (modified version of Wittwer,1986) in chronological order.

Year	*Observations*	*Observers*
1648	Major increase in mass of a willow from the atmosphere	Van Helmont
1804	First observations of CO_2 enrichment on plant growth	de Sassure
1894-1902	Positive effects of CO_2 enrichment on plant growth	de moussy
1902	Negative effects of CO_2 enhancement on plant growth	Brown and Escombe
1918	Positive effects of CO_2 on plant growth	Cummings and Jones
1931	600 nurseries reported using CO_2 in Germany	Reinau
1959	Basic studies on CO_2 and light responses in plants	Gaastra
1961	Dutch growers add CO_2 for improving yields of 400 acres lettuce	Anon
1962	Response of cucumber reported to the complimentary effects of CO_2 and light	Hopes & Ries
1962-1966	Response of flower crops reported	Goldberry and Holley
1964	Comprehensive studies on tomato and cucumber	Wittwer and Robh
1976	Positive effects noted for the growth of tree seedlings	Hannover *et al.*
1987	Reduction in transpirational cooling in leaves of cotton	Idso *et al.*
1983-1986	Observed 31 % increase in agricultural plant productivity	Kimball
1983-1987	64% increase in cotton lint and seed yield at 650 ppm CO_2	Klmball *et al.*
1995-2004	South Asian studies demonstrated amelioration of adverse moisture stress effect in *Brassica* plants. Demonstrated the transfer of CO_2 responsive characters from the parent *B.campestris* to hybrid *B.oxycamp.*	Uprety *et al.*

The increase in the concentration of CO_2 in the atmosphere cause global warming by absorbing long wave heat radiations from earth's surface and may also affect cloudiness and precipitation. This is referred as Climatic effect. However, according to general circulation models (GCM) temperature changes is not uniform. The temperature rise in polar region may range from 4 to 5° C and near equator less than 1° C. The cumulative value for surface temperature change without feedback for the period 1850-1980 is 0.56° C. CO_2 increase contributed 0.40° C and other trace gases contributed the balance of 0.16° C. Several theoretical models have predicted that the mean global temperature will increase 3 to 4° C, with doubling of CO_2 (IPCC 2001). Long term temperature records have concluded that global temperature has increased about 0.5° C over the last 100 years (Hurrell, 1998).

Plants have been directly affected by rising atmospheric CO_2 concentration because it is the first molecular link between atmosphere and biosphere. It serves as substrate in the photosynthetic carbon assimilation. There is concomitant decline in photorespiratory activity and alteration in stomatal activity (Idso, 1995). C_3 plants (wheat, rice, oilseed, pulses, etc.) respond to elevated CO_2, since it reduces the oxygenase activity of RuBP carboxylase oxygenase enzymes in plants. C_4 plants (sorghum, maize, sugarcane, etc.) show little or no photosynthetic response to elevated CO_2 because C_4 pathway is not competitively inhibited by O_2 and is completely CO_2 saturated. The differing responses of C_3 and C_4 species to CO_2 will likely to change their competitiveness and C_3 weeds are likely to become problem in C_4 crops while C_3 crops may get advantage over C_4 weeds. This could affect the complexion of agroecosystems. The natural vegetation is also expected to show significant response to rising CO_2, since 95% of the species of higher plants and 100% of the trees are C_3 plants (Drake *et al.*, 1992). C_4 system represented an evolutionary adaptation to low CO_2 environment (Hall and Allen, 1993) and has lower CO_2 compensation point. Plants with C_4 photosynthetic pathway have negligible photosynthetic response to elevated CO_2 because the C_4 cycle increase the CO_2 concentration in bundle sheath cells to the point where very little photorespiration occurs and the Calvin cycle is nearly saturated with CO_2 (Conroy; 1992; Ziska *et al.*, 1999). However, there is no consensus on the quantitative effects of increased CO_2 on plant physiology and growth due to differences in response at different stages of growth, species of crops and growth limiting environmental factors.

TECHNOLOGIES FOR MEASURING EFFECTS OF CO_2

Most of the research on plant responses to high CO_2 have been conducted in the past under laboratory, greenhouse or controlled field condition (Bazzaz and Garbutt, 1988). With the development of temperature control in greenhouse, the studies on interactive effect of CO_2 and temperature on plants became possible (Mitchell *et al.*, 1995). Agricultural scientists have used controlled environments and field based facilities to study the crop responses to individual factors such as radiation, temperature, humidity and water availability. Technologies, used earlier, were based on single component approach and have been successfully applied to study the responses of plants to increased CO_2. Although this single component approach is the highly effective way of discovering how one atmospheric factor influences crops, in practice several interactions will affect the responses. Results obtained with isolated plants growing under controlled (mostly optimal) growth conditions, life phase synchrony, absence of organismic interactions and provision of luxurious amounts of mineral nutrients appear to be the major limitations for the application of results of a large number of CO_2 enrichment studies. Even in single species experiments plant to plant interactions may have an effect. Thus, population should become the common experimental unit rather than isolated individuals.

Innovative approaches have been developed for conducting long term experiments to investigate the impact of rising atmospheric CO_2 on crops. Several systems are used to study the responses of plant communities to elevated CO_2. Before 1980, most studies on the effects of elevated CO_2 were carried out in controlled environmental facilities including leaf cuvettes, whole plant growth chambers and greenhouses. There are extensive precisely controlled closed systems that continuously recondition and recirculate the air in controlled environment of growth cabinets, phytotrons, and open field exposure systems such as Free Air CO_2 Enrichment (FACE), which is moderately more expensive and less precise than closed systems (Allen *et al.*, 1992). Each of these research systems has its advantages and disadvantages. New methods and facilities have been devised and field exposure techniques that do not modify microclimate are being developed and improved for exposure of large areas of plants to elevated CO_2 and other trace gases. Though the cost and difficulties of studying the CO_2 effects in the field are considerable but such research must be done to give confidence in the conclusions arising from the simpler controlled environmental studies. During the current decade, there have been new approaches to CO_2 enrichment technology under field condition, which have potentially reduced the costs. Further, the computer models which make predictions for the field conditions should at least partially be based on the data collected at the field (Lawlor and Mitchell, 1991).

Some of these CO_2 enrichment technologies are described below:

(a) Leaf Cuvettes

They are designed for single leaf gas exchange measurements. These can be used for measuring gas exchange as a part of system for studying the short time effect of elevated CO_2 levels on the carbon dioxide exchange processes in leaves. Allen *et al.* (1996) used this system for studying short-term responses of soybean leaves to elevated CO_2.

(b) Sunlit controlled environment chambers

These chambers usually had Mylar polyester film walls. Non-dispersive IRGA allowed rapid measurement of CO_2 concentration, and these analyzers in combination with metered CO_2 allowed direct and continuous measurement of photosynthetic CO_2 exchange rate. Sunlit controlled environment chambers were used successfully for measuring photosynthesis and transpiration as a function of CO_2 concentration, light, temperature and soil moisture conditions.

(c) Soil-plant-atmosphere research (SPAR)

SPAR chambers were designed to provide accurate, flexible control of dry bulb temperature, CO_2 concentration and humidity of the canopy air, as well as control and measurement of soil water and root conditions. Jones *et al.* (1984) designed the SPAR system units. SPAR system consists of (1) Sensors for measuring temperature (Copper-constantan thermocouple), (2) Infrared gas analyzer (IRGA) for sensing CO_2, and (3) Dew point hygrometers for measuring humidity. Feedback mechanisms such as thermostats on loops in computer logic that compares sensed with desired conditions. Control devices such as heaters that were regulated to produce the desired temperature treatment conditions, CO_2 injection valves for replacing CO_2 as it was used in photosynthetic and cooling coil flow for regulating dew point temperature.

Air in these SPAR chambers was circulated through the canopy from top to bottom and than goes out through ducts, where the air was reconditioned before flown back to the canopy chamber. Sensors, air sampling ports and control devices were located within the ducts so that the air circulated to the top of the canopy had the experimentally prescribed set point of temperature, CO_2 concentration and humidity level. Changes in CO_2 and humidity levels within the chamber were driven by canopy CO_2 and H_2O gas exchange process. Thus, except during occasional operational problems such as excessive condensation in sampling lines/on chamber walls (during cold winter events) or, failure of air sampling pumps, successful control actions provided a mirror image of canopy transpiration.

Advantages of the SPAR system include (A) High light, similar to natural irradiance, (B) Variable condition, (C) Provides continuous integrated measurements of CO_2 and H_2O balance, (D) Root zone similar to field, and (E) Control to specified set points or, track the ambient environment, temperature and humidity.

Disadvantages of this system include (A) Complex control, chamber effects (humidity, wind gradients), (B) Limited replications, and (C) Expensive cost per unit.

(d) Open Top Chamber

Open top chamber (OTC), as described by Heagle *et al.* (1973), were extensively used as plant exposure units both in air pollution and in CO_2 response studies in the field. These systems have been used to expose potted plants, annual crops and trees to a variety of aerial pollutants and CO_2. They are currently in use at a number of laboratories in North America, Europe and South Asian countries.

Hardy and Havelka (1975) first used a square walled open top enclosure to expose soybeans to the atmosphere enriched with CO_2 for the purpose of studying the effect of increased photosynthate production on symbiotic nitrogen fixation. Rogers *et al.* (1983) adapted the basic cylindrical open top chamber system to generate large scale CO_2 test atmospheres in the field. Nakayama and Kimball (1988) used a square wall open top chamber with 0.2 m diameter perforated polythene rows in cotton. Drake *et al.* (1989) developed an open top chamber system for CO_2 enrichment of salt marsh vegetation that re-circulated part of input air. Uprety *et al.* (1995) designed and developed open top chambers for South Asian climatic conditions. These chambers with frustum are being used by South Asian CO_2 research network for multi-country multi-disciplinary CO_2 crop response studies.

Uprety (1998) described the South Asian OTC technology with five sub systems (1) Supply of pure and high concentration of CO_2, (2) System of valves, regulators and flow meters, (3) Chambers to be CO_2 controlled, (4) Appropriate gas analyzer with feedback control, and (5) Computed data acquisition and programming (Fig. 21.4). These open top chambers were made of an aluminum

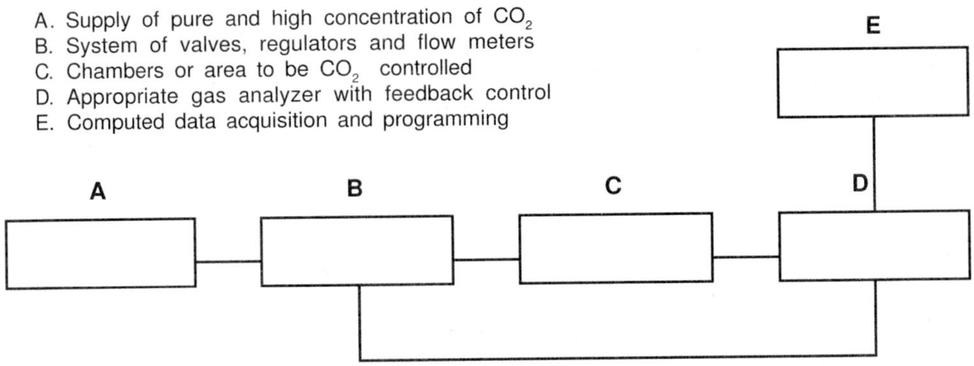

Fig. 21.4. Basic scheme of measurements and control system for CO_2 enhancement

frame installed on the ground. Each chamber covered a circular area of 3m diameter. The vertical sidewall was also 3m in height and top of chamber was kept open to provide the near natural conditions. These chambers had a characteristic slender shape and were covered with a 0.15 mm thick transparent polyvinyl chloride (PVC) sheet. Chambers equipped with a frustum at the top to deflect air and prevent dilution of the desired CO_2 concentrations within the chamber. There was a cylindrical double walled plenum around the base for uniform CO_2 circulation. The inner side of the plenum was having numerous gas outlets of different sizes, small being nearest to the gas inlet pipe. Commercial CO_2 was added in the upstream of the frame to ensure adequate mixing and to create the desired concentration. Blowers were used to distribute CO_2 enriched air uniformly throughout the plant canopy. The CO_2 gas was supplied to the chamber from gas cylinders using a manifold, gas regulators and pressure gauze pipelines. Gas cylinders were fitted in a row with thick copper outlet pipe. CO_2 gas was released from the cylinders to the chamber through a manifold and under ground pipe and injected into the chamber along with the ambient air using a blower. The air blowers (12 inch diameter) are used for maintaining desired concentration of CO_2 into the chamber tunnel. The blower also helped in maintaining the inside air temperature closure to that of outside ambient atmosphere. The relative photosynthetic photon flux density (PPFD) in the chambers was maintained about 95% of full sunlight by frequent and gentle washing of the transparent polythene sheet. The daily observations on temperature and relative humidity were monitored in the chambers and recorded in thermo hygrograph (Lawrence and Mayo, India model W/1712/Z). Weekly changes in chart papers of thermo-hygrograph are done during the cropping seasons. The diurnal variations in temperature, light intensity and CO_2 concentration in side and out side the chambers are recorded regularly.

The concentration of CO_2 in OTC was regulated by three step operation: (A) Standardization of the gas release from the source to the manifold with the help of pressure gauze, (B) Opening two successive finer metered valves for controlled release of gas from manifold to the under ground pipe, (C) Adjusting the solenoid valve at the end of pipe with the help of pressure meter, and (D) The monitoring of CO_2 concentration was done by IRGA by drawing air from different positions of crop canopy (Uprety, 1998).

There are several research needs to improve the open top chamber as a field experimental tool. Fluid dynamics and heat exchange properties, particularly the effects of turbulence and altered flow path of wind within the canopy, should be studied to determine potential effects of increased CO_2 on plant growth in the field.

Open top chamber is one of the most widely used techniques for studying elevated CO_2 impact on crops and ecosystems. Climate in OTC tracks the dynamic change in temperature, light, rainfall as experienced in unclosed area without costly and complex environmental control characteristics of closed chambers. OTC subjects the whole plant to the treatment without restricting the root growth. The principal advantage of OTC is the ability to provide control over the atmospheric variations to which plants are exposed while maintaining more or less the same soil conditions of the field setting.

(e) Free Air CO_2 Enrichment Technology (FACE);

FACE system provides a unique opportunity to examine the effects on growth at elevated CO_2 under other environmental conditions without any direct perturbation of microclimate (Hendry and Kimball, 1994). The development of FACE program had following objectives: (A) To maintain sufficiently steady CO_2 concentration so that fluctuations in CO_2 concentration cause no perceptible effect on plants grown within the environment, (B) To provide reliable field data on the response of crops and other plant communities to develop and evaluate mechanistic plant growth models which can contribute to predict the effects of future CO_2 concentration and climate changes on plant growth and water use, and (C) To contribute for an evaluation of global CO_2 fluxes and carbon balances between vegetation and rising atmospheric CO_2.

Maricopa FACE system had arrays of 32 vertical vent pipes, with each vent pipe containing multiple gas emitter ports. Each vent pipe in an array is connected to a common 22 m diameter torridly distribution plenum through an individually controlled valve. High volume blower draws ambient air through 60x60x5 cm fiber filter packs to remove coarse particles and inject pure CO_2 to mix with air by placing the outlet from CO_2 flow controller directly in front of the blower. The amount of CO_2 metered in to air stream entering the plenum is based on wind speed and gas contact sampled at the center of the array. The proportional integrative differential (PID) adjust the amount of gas into the plenum and another algorithm considers both wind direction and speed and controls the vent pipes that emit CO_2 enriched air. The CO_2 gas is stored in liquid state at a pressure of 1725 k pa. The outlet of the flow controller is positioned just in front of blower, which draws air at constant rate and CO_2 concentration varies in direct demand for CO_2. The wind direction and wind speed controls CO_2 concentration by measuring them by Climatronics model 101871 and cup anemometer. A non-dispersive infrared CO_2 analyzer (NDIR) is kept adjacent each array for measuring gas concentration in each plot. Air is sampled from center of the array. The height of sampling port is 5 to 25 cm above the canopy.

The south Asian FACE (Uprety *et al.*, 2000) is also based on the principle of injecting additional CO_2 gas in open field suitably so as to attain a predetermined elevated level of gas concentration with uniform distribution inside the FACE ring under varying metrological conditions of wind, temperature and humidity. The main components of this facility are (1) CO_2 gas storage and distribution system, (2) The FACE ring (Plenum), through which CO_2 enriched air is injected into the open field, (3) Different sensors and actuators used for monitoring the environment and controlling the operations, and (4) An electronic system for controlling these operations.

The system design of the Mid-FACE ring was octagonal in shape having a diameter of 8 m as shown in block schematic diagram. The eight arms of the FACE ring were made of rigid PVC pipes of 200 mm diameter and wall thickness of 4.5 mm. Each arm of the octagon was fitted with a centrifugal air blower at one end for blowing air into the pipe and was closed at the other end. Four rows of holes (4 mm diameter) were drilled on each arm for the outlet of CO_2 enriched air. The CO_2 was injected at the input of each arm blower for pre-mixing with air within the pipe before its injection in the fields. The air blower pressure and holes on the pipe were arranged in such a way that the air outlets from the holes along the whole length of pipe remained at uniform pressure. All the eight arms of the octagon had independent supply of CO_2 controlled by 8 on/off valves. A common computer controlled proportional integral differential (PID) valve controlled the quantity of CO_2 to be released into the arms.

The CO_2 was released at a time into the field from 3 adjacent arms of the FACE ring. The CO_2 concentration was measured at the center of the array with the help of an IRGA. Sampling of gas was done continuously through a thin polythene tube (4.3 mm ID) with the inlet height adjusted at about 5-25 cm above the crop canopy. The output from the IRGA was in DC voltage of 0.5 V corresponding to 0-2000 μmol mol^{-1} of CO_2. The sensors for wind speed (anemometer), wind direction (wind vane) and temperature were located near the center of the FACE ring at plant canopy height. Their output signals were continuously measured in real time by the system controller. The output from the wind speed and wind vane sensors was a DC voltage in the range of 0-5 V corresponding to wind velocity of 0-33 m/s and wind direction of 0-359^0, with reference to the north, respectively. The PID valve controlled the quantity of CO_2 to be released in to plenum arms, which depends on the control voltage, applied to it (0-10 V). To evaluate the proper operation of the PID valve, a feedback signal was also monitored to know the actual status of the valve. This signal was in the range of 0-5 V corresponding to the fully closed and fully open conditions of the PID valve.

The fumigation of CO_2 gas into the FACE ring from the plenum was made at the crop canopy height. The height of the plenum could be adjusted from time to time to match with the height of the crop canopy by resting the plenum on height adjustable stands. Plenum could be raised up to 5 feet from the ground level in steps of 2 inches. Two banks of 25 cylinders in parallel were used for CO_2 storage and dissemination under constant pressure. A buffer tank was installed after the cylinder bank, which took care of the pressure variations in supply line due to decreasing level of CO_2 in the cylinders during Mid - FACE operation. Apart from maintaining the CO_2 concentration at the target value, the system controller was also used to record all the data, make computations and display the data in suitable tabular and graphic format. The data were recorded in the forms of files.

The data from the meteorological sensors (wind speed and wind direction) along with the CO_2 concentration were measured in every second by the control system. Based on these inputs, the control system, through a specially developed PID algorithm, controlled the flow rate of CO_2 gas in the plenum with the help of PID controller valve. The direction of wind decides which on/off valve was to be opened to release CO_2. In case, the wind velocity was less than 0.5 m/s (which is taken as the case of calm air), the valves were opened in a cyclic order as no dispersion of CO_2 occurs in such conditions. Thus a predetermined constant level of enriched CO_2 was maintained at the center of the plenum. The South Asian FACE is expected to include temperature enrichment device in future. (Figs. 21.5, 21.6 and 21.7).

EFFECTS OF ELEVATED CO_2 ON CROPS

Crops

The major change in the earth's atmosphere is the exponential rise in the concentration of CO_2, which has direct effect on the crops of major economic consequence. Although many reviews have been published on the responses of natural vegetations to the elevated CO_2, it must however be emphasized that most of the findings on the physiological response to CO_2 were first discovered

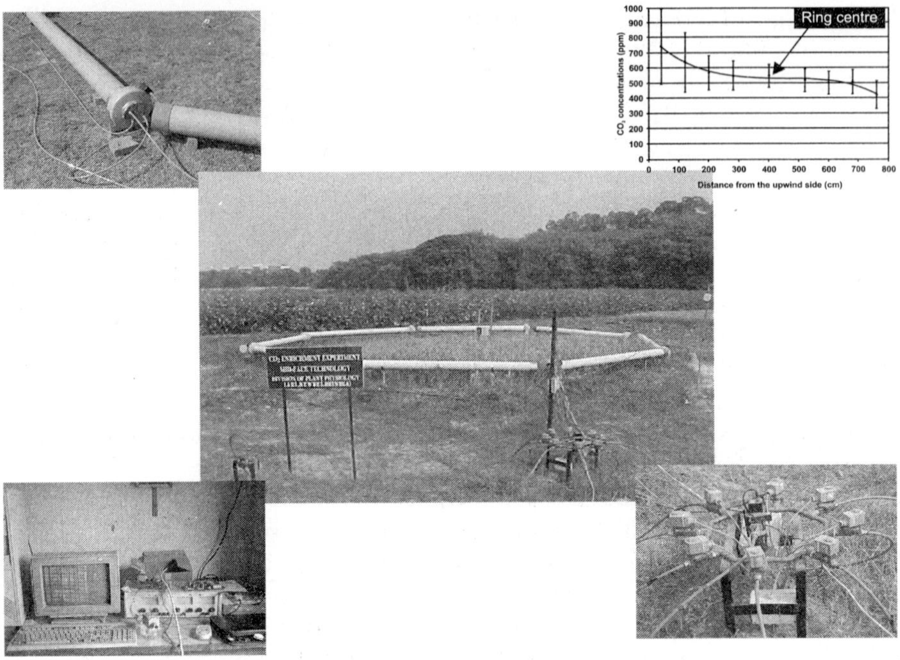

Fig. 21.5. Free air CO_2 enrichment technology (FACE)

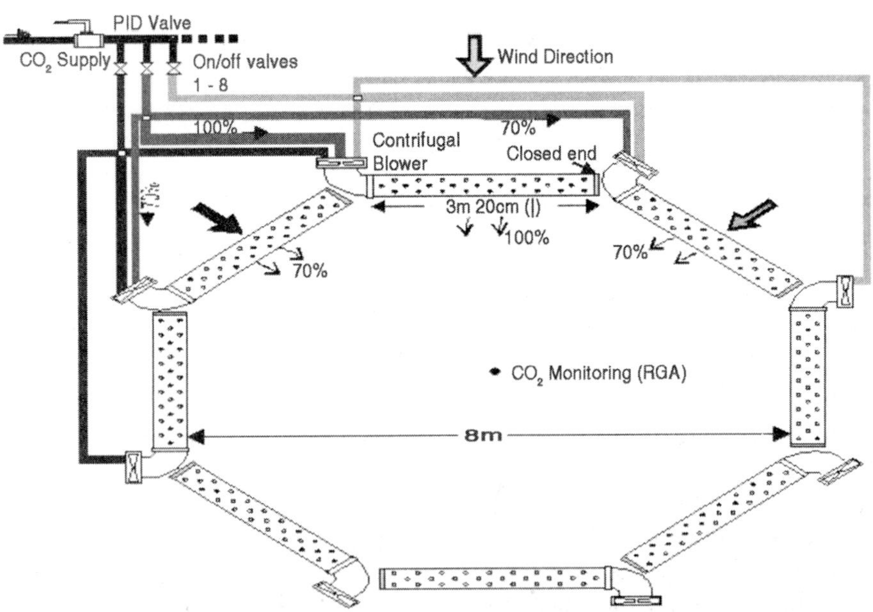

Fig. 21.6. Structural diagram of FACE ring of South Asia

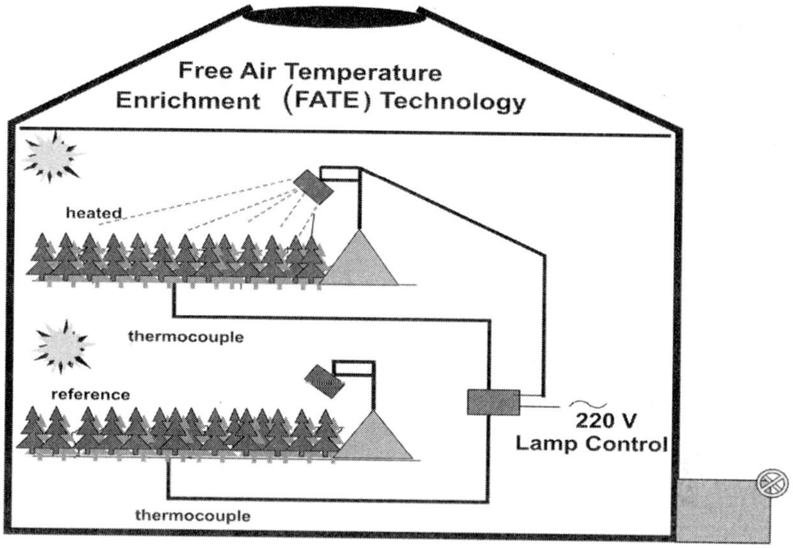

Fig. 21.7. Temperature regulating system in Open top chambers and FACE (Proposed Uprety, 2004)

in agricultural crops. Thus it was considered important to review various studies made on the responses of crop plants to the rising atmospheric CO_2 so that plant types could be developed for changed climatic conditions on the basis of realistic biological data obtained using facilities close to near natural conditions. The structural modeling for crops to influence farmer's decisions and spatial analogue model to exploit agricultural production based on experimental data will help in altering cultural practices, engineering techniques, and crop nutrient and cultivation management for future climatic conditions. The importance of developing cultivars to the changed climatic conditions will increase the use of both transgenic and conventional breeding procedures.

CEREALS

Wheat

The major concerns of studies of the impacts of global environmental changes on wheat include identifying the effects on growth and yield due to changes in the CO_2 and temperature. Watson *et al.* (1996) emphasized considerable uncertainty in estimating impacts of doubling CO_2 and increasing temperatures by 4° C. Modeling studies predict that wheat will be grown at higher latitudes and the production in region closure to the equator will decrease (Rosenzweig and Parry, 1994). Rosenzweig and Parry's study of the impacts of 555 µmol CO_2 mol^{-1} and approximately 4°C, increase in temperature by 2060, simulated global agricultural production of major cereal crops declined, despite assuming farm-level adaptation and future technological improvements. However, according to Leemans and Solomon (1993) the 2°C warming might improve yield by 12%. Methods for studying the responses of wheat cultivars to CO_2 and temperature have been discussed (Morison and Lawlor, 1999). Hendry and Kimball (1994) studied the agronomical changes in wheat cultivars using FACE technology under field conditions. However, the major limitation was the current inability to test the interaction of CO_2 and temperature. Elevated CO_2 increases the rate of photosynthesis and decreases the photorespiration in C_3 plants like wheat. The enzyme responsible for the reaction of CO_2 with ribulose bisphosphate is ribulose bisphosphate carboxylase-oxygenase (Rubisco). The reaction is not saturated at current CO_2; also oxygen reacts with RUBP, producing phosphoglycolate, which is metabolized, ultimately releasing CO_2 in photorespiration. Doubling CO_2 saturates Rubisco and greatly decreases photorespiration, thus increasing net photosynthesis and the rate of carbon assimilates supply to the plant (Long, 1991). Although elevated CO_2 decreases the stomatal conductance with potentially large reductions in water loss, the increased atmospheric CO_2 is more than sufficient to compensate so that the internal CO_2 concentration, is greatly increased.

Bloom *et al.* (2002) indicated that short term exposure to elevated CO_2 concentration diverted photosynthetic reductant from NO_3 or NO_2 reduction to CO_2 fixation, whereas with longer exposures to elevated CO_2, wheat leaves showed a diminished capacity for NO_3 photoassimilation at any CO_2 concentration. Moreover, high bicarbonate levels impeded NO_2 translocation into chloroplast isolated from wheat and pea leaves. Thus, elevated CO_2 inhibits NO_3 photoassimilation and when wheat plants received NO_3 rather than NH_4 as a nitrogen source, CO_2 enhancement of shoot growth halved and CO_2 inhibition of shoot protein doubled. This study will have implications for the ability of wheat to use NO_3 as a nitrogen source under elevated CO_2.

A range of different responses to elevated CO_2 were observed in wheat. Under controlled environmental condition reduction in photosynthetic capacity with decrease in the amount and activation states of Rubisco in individual wheat plants were observed (McKee and Woodward, 1994). However, Delgado *et al.* (1994) and Garcia *et al.* (1998) did not find any acclimation due to growth and observed stimulation of photosynthesis under FACE conditions. Mitchell *et al.* (1999) did not find any increase in photosynthesis in the flag leaf of a European spring wheat during grain filling with less applied nitrogen. Increased photosynthetic capacity in response to elevated CO_2 have been observed in young plants grown with warm temperature and unlimited N supply (Habash *et al.*, 1995). Such conditions stimulate extra root growth and N uptake. There is little evidence of down regulation of photosynthetic capacity at high light (Delgado *et al.*, 1994, Lawlor *et al.*, 1995, Nie *et al.*, 1995 and Theabald *et al.*, 1998). McKee and Woodward (1994) explained the down regulation of Rubisco to occur when it is in excess compared with some other limitations on net photosynthesis. Changes in amounts of photosynthetic components, and Rubisco in particular are responsible to the reduction in photosynthesis. Morison and Lawlor (1999) explained the observed effect of elevated CO_2 on photosynthetic capacity in wheat in terms of changes in the sink-source balance. Theobald *et al.* (1998) however, considered decrease in Rubisco as a CO_2 specific response. van Oosten and Besford (1996) consider the changes in photosynthetic capacity was the consequence of the feedback regulation caused by increased carbohydrates on gene expression. Nie *et al.* (1995) observed a poor correlation of the mRNA transcripts of various photosynthetic components such as large and small subunits of Rubisco with the concentration of carbohydrates, which were different between the CO_2 treatments in a FACE experiment. Thus, the sugar repression model of photosynthetic acclimation does not operate statically in wheat grown in elevated CO_2. Prasad *et al.* (2003) demonstrated that the regulation of Rubisco often leads to upregulation of carbohydrate synthesizing enzymes such as SPS and ADPG.

The stomatal component of photosynthesis is directly affected by elevated CO_2. Samarakoon *et al.* (1995) observed 30–40% reduction in stomatal conductance in wheat by doubling CO_2. Estiarte *et al.* (1994) attributed it to changes in aperture since stomatal density (stomata/unit leaf area) and stomatal index (ratio of stomata/leaf epidermal cell) were unaffected within a single season. There is no evidence of acclimation of stomatal conductance to elevated CO_2 both in phytotrons as well as FACE studies in wheat (Wall *et al.*, 1997).

The characterstic decrease in respiration was often observed in plants grown in elevated CO_2 (Poorter *et al.*, 1992). The decreased respiration is probably due to the decreased proportion of metabolic components per unit dry matter as carbohydrates accumulate (Drake *et al.*, 1997). However, this can not explain the reduction in the respiration rate due to elevated CO_2 of young wheat seedlings in darkness (Natr *et al.*, 1996). Gifford (1995) observed a constant respiration: photosynthesis ratio regardless of temperature and CO_2.

It is difficult to separate the direct and indirect effects of CO_2 on crop development. Slightly faster rates of leaf, ear, spike and tiller development are frequently observed in wheat grown at elevated CO_2 and are probably related to increased tissue or canopy temperature due to partial closure of stomata. The complete absence of an effect of elevated CO_2 on wheat phenology in conditions of low radiations also supports this conclusion (Wheeler *et al.*, 1996). In the FACE experiments, N fertilizer altered crop architecture but elevated CO_2 had a minimal effect on the rates of overall canopy development and senescence (Wall *et al.*, 1997). This was in contrast to accelerated development seen in earlier studies when temperatures differed (Garcia *et al.*, 1998). Kimball *et al.* (2002) observed that crop such as wheat with modest growth response to elevated CO_2 with ample water (e.g.+ 8 %) had a water saving of 7 % compared to + 44 % response in cotton with near zero water saving.

Effect of CO_2 and temperature interaction on wheat biomass and grain yield has been studied in several experiments in UK (Mitchell *et al.*,1993, 1995). Early field studies indicated responses ranging from 30% increase in yield for a doubling of CO_2. Similarly, doubling CO_2 caused 15-27% increase in biomass in experiments simulating the field temperature (Batts *et al.*, 1998) and a 6-34% increase when CO_2 was doubled in temperature gradients tunnel. Rawson (1995) obtained increase ranging 7-36% depending on temperature. Kimball *et al.* (1997) reported 10 to 15% increase in biomass in 2 years with increase of a 200μmol mol^{-1} CO_2 under the FACE condition. Tiller production and survival and root growth are stimulated most, but leaf size, grains per ear and grain size can all be increased depending on environment and cultivar. Grain yield is more sensitive to biomass production during the reproductive growth. It was also seen that increase in biomass and yield was more in chamber studies than FACE.

The effect of elevated CO_2 on stomatal conductance is consistent. Under water limiting conditions in the FACE studies, a 200µmol mol^{-1} increase in CO_2 reduced evapotranspiration by 5% under irrigated condition, whereas with drought it increased by 3% (Hunsaker *et al.*, 1997). This may have resulted from the larger root system exploiting more water, greater leaf area and slightly greater temperatures. Despite relatively small effect of CO_2 on evapotranspiration, elevated CO_2 increased biomass substantially and so increased water use efficiency (WUE) by 45% under drought condition and 21% in wet condition in FACE. WUE was increased by about 64% by doubling CO_2 under drought (Samarakoon *et al.*, 1995). The relative increase in biomass due to doubled CO_2 was much greater under dry than wet condition. In the FACE studies, a greater effect of CO_2 enrichment on biomass occurred under drought than under irrigated conditions, which was reflected in a 20% increase in grain yield in drought and a 10% increase in irrigated plots.

Rao *et a*l. (1995), while studying the effect of high CO_2 on the ozone induced oxidative damage in wheat plants observed that O_3 exposure under ambient CO_2 decreased photosynthetic pigments, soluble proteins and Rubisco protein and enhanced oxidative damage to proteins, but these effects were not observed in plants exposed to O_3 under high CO_2.The higher growth under elevated CO_2 and O_3 led to the synthesis of one new isoform of glutathione reductase. This could explain why plants grown under a combination of high CO_2 and O_3 are capable of resisting O_3^- induced damage to growth and proteins.

Elevated CO_2 tends to increase mass per grain and decrease percentage nitrogen because of the increased supply of carbohydrate from photosynthesis either during grain fill or from reserves (Mitchell *et al.*, 1993 and Rogers *et al.*, 1998). Elevated CO_2 does not greatly alter the composition of carbohydrates, the types of sugars or starch or sugar: starch ratio (Rogers *et al.*, 1998). It also does not affect protein composition (Shewry *et al.*, 1994). Complex changes in lipid composition have been identified (Williams *et al.*, 1995). The non-polar neutral lipids mono and di-galactosyldiacylglycerol increased where temperature was 4°C above ambient and nitrogen was deficient at elevated CO_2 but decreased in ambient temperature conditions. With elevated CO_2 and temperature only starch lipids were affected. Fatty acid composition was affected by warmer conditions, which decreased oleate but increased linoleate of non-polar lipid. Warmer temperature, however, substantially decreased mass per grain and starch content, due to fewer and smaller type starch granules and fewer amyloplast per endosperm. Also, total and lipid free amylose increased, but amylolipid and lysophospholipids were not affected by high temperature. Given the dynamic nature of the processes and conditions, there is considerable scope for variation in final grain composition. Such alterations may affect the nutritional quality and industrial uses of wheat grain. It is surprising; therefore, that more attention has not been given in assessing the effect of global climate changes on wheat grain quality and needs attention.

Rice

Several studies predict positive and negative effects of global climate changes on rice production (Horie *et al.*, 1995; Singh and Padilla, 1995). However, these modeling studies point to the need for further research in order to test and improve our understanding of environmental effects on rice development, growth and yield. More specifically, research is needed to quantify further interactive effects of CO_2, air temperature and other environmental variables on rice growth and yield formation processes.

Mostly the experimental systems used to investigate rice responses to CO_2 include closed chamber systems (Morison and Gifford, 1984; Imai *et al.*, 1985; Baker *et al.*, 1990), open top chambers (Moya *et al.*, 1997; Uprety *et al.*, 2000a) and temperature gradient chambers (TGCs) (Horie *et al.*, 1995). Recently, free air CO_2 enrichment system was used to study rice responses to CO_2, (Kobayashi, 1999, Uprety *et al.*, 2000b).

Rice being a C_3 plant responds significantly to increased CO_2. Several long and short term CO_2 enrichment studies show that doubling the current ambient CO_2 increased leaf photosynthesis rate by 30 to 70%. The magnitude of this increase depends on the particular rice cultivar, growth stage and environment (Morrison and Gifford 1983, Akita 1980). Ziska *et al.* (1997) observed that the effect of CO_2 enrichment on rice leaf area was small, the canopy photosynthetic response to CO_2 enrichment can be mainly attributed to responses at unit leaf area level. Lin *et al.* (1997) and Nakagawa *et al.* (1997) found that higher temperature stimulated single leaf photosynthesis of rice subjected to long term CO_2 treatments during the vegetative stages. However, Baker and Allen (1993) reported that rice canopy photosynthesis was relatively unaffected by a wide range of temperatures. Since one of the major effects of elevated CO_2 on net photosynthesis is through the suppression of photorespiration, it could be expected that optimum temperature for photosynthesis shifts upward as CO_2 increases. This type of interaction of temperature and elevated CO_2 has been explained by Long (1991) for some C_3 plants. Uprety *et al.* (2003) in a study on the effect of elevated CO_2 on rice leaf protein profile using electrophoresis, demonstrated depression in the bands of low molecular mass proteins including larger and smaller subunits of Rubisco and this reduction has not affected the CO_2 induced enhancement in the photosynthetic rate of rice cultivars. The decreased level of low molecular mass proteins may have been resulted from the accumulation of sugars. However, the increase in higher molecular mass proteins observed in this study was found parallel to the net photosynthesis in rice cultivars at elevated CO_2 (Fig. 21.8). Conversely very small interactive effect of temperature and CO_2 on photosynthesis occurred at canopy level. This is consistent with the temperature effects on rice responses to CO_2 in term of biomass production under field conditions (Baker *et al.*, 1992; Nakagawa *et al.*, 2000 and Uprety *et al.*, 2000). CO_2 enrichment causes partial stomatal closure, increased stomatal resistance (Uprety *et al.*, 2000 b and 2002), reduced leaf and whole canopy temperature (Baker and Allen, 1993). Vu *et al.* (1997) reported that leaf photosynthetic rates of rice increased by CO_2 enrichment, but decreased by supraoptimal temperatures.

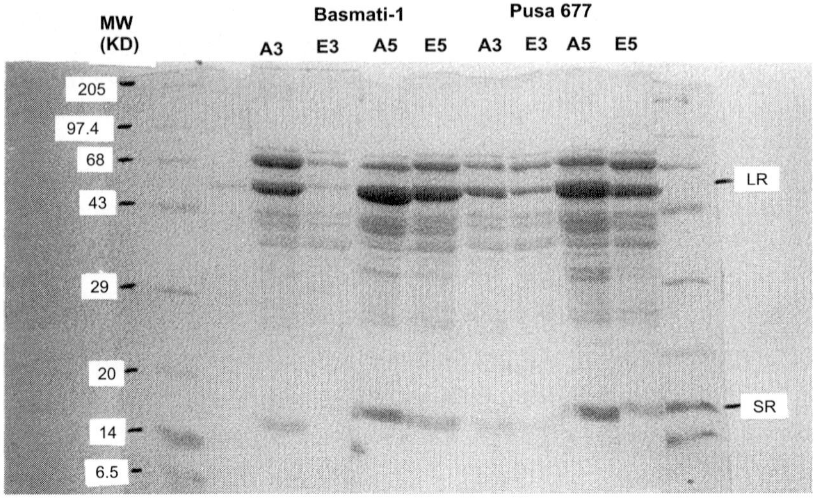

Fig. 21.8. Effect of elevated CO_2 on the protein profile in the leaves of rice cultivars (Uprety *et al.*, 2003).

However, CO_2 enrichment compensated the temperature induced decline in photosynthesis. The CO_2 enrichment and high temperature reduced the Rubisco content of rice by 22 and 23%. Total activity and activation of Rubisco were reduced.

The photosynthetic acclimation to elevated CO_2 is not likely to occur for rice grown under field conditions within the CO_2 range upto twice the current ambient level (Horie *et al.*, 2000). Conversely, photosynthetic acclimation to enriched CO_2 could result from plant responses to much higher CO_2 (above 1000 μmol mol^{-1}) as shown by Imai and Murata (1978) or from reduced sink size caused either by restricted root growth in the case of potted plants (Arp, 1991) or by high temperature induced spikelet sterility (Lin *et al*, 1997).

Baker *et al.* (1992) demonstrated that dark respiration rate of rice expressed on a land area basis, increased with increasing CO_2 due to increased biomass but specific respiration rate per unit biomass decreased. Since elevated CO_2 can reduce plant tissue nitrogen concentration (Baker *et al.*, 1992; Ziska *et al.*, 1996; Uprety *et al.*, 2000 and 2003) both the growth and maintenance components of respiration can be affected by tissue-N concentration (Amthor, 1994).

It is established that increase in CO_2 increases stomatal resistance through a reduction in stomatal aperture (Homma *et al.*, 1999 and Uprety *et al.*, 2002). In a recent study on rice cultivars Uprety *et al.* (2002) demonstrated that the stomata of lower epidermis were more sensitive to CO_2 enrichment than those of the upper epidermis (Table 21.4). They have attributed the reduction in stomatal conductance to balance the resource use and to maintain Ci : Ca ratio under elevated CO_2. Such changes occurred in rice cultivars through its effect on stomatal size and numbers. Uprety *et al.* (2002) also indicated that CO_2 induced increase in leaf size affected the stomatal density by reducing the penetration of radiation and spreading the stomata differentially. The reduction in stomatal density may be a consequence of the increase in the size of epidermal cells although there was decrease in epidermal cell density indicating greater effect on cell expansion than on cell division. This increase in cell expansion has been attributed to the increased osmotic potential of leaf cells with their high saccharide contents, which cause the cells to absorb more water and thus enlarge (Delucia *et al.*, 1985). These changes in the stomatal parameters may help rice cultivars to regulate the gas exchange processes and adjust under elevated CO_2 conditions.

Table 21.4. Effect of elevated CO_2 concentration (EC) on the leaf, stomatal, and photosynthetic characteristics of four rice cultivars. g_s- stomatal conductance, P_n- net photosynthetic rate n.s.- not significant (Uprety *et al.*, 2002)

		(CO_2)	Pusa Bac mati -1	P-677	P-834	P2503-6-893	C.D. 5% culti-vation	treat-ment	Inte-raction
g_s		AC	0.187	0.208	0.170	0.187	0.013	0.013	n.s
[mmol $m^{-2}s^{-1}$]		EC	0.072	0.087	0.075	0.082			
Stomatal	Abaxial	AC	93.56	128.65	105.26	93.56	3.75	3.75	n.s
Density		EC	81.87	105.26	119.88	87.74			
[n mm^{-2}]	Adaxial	AC	116.95	149.12	116.95	113.50	4.20	4.20	n.s
		EC	111.11	132.42	131.57	149.12			
Stomatal	Abaxial	AC	28.6	31.9	23.9	24.4	1.20	1.20	n.s
Index		EC	35.0	35.8	29.5	31.6			
[%]	Adaxial	AC	29.8	29.8	28.1	26.6	1.60	1.60	n.s
		EC	35.8	31.8	34.6	40.1			
Stomatal size	Length	AC	20.16	17.52	20.19	19.07	2.62	2.62	n.s
[µm]		EC	21.07	24.00	22.58	19.29			
(abaxial)	Width	AC	10.38	8.87	9.18	8.39	n.s	1.35	n.s
		EC	10.95	11.69	11.12	11.51			
(adaxial)	Length	AC	19.80	17.29	18.11	18.14	1.25	1.25	n.s
		EC	20.29	19.31	18.80	21.71			
	Width	AC	11.30	7.04	7.51	9.61	1.10	1.10	n.s
		EC	12.06	7.66	11.46	11.75			
Stroma [µm]	Length	AC	1.55	0.975	1.98	1.03	0.35	0.35	n.s
(abaxial)		EC	2.09	2.240	2.18	1.66			
(adaxial)		AC	0.876	1.440	0.923	0.696	0.018	0.018	n.s
		EC	1.860	1.720	0.976	1.660			
Epidermal cell	Abaxial	AC	233.91	214.85	336.25	290.63	7.25	7.25	9.61
density [mm^{-2}]		EC	152.04	192.98	286.54	190.05			
	Adaxial	AC	274.85	350.87	298.24	312.86	8.14	8.14	10.21
		EC	198.82	283.62	248.53	222.22			
Epidermal	Length	AC	82.3	81.30	80.00	102.00	4.85	4.85	n.s
cell [µm]		EC	106.6	114.60	86.60	104.00			
	Width	AC	22.30	22.60	24.00	25.30	1.92	1.92	n.s
		EC	26.30	27.30	24.00	29.30			
Leaf area		AC	781.43	708.53	897.10	88.80	52.06	52.06	73.65
[cm^2 $plant^{-1}$]		EC	980.49	775.25	960.20	104.70			
Leaf dry mass		AC	12.17	14.27	17.35	18.00	3.24	3.24	4.56
[g $plant^{-1}$]		EC	18.38	16.54	19.80	22.85			
Leaf area per		AC	6.420	4.965	5.170	4.548	4.86	4.86	n.s
unit mass [m^2 kg^{-1}]			EC	5.334	4.687	4.849	4.440		
Leaf mass		AC	15.5	20.1	19.3	22.0	2.1	2.1	n.s
per unit area [g m^{-2}]			EC	18.7	21.3	20.6	22.5		
P_N		AC	18.07	13.55	15.26	21.75	4.54	4.54	6.42
[µmol $m^{-2}s^{-2}$]		EC	26.45	16.92	19.40	28.50			

Rice displays higher water use efficiency under elevated CO_2 primarily due to increased biomass production and partly due to reduced transpiration. Morrison and Gifford (1984) found CO_2 induced increase in WUE by 53 to 63% in potted rice plants, whereas under TGC with CO_2 enrichment at 24-26^{0}C, WUE increased by 40-50 %. Nakagawa *et al.* (1997) found that leaf WUE at ambient CO_2 was not affected by growing temperature, but at elevated CO_2, WUE declined with increasing temperature.

Elevated CO_2 increases phenological development rate and hence reduces the number of days to heading (Nakagawa *et al.*, 1993; Kim *et al.*, 1996 and Hamid *et al.*, 2003). This CO_2 effect on rice development rate has been shown to be temperature dependent. The CO_2 induced accelerated development rate may be due to increase in stomatal resistance, decreasing the transpiration cooling which increases the tissue temperature (Kim *et al.*, 1996; De Costa *et al.*, 2003) and or due to high C/N ratio (Ziska *et al.*, 1996). Baker *et al.* (1990) observed reduction in final main stem leaf numbers with elevated CO_2 due to an acceleration in the phenological development. Ziska *et al.* (1997) demonstrated that under planting densities followed in Asian rice cultures, elevated CO_2 may substantially promote tiller production. Elevated CO_2 has minor effects on individual leaf size and also on total leaf area of rice except during initial growth stage (Morrison and Gifford, 1984). Elevated CO_2 increased specific leaf weight of rice during early growth but less at later growth stages (Baker *et al.*, 1990). Similarly, the root/shoot ratio was significantly increased during early growth stages (Ziska *et al.*, 1996). Elevated CO_2 markedly increased rice spikelet number per unit area, through increases in number of productive tillers per unit area and spikelets per tiller (Kim *et al.*, 1996). The fact that elevated CO_2 increased the spikelet number despite a reduction in plant nitrogen concentration suggests that elevated CO_2 promoted spikelet production efficiency per unit of plant nitrogen. Elevated CO_2 increases sink size by increasing spikelet number and this generally leads to increased yield provided successful fertilization and grain fill follows.

Horie (1993) observed that under field conditions CO_2 enrichment increases biomass production by 24% across a wide range of air temperature changes. Generally the pot grown plants showed a strong CO_2 by temperature interaction for biomass production may be that pot grown plants continue to produce tillers for a longer time than does field grown rice. This is likely due to reduced mutual shading in potted plants compared with field grown plants and is similar to the light interception effects on dry matter production and tiller development. Further studies also showed that the relative enhancement of rice biomass production with CO_2 depends strongly on plant nitrogen content. For pot grown rice plants, CO_2 enrichment, under drought stress conditions resulted in increased relative biomass production compared with drought stress ambient controls (Rogers *et al.*, 1984). This effect was attributed to higher leaf water potentials for the CO_2 enriched plants under drought conditions. Baker *et al.* (1997) reported a 10% reduction in crop water use due to anti-transpirant effect of CO_2 enrichment. Weerakoon *et al.* (2000) observed that the plant nitrogen requirement for radiation interception was similar for rice grown under ambient or elevated CO_2 conditions. The conversion efficiency of intercepted radiation to dry matter, radiation use efficiency (RUE) was about 35% greater at elevated CO_2. At ambient CO_2, RUE was fairly stable across the level of nitrogen but N less than 2.5% lowers RUE under elevated CO_2. Decreased leaf N with increased CO_2, therefore, decreased RUE of rice plants grown under elevated CO_2. RUE should be adjusted with a decrease in leaf N at the time of predicting responses of rice to elevated CO_2 (Table 21.5). Anten *et al.* (2003) demonstrated that LAI growth in rice is strongly determined by N availability, more or less independent of CO_2 concentration. When N availability is low, the canopy is relatively open and most leaves receive relatively high light. Under these conditions, the effect of elevated CO_2 on canopy photosynthesis will be predominantly through its effect on photosynthetic capacity. But as N availability increases, the canopy becomes denser and lower leaves become increasingly shaded. Under these conditions the enhanced quantum yield under elevated CO_2 will have an increasingly positive effect on canopy photosynthesis.

Table 21.5. Radiation Use Efficiency (RUE) of IR 72 rice for different growth periods when grown under ambient or enriched atmospheric CO_2 and different rates of fertilizer N (Weerakoon *et al.*, 2000)

Growth period	Fertilizer	Ambient CO_2 Nitrogen		Enriched CO_2	
Days after transplanting	*(Kgha^{-1})*	*RUE (g MJ^{-1})*	r^2	*RUE(g MJ^{-1})*	r^2
22-42	0	1.30 ± 0.10	0.93	1.82 ± 0.08	0.96
42-55	0	1.24 ± 0.06	0.73	1.55 ± 0.06	0.36
55-67	0	1.27 ± 0.15	0.93	1.26 ± 0.05	0.65
22-42	90	1.47 ± 0.11	0.95	1.74 ± 0.04	0.99
42-55	90	1.21 ± 0.11	0.45	1.58 ± 0.06	0.67
55-67	90	1.32 ± 0.11	0.77	1.72 ± 0.09	0.87
22-42	200	1.52 ± 0.07	0.97	1.98 ± 0.11	0.93
42-55	200	1.56 ± 0.05	0.89	1.92 ± 0.10	0.69
55-67	200	1.33 ± 0.08	0.82	1.74 ± 0.07	0.62

Allen *et al.* (1993) observed 44% increase in the yield of Indica type rice due to doubling CO_2. Ziska *et al.* (1997) obtained a 27% increase in yield with CO_2 enrichment for the cultivar IR 72 and in Japonica rice it increases from 20-40% (Kim *et al.*, 1996). It is important to examine the effect of elevated CO_2 in association with temperature effects as it is established that spikelet sterility in rice occurs below 20^0 C and above 35^0 C, which inturn reduces the yield. Ziska *et al.* (1997) reported that larger yield reduction caused by high temperature occurred at elevated CO_2 compared to ambient CO_2 in dry season. However, Matsui *et al.* (1997) and Kim *et al.* (1996) suggest a negative interaction between CO_2 and temperature on seed set and yield.

Gesch *et al.* (2003) studied the Rubisco expression in rice leaves in relation to genetic variation of photosynthesis under elevated CO_2 and temperature using an Indica and Japonica cultivars. They observed that Rubisco total activity and protein content were negatively affected in both genotypes by high temperature and elevated CO_2. IR 72 (Indica type) leaves had 71 and 39% more rbcS transcripts than M103 (Japonica type) under ambient and elevated CO_2, respectively, and had greater Rubisco activity and protein content. Expression of psbA (D1 protein of PS II) in IR 72 leaves increased with temperature but it remained constant for M 103 under elevated CO_2. PS II photochemical activity was not impaired in both the genotypes. Such differences for carboxylation responses to elevated CO_2 and temperatures will be useful in developing genotypes suited to cope with global climate changes (Table 21.6). Furthermore, Baker (2004) indicated that Southern US rice cultivars were more sensitive to high temperature stresses during reproductive development than Asian cultivars. These observations also point to the possibility of selecting and breeding rice cultivars with enhanced capability to take advantage of future global increases in CO_2 and temperature.

Table 21.6. Effects of growth temperature treatments under ambient and elevated CO_2 on leaf maximum PSII quantum yield (Fv/Fm), PSII electron transport (*J* PSII) and light adapted quantum yield (o/PSII) for IR72 and M103 rice cultivars (Gesch *et al.*, 2003).

Attributes	*Max. growth temperature (°C)*	*CO_2 (350 μmol mol^{-1})* M 103	*IR 72*	*CO_2 (700 μmol mol^{-1})* M103	*IR 72*
Fv/Fm		0.74 ± 0.02	0.70 ± 0.01	0.75 ± 0.01	0.70 ± 0.02
	40	0.71 ± 0.04	0.67 ± 0.02	0.70 ± 0.01	0.72 ± 0.02
J PSII (μmol m^2 s^1)	28	151 ± 9	133 ± 7	129 ±15	115± 6
	40	135 ± 16	123 ± 12	120 ± 9	138± 6
O/PSII	28	0.31 ± 0.02	0.31 ± 0.03	0.32 ± 0.05	0.28± 0.02
	40	0.31 ± 0.05	0.29 ± 0.04	0.31 ± 0.02	0.36± 0.03

More experimental studies are required to quantify the effects and interactions of CO_2 and temperature on biomass production, stomatal conductance, water use efficiency, spikelet sterility and grain yield. Studies are also needed to understand the mechanisms controlling the adverse effect of elevated CO_2 on high temperature induced spikelet sterility. Variability among rice genotypes in response to CO_2 and temperature in the field of crop water use, temperature induced spikelet sterility needs to be studied. This will help in identifying rice genotypes that are better adapted for future climate change.

C_4 PLANT

Maize

Maize is the third important food crop of the world in terms of sources of energy and protein in human nutrition. It is also a major food crop that has C_4 photosynthesis. During last few years its cropping area has increased 40% and yield gone up from 2 tha^{-1} to 4 tha^{-1} even than its demand is predicted to increase 45% from 1997 to 2020 (Rosengrant *et al.*, 2001). Its flowering is temperature and day length dependent. The unique combination of the photosynthetic metabolism and leaf anatomy of this C_4 plant is the elevation of CO_2 concentration at the site of Rubisco in bundle sheath. This elevated concentration of CO_2 has two effects.

Firstly, competitive inhibition of the oxygenase reaction of Rubisco eliminates most of the C_2 - photosynthetic oxidative or photorespiratory pathway activity and its expenditure of energy associated with photorespiration. Secondly, it allows Rubisco to approach its maximum rate of catalysis despite its low affinity for CO_2. The PEP carboxylase from C_4 photosynthetic tissues has a high affinity for CO_2, and photosynthesis is saturated at low CO_2 concentration (Long, 1999). Although no direct effect of the rise in atmospheric CO_2 in this C_4 species (Drake *et al.*, 1997) but at arid and low nitrogen conditions leaking of CO_2 from bundle sheath cells may lead to a direct response of photosynthesis to elevated CO_2 (Long, 1999). Uprety *et al.* (1996) reported a 10% increase in the productivity of maize cultivars compared to 28% increase in mung bean (C_3 cultivar) under elevated CO_2 and this improvement in maize productivity has been attributed to the CO_2 induced increase in stomatal resistance, greater root growth and improved water status. Drake *et al.* (1999) reported significant inhibition of mitochondrial respiration in maize.

There are still arguments about that whether or not the elevated CO_2 enhances plant growth in C_4 species. Increased growth under elevated CO_2 has been reported for a number of C_4 species including maize and sorghum (Wong, 1979; Poorter, 1993; Amthor *et al.*, 1994). However, others found no effect (Hocking and Meyer, 1991; Ziska *et al.*, 1991; Ellis *et al.*, 1995) or an effect when CO_2 enrichment was done under drought condition (Samarakoon and Gifford, 1996). Samarakoon and Gifford (1996) observed 35% more leaf area and 50% more dry matter due to the interactive effect of elevated CO_2 and soil drought in maize.The well watered plants showed no significant growth effect . However, Wong (1979) also demonstrated a 20% increase in dry matter production of maize grown at elevated CO_2, under irrigated condition.

Reported enhanced growth under elevated CO_2 in C_4 plants like maize has generally been attributed to increased water use efficiency. However, there are reports of photosynthetic stimulation of C_4 under elevated CO_2, although this enhancement may be limited to conditions where elevated CO_2 ameliorates the mild drought stress effect (Samarakoon and Gifford, 1996). Brown and Rosenberg (1997) while modeling for yield prediction, observed that high temperature induced reduction in grain yield in five crops including maize and sorghum, were to some extent mitigated by elevated CO_2 and increased precipitation. The mitigation was most pronounced in sorghum. Singh *et al.* (1998) suggested that the doubling CO_2 climate would increase the maize and sorghum yield in Quebec by 20%. Despite various differences on the impact of climate change on maize yield as predicted by various models, Young and Long (2000), suggested a reduction in maize yield for all climate change scenarios. On the contrary, Leakey *et al.* (2004) observed that the growth at elevated CO_2 significantly increased leaf photosynthetic CO_2 uptake rate by up to 41% and accounted it with greater intercellular CO_2, lower stomatal conductance and lower transpiration. These results call for a reassessment of the established view that C_4 photosynthesis is insensitive to elevated CO_2 under favorable growing conditions and the production potential of corn will not be affected by global rise in CO_2.

OILSEED CROPS

Brassica

Among oilseed crops, rapeseed and mustard group is the second most important after groundnut contributing nearly 30% of the total oil production in India. They play an important role in agriculture industries and foreign trade. India ranks first in acreage of mustard in the world occupying an area of 5.83 million hectares of land and a total production of 5.34 million tones with an average yield of 1106 kg ha^{-1} (Chakravarty and Gautam, 2002). The annual requirement of oilseeds in India is much higher than the production. The reasons for its low productivity are: (1) The large-scale rainfed cultivation (76 %) of *Brassica* grown area is rainfed, (2) Minimal resource capacity of the small and marginal farmers, (3) Low technology of production for this crop, and (4) Vulnerability of this crop to pests and diseases. Response studies to elevated CO_2 with *Brassica* crop in India was done using OTC (Uprety, 1993) and FACE facilities (Uprety *et al.*, 2000). *Brassica* experiences intermittent drought during its growth period from vegetative to siliqua formation stage and there was variability in the responses of these cultivars to the moisture stress (Lawlor and Uprety, 1993; Uprety and Tomar, 1993). Uprety *et al.* (1995) studied the interactive effect of elevated CO_2 and moisture stress on the morpho-physiological processes of four *Brassica* species namely, *B. campestris, B. juncea, B. carinata and B. nigra*, using OTC. It was observed that the rate of photosynthesis, stomatal resistance and plant growth was considerably increased under elevated CO_2. This effect was highly significant under moisture stress conditions mitigating the adverse effect on these components in plants. However, relatively drought susceptible species *B. campestris* and *B. nigra* responded to CO_2 markedly compared to less sensitive *B. carinata* and *B. juncea*. The stomatal conductance of plants exposed to high CO_2 does not decrease as rapidly during stress period as under ambient conditions. This is an adaptive advantage in maintaining the rate of CO_2 uptake. The water status of plant was significantly improved under elevated CO_2. This was reflected in both water potential and relative water content in leaves. This might be caused by either increased root growth and/or increased stomatal resistance. One may bring more water to the plant and other may reduce water loss by reducing transpiration. Here again those species (*B. campestris* and *B. nigra*), which were more susceptible to moisture stress responded significantly to higher CO_2.The mitigating effect of CO_2 was observed in both water potential and relative water content. The quantitative increase in apoplastic water and the positive effect on bulk modulus of elasticity of cell walls possibly helped to maintain cell turgor for effective physiological functions. These results correlate well with changes in seed yields of these species. Furthermore, a large increase in the C: N ratio was observed in *Brassica* leaves. The accumulation of carbohydrate storage or shifting in the carbon-nitrogen balance is the response to increased carbon availability. Elevated CO_2 also contributed to investment in roots without losses in capability for carbon acquisition and utilization. Morison and Gifford (1984) reported that in oil seed rape (*B. napus*) high CO_2 increased the leaf area by 40%, dry weight by 48% and WUE by 50%. Frick *et al.* (1994) observed a delayed onset of flowering and senescence under elevated CO_2. Marina *et al.* (2002) observed varietal difference on the response of *Brassica napus* to the elevated CO_2 and reported that old cultivars with clustered growth responded and new varieties did not. Andreeva *et al.* (1989) observed that mustard (*B. juncea*) under high CO_2 had greater net assimilation and photosynthetic rates without additional synthesis of fraction 1 protein and with the concurrent enhancement of N metabolism, plant growth and biomass accumulation during vegetative phase. They also observed that high starch accumulation in leaves does not cause suppression of photosynthesis unless it damages chloroplast structure.

Rabha and Uprety (1998) in a study in *Brassica juncea* leaves observed that elevated CO_2 brought about reduction in the fluorescence value of Fv, Fm and Fv/Fm, though the reduction in Fv/Fm was only 10% compared to plants grown under ambient CO_2. The electron transport capacity, as studied from Fv/Fm, did not change much with CO_2 enrichment but was sufficient to sustain high rate of carbon assimilation. It was demonstrated that there was relatively higher capacity of carbon assimilation at a given efficiency of PSII photochemistry in leaves grown under elevated CO_2, they concluded that PS II photochemistry was more efficient in leaves grown under elevated CO_2 conditions. The greater proportion of electrons helped to compensate for the stress effect on the photosynthesis in leaves. Much of the reduction in Fv/Fm without a decline in the rate of photosynthesis in moisture stressed plants grown under elevated CO_2 may be plant's strategy of photo-protection.

It appeared that the CO_2 induced dry matter accumulation in roots was mainly used to increase root length and dry weight, whereas in shoots it caused an increase in thickness and surface area. The continuous translocation of photosynthates to the growth of roots and leaves under higher CO_2 attributed to their greater growth. The impact of CO_2 on the production of new sinks before the stress effect lead to the greater leaf area and better root growth. Furthermore, there was remobilization of the additional assimilates for the development of new sinks. The continuous higher source activity under elevated CO_2 conditions relieved the source limitations for photoassimilate transport to siliqua and seed growth, thereby, reducing the stress effect on productivity. Uprety and Rabha (1999) in an another experiment on *B. juncea* noted that the ameliorating effect of CO_2 to water stress induced changes can be attributed to the accumulation of sugars in the cell sap. This may possibly help in osmo regulation. In these studies, Uprety *et al.* (2000) contended that at an elevated CO_2, plants accumulate increased metabolic reserves, which might possibly enhance metabolic flexibility to respond to moisture stress. This flexibility would give plants an advantage economically and ecologically. The *Brassica* plants would be able to survive stressful conditions better in the future than currently regarding resistance to drought.

Uprety *et al.* (1998) characterized the transfer of CO_2 responsiveness between *B.campestris* and hybrid *B. oxycamp* using *B. oxyrrhina* as a female parent. It was shown that the characterstic CO_2 responsive nature of the *B.oxycamp* hybrid was inherited from the responsive characters of *B. campestris*. The variation in the response for photosynthesis and root and leaf growth to elevated CO_2 between the parents and the hybrid was attributed to their sink potential and the photosynthetic changes under elevated CO_2 were attributed to stomatal behavior and to some of the fluorescence parameters. It is likely that the empirical selection method for CO_2 responsive characters in *Brassica* species may be inefficient because of the steady increase in atmospheric CO_2. However, a possibility of transferring some of the CO_2 responsive characters of *B. campestris* to the hybrid *B. oxycamp* was observed in this study (Table 21.7 and Fig. 21.9). Bunce and Sicher (2003), studied the down regulation of photosynthesis at elevated CO_2 in two cultivars of *B.oleracea* and observed that down regulation was not occurred in kohlrabi because of larger sink, however, in collard which lack sink it occurred. The degree of down regulation was associated with leaf glucose, fructose and sucrose and less with starch and day-to-day variation in the occurrence of feedback inhibition of photosynthesis at elevated CO_2 was predictable from PAR.

Table 21.7. Average net assimilation rate (NAR), leaf area ratio (LAR) and relative growth rate (RGR) for maize grown at ambient and twice–ambient pCO_2 in wet soil (moisture content held close to field capacity) and in dry soil (allowed to dry from field capacity). Drying commenced after germination. Growth rates are for 26-37 days after germination. (Samarakoon and Gifford, 1996).

Soil /CO_2 treatment	*NAR(g $m^{-2}day^{-1}$)*	*LAR (m^2kg^{-1})*	*RGR (mg g^{-1} day^{-1})*
Wet soil Ambient CO_2	13.4	10.8	135.8
Twice ambient CO_2	13.8	10.4	133.8
Change Dry soil	+3%	-4%	-1%
Ambient CO_2	5.1	10.2	50.6
Twice ambient CO_2	7.9	9.6	73.3
Change	+56%	-6%	+45%

Uprety and Mahalaxmi (2000) studied the response of *B. juncea* plants to elevated CO_2 and nitrogen nutrition and concluded that the additional carbohydrates, sequestering of carbon and greater water status in plant systems made it flexible and responsive to nitrogen nutrition, thus sustaining the responses of plants to changed CO_2 in the atmosphere through a balanced C/N ratio. However, it is true that to sustain greater productivity of plants resulting from increased CO_2, nutrient availability needs to be maintained and may require changes in soil, plant, and fertilizer management.

Uprety *et al.* (2004) observed that moisture stress lead to the oxidative stress in *Brassica* species by accumulation of substrates like H_2O_2 and MDA resulting in to the production of potentially toxic radicals such as O_2^{-1} and H_2O_2. The elevated CO_2 significantly reduced the activity of anti-oxidative enzymes because of the reduction in the cellular reactive oxy species (ROH) production. This may possibly be due to increase in CO_2: O_2 ratio within chloroplast, which may decrease electron leakage of PSI to O_2. The possibility of attributing it to the reduction caused by the effect of CO_2 enrichment on photorespiration (Causin *et al.*, 2001) and respiration (Uprety *et al.*, 2001) and resultant cellular H_2O_2 radicals is

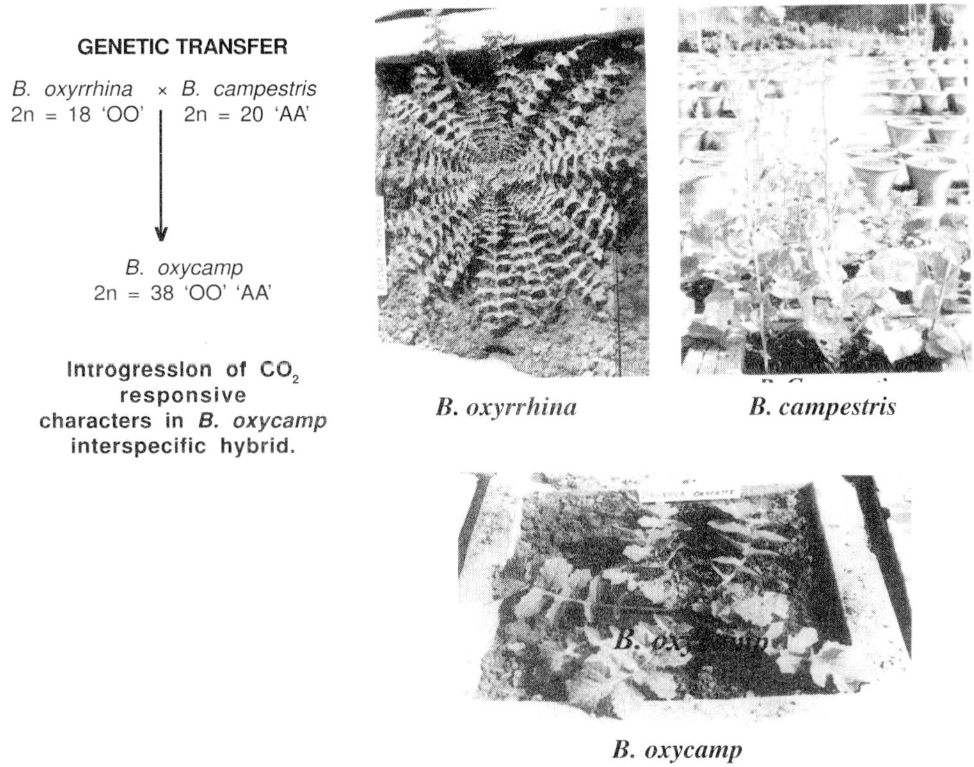

Fig. 21.9. Introgression of CO_2 responsive characters in *B. oxycamp* interspecific hybrid (Uprety *et al.*, 1998),

indicated. Foyer and Harbinson (1994) demonstrated increase in the production of ROS at low chloroplast concentration of CO_2. However, sufficient intercellular CO_2 was available to utilize the reducing power and for the generation of ROS in *Brassica* cultivars. It was clearly indicated that the elevated CO_2 decreased the whole antioxidant system in *Brassica* cultivars. However, significant increase was observed in all anti-oxidant enzymes under water stress condition. The high CO_2 helped for quick induction of the anti-oxidative system to scavenge the stress induced ROS. It was observed that *Brassica* plants grown at an elevated CO_2 contain increased metabolic reserves (Uprety and Mahalaxmi, 2000) to respond to oxidative stress compared to plants grown in ambient air. The stress induced enhancement of H_2O_2 and malondialdehyde (MDA) significantly depressed under elevated CO_2 indicating the enhancement in antioxidant enzyme pool in these plants. Thus the *Brassica* cultivars developed strategy to combat the stress induced oxidative damage on cell membrane both by reducing ROS and increasing antioxidant enzymes. The plants grown with the elevated CO_2 were better protected from drought stress by an increased capacity for detoxification and repair (Fig. 21.10).

The investigations on anatomical changes caused by elevated CO_2 are limited. Such studies were required as additional evidences to explain the physiological

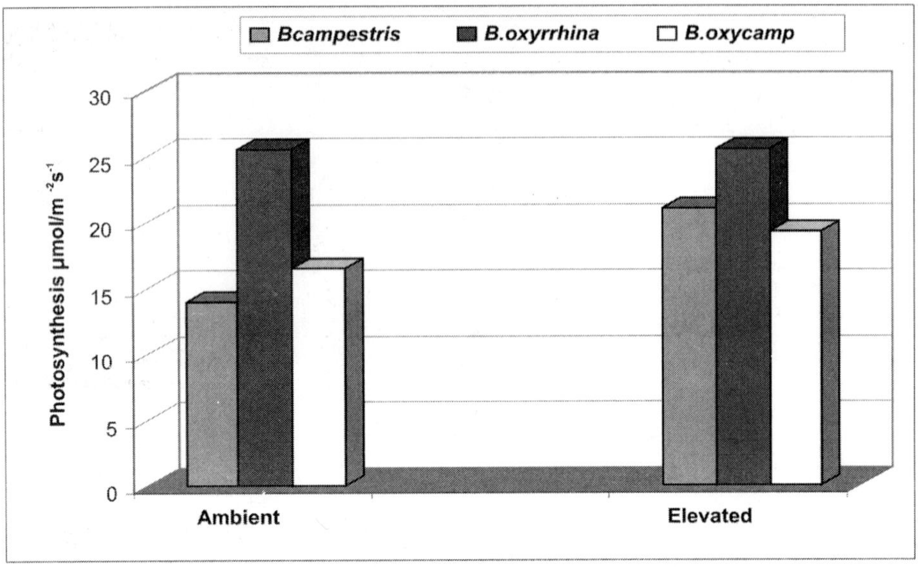

Fig. 21.10. Effect of elevated CO_2 on the photosynthesis of *B. oxycamp* and its parents *B. oxyrrhina* and *B. campestris* (Uprety *et al.*, 1998)

changes in plants, particularly in relation to water stress. Uprety *et al.* (2001, 2004) used transmission electron microscopy (TEM) to identify such anatomical changes in *Brassica* cultivars. They demonstrated that leaf tissues showed a significant increase in thickness of epidermis, size of mesophyll cells, accumulation of starch and size and number of starch granules per chloroplast in *B. juncea*. It appears that by increasing the number and size of palisade cells and concomitantly the number of chloroplast per leaf; more storage sites for starch were made available under stress condition which was otherwise lacking such sites and sink capacity for starch accumulation. The adverse effect on chloroplast structures both due to moisture stress and starch overloading was checked by elevated CO_2. The production of photo assimilates under elevated CO_2 possibly increased the sink capacity of leaf blade and consequently reduce the moisture stress effect. The feedback inhibition did not occur because of the larger palisade and epidermal cells, which compensate the adverse water stress effect on these components. However, there was no significant increase in the number of epidermal cells per unit leaf area in high CO_2 treatment. The leaf thickness was affected by increase in the length of epidermal cells and increase in number and elongation of mesophyll cells. It seems that cell enlargement was more sensitive to CO_2 than cell division and this may be related to changes in leaf water relations (Uprety *et al.*, 1995). The leaf anatomy reflected an optimization strategy due to high CO_2 that allowed the loading of chloroplast with excessive oval starch grains to meet the adverse effect of drought on these components in *B. juncea* leaves (Fig. 21.11).

Uprety (2004) observed that elevated CO_2 altered the vessel characters of shoot and root of *Brassica* species. The increase in radial width of xylems

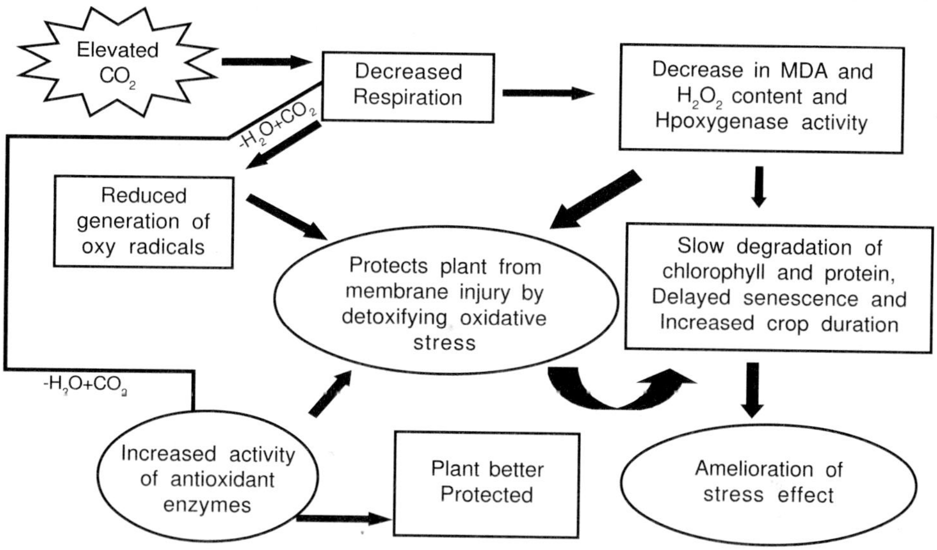

Fig. 21.11. CO_2 induced amelioration of oxidative stress effect in *Brassica* species (Uprety, 2004).

(meta+proto) under elevated CO_2 indicated more transfer of water and nutrients from soil to plants resulting in higher water status maintained throughout the growth period, whereas, the increase in phloem size helped to transfer the excessive solute material from source to sink. The proto-phloem sieve elements solely imports solute for growth (Esau, 1978). Meta-phloem functions primarily to transport solute to other developing regions of plant (Wood *et al.*, 1994). The coexistence of two ontogenetically different phloem sieve element types in *Brassica* plants under elevated CO_2 indicated the possibility of two different transport functions at the same time one may be involved in supplying for the structural development and another may cater to the need of increased new sinks. This indicated that metabolite adjustment through phloem might be one of the important area for future studies. Another important consequences of greater phloem size and number in roots, was to channelize basipetal movement of the larger amount of carbohydrates to the roots for the development of more number of secondary roots, increase in root length and root volume in *Brassica* species under elevated CO_2.

Based on South Asian studies on the characterization of the responses of *Brassica* species to the interaction of elevated CO_2 and drought stress, Uprety (2004) outlined various adaptation strategies developed by *Brassica* plants to ameliorate the adverse drought stress effect under elevated CO_2 conditions. They include (A) Maintenance of water status by greater root growth, reduction in transpiration, increase in content and mobility of water, improvement in cell wall elasticity and greater apoplastic water and accumulation of non structural carbohydrates for osmoregulation. (B) Metabolic flexibility by enhancement in photosynthesis, reduction in respiration, enhancement of stomatal resistance and stomatal index and reduction in stomatal density (C) Detoxification of oxidative stress by reduction in oxy radical production, increase in antioxidant enzymes and accumulation of metabolic reserves to combat toxic effects of oxy radicals (D) Growth adaptation by increase in plant growth and leaf area development, reduction in senescence, increase in leaf area duration and improvement in source sink balance, and (E) Anatomical modifications by increasing the size of xylem vessels for easy transport of water, compact phloem for smooth movement of photosynthates to sink and protection of chloroplast in leaves by increasing their size, number and starch granules to strengthen thylakoid membrane. However, more studies are in progress to fill in various gaps to explain these strategies (Fig. 21.12).

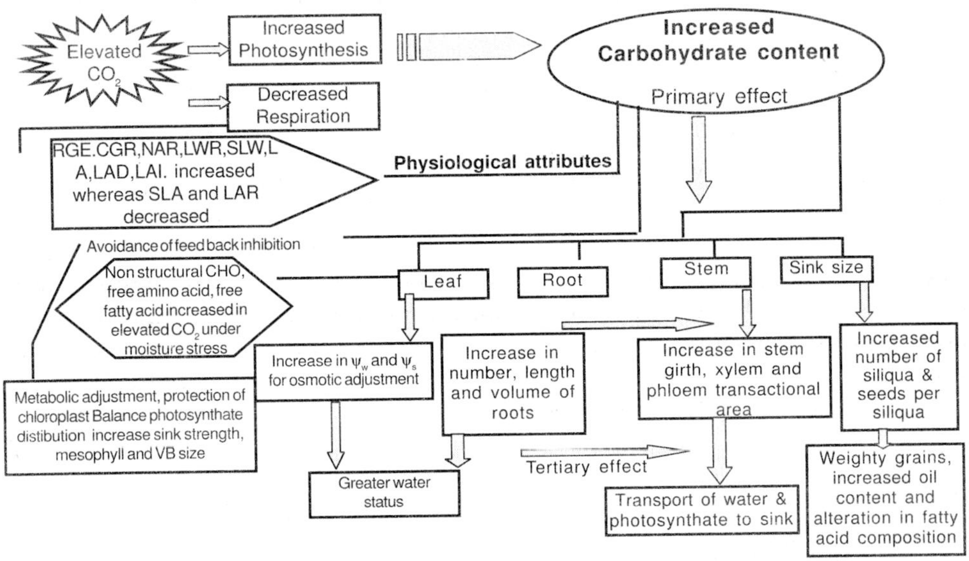

Fig: 21.12. Amelioration of water stress effect by elevated CO_2 in *Brassica species* (Uprety, 2004)

Reddy *et al.* (2004) studied the effect of elevated CO_2 on the chemical quality of leaves of *B. oleracea* and *B. rapa* plants and observed that elevated CO_2 significantly decreased the concentrations of leaf phytochemical constituents in oilseed rape, i.e., total phenolics and total nitrogen, but not in cabbage. However, the effect on the leaf glucosinolate concentrations of both plant species was marginal.

Uprety *et al.* (1997) also demonstrated the sustenance of the stimulatory effect of elevated CO_2 up to seed formation stage in *Brassica*. The sequestration of carbon in plant parts leads to the accumulation of greater carbohydrate components in *Brassica* seeds. The maximum increase was in starch content followed by reducing and nonreducing sugars. The stress induced decrease was ameliorated by CO_2 enrichment from 19.9% to 1.1% for starch, from 24.9% to 1.4 % to reducing sugars, from 17.3 to 3.5% for nonreducing sugar and from 20.12% to 2.76% for total sugars, respectively. Such a marked recovery among carbohydrate components may be due to greater photosynthate transfer to the seeds. Another carbon component which was significantly boosted by increased CO_2 in *Brassica* was that of its oil content. However, protein content of seeds was adversly affected by CO_2 enrichment. It was clear from this study that stress induced adverse effects on grain composition, particularly on carbohydrate and oil content were significantly ameliorated by sequestration of carbon and improved water status. Participation of additional carbon to lipid synthesis at the acetyl-coenzyme A level (Williams *et al.*, 1995) must have helped in the oil synthesis in *Brassica* seeds. Uprety (2004) demonstrated that besides increasing oil content, elevated CO_2 also altered the fatty acid composition of oil in seeds. He observed that palmitic, stearic, linolenic and erucic acid contents were decreased and linoleic acid content was increased significantly due to CO_2 enrichment. There was no significant difference between cultivars for their fatty acid concentration. The decrease in the saturated fatty acid concentration was attributed to the lower O_2:CO_2 ratio. However, the presence of higher linoleic acid in seeds of CO_2 enriched plants was due to the fact that erucic acid synthesis and conversion of linoleic acid to linolenic acid requires sufficient quantity of O_2 that was not available due to higher intercellular concentration of CO_2. Another possibility of reduction of linolenic acid was due to its use for chloroplast membrane organization (Thies and Nitsch, 1974). A general decrease in protein level due to elevated CO_2 can be attributed to the dilution effect caused by excess carbohydrates. Amelioration of this effect by increased fertilizer treatment was not ruled out and needs further investigation. Such an analysis of grain quality will have far reaching implications in breeding of *Brassica* for oil content and cattle feed and in fertilizer demand in a future high CO_2 world (Fig. 21.13 and Table 21.8).

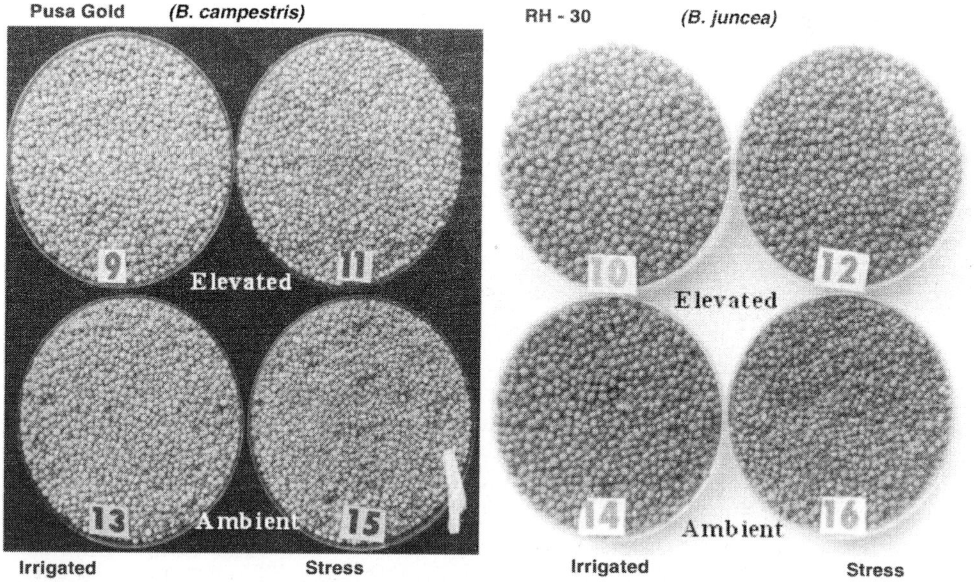

Fig. 21.13. Interactive effect of carbon dioxide and water stress on the grain size of *Brassica species* in FACE

Table 21.8. Interactive effect of elevated CO_2 and water stress on the fatty acid composition of *Brassica* seeds

Treatment	Stearic acid (%)		Palmitic acid (%)		Linoleic acid (%)		Linolenic acid (%)		Erucic acid (%)	
	V1	V2	V1	V2	V1	V2	V1	V2	V1	V2
Ambient CO_2										
Control	1.40	1.30	2.80	2.79	12.20	10.7	6.37	7.99	52.44	55.39
Moisture stress	1.20	1.20	3.30	2.74	11.90	10.4	6.29	7.84	51.77	55.24
Elevated CO_2										
Control	0.98	0.92	2.00	2.70	14.10	12.90	5.24	6.66	47.75	47.32
Moisture Stress	0.92	0.94	2.10	1.94	13.80	12.70	5.20	6.70	47.19	48.21
CD at 5 %										
Variety	N.S		N.S		1.63		1.43		N.S	
Treatment CO_2	1.19		0.54		1.02		0.58		4.44	
Var x Treat	N.S		N.S		N.S		N.S		N.S	
Interaction	N.S		N.S		N.S		N.S		N.S	

Variety x treatment x moisture stress
V1-*Brassica campestris*, "Pusa gold"
V2-*Brassica juncea*, RH-30

CAM CROPS

CAM plants (Crassulacean acid metabolism) such as pineapple (*Ananas comosus*), *Agave, Hylocereus* and *Opuntia ficus-indica* uptake the substantial nocturnal CO_2 by their photosynthetic organs (Joshi *et al.*, 1965). Because photosynthesis cannot occur without light, the CO_2 taken up at night cannot be immediately fixed in to photosynthetic products such as glucose and sucrose. Rather, the CO_2 is incorporated in to phosphenolpyruvate (PEP) by the enzyme PEP carboxylase, leading to the formation of an organic acid such as malate. Although the long-term effects of elevated CO_2 are relatively unstudied for commercial CAM species, such effects need to be evaluated in view of the rising atmospheric CO_2. Zhu *et al.* (1997) observed 23% more dry mass of *Ananas comosus* with 730μ mol mol^{-1} CO_2 treatment for 4 months. Its nocturnal increase in leaf acidity was 42% higher for higher CO_2. Graham and Nobel (1996) reported 49% increase in photosynthesis in *Agave deserti* plants grown under elevated CO_2. More leaves were produced, increased total leaf surface area, net CO_2 up take per unit leaf area and enhanced dry mass accumulation by 88%. Nobel (2000) summarized the results for all of the cultivated CAM species examined for months under elevated CO_2, concluded that the daily net CO_2 uptake and biomass productivity increases about 1% for each increase of 10μ mol mol^{-1} CO_2 in the atmosphere.

Future research will inevitably be devoted to biotechnological improvements in the yield and general performance of cultivated CAM species because Agave and Cacti are currently important crops in Mexico and other Latin American countries. The increased CO_2 accompanying global climate change will be advantageous for net CO_2 uptake and productivity by CAM crops. The increased temperature will favor cultivation at high altitudes. The high productivity of certain cultivated CAM species under optimal conditions and the generally high WUE of CAM plants warrants increased agronomic exploitation of such plants in the future.

Role of breeding and biotechnology for crop productivity in rising atmospheric carbon dioxide:

Crop productivity potential may be increased by breeding or selecting cultivars that are more responsive to elevated CO_2. Climatic changes such as increase in temperature also may occur in 21st century (Houghton *et al.*, 1996), which may decrease crop productivity in some cases. However, the projected exponential rise in atmospheric CO_2 should result in increased productivity of crop species. Grain yield of some of the cereal crops may have increased by 7-25 % over the last 100 years, solely as a result of increase in CO_2 during this period (Goudriaan and Unsworth, 1990; Mayeaux *et al.*, 1997). The changes in atmospheric CO_2 and climate are rapid, consequently it is appropriate to begin designing new breeding strategies and cultivars for 21st century. This can be facilitated by considering past, present and possible future climate.

Plants with C_4 photosynthetic system evolved during the early period after the atmospheric CO_2 became low and this system represents a specific adaptation to the low CO_2 environments of the last 200,000 years (Bowes, 1993). The growth responses of C_4 and CAM plants to elevated CO_2 are smaller than those of C_3 species (Porter, 1993).

Empirical selection for yield under field conditions may indirectly select plants that are responsive to the continually increasing levels of CO_2 (Kimball, 1985). However, it is likely that indirect selection will be inefficient because the atmospheric CO_2 is increasing rapidly and yield is dependent upon many abiotic and biotic factors. Direct selection in breeding nurseries that are subjected to elevated CO_2 would not appear feasible in commercial programs due to the large number of plants that must be screened and the high costs of nursery environments that have elevated CO_2 compared with conventional field nurseries. Maxon Smith (1977) with elevated CO_2 environment found selection was effective in improving agronomic traits but did not enhance yield. He attributes the lack of success in increasing yield to small number of plants that were screened. However, if the biochemical, physiological and morphological traits that contributes to responsiveness to elevated CO_2 could be identified, they could be used to supplement empirical selection based on yield under field conditions and thereby enhance breeding programs.

At optimum temperature, doubling atmospheric CO_2 increased the grain yield of various small grain cereals by 32% and grain legumes by 54% (Kimball, 1983), but the increases often were less than the increase in photosynthesis that occurred in short-term doubling of CO_2 in the same temperature (Poorter 1993; Allen 1994). There are several explanations: A major factor is the downregulation of photosynthetic capacity under long term exposure to elevated CO_2. Downregulation has been attributed to either to be a consequence of artificial growth conditions or it may indicate that current cultivars of C_3 plants are not well adapted to elevated CO_2. Downregulation could either be a consequence of feedback mechanisms that operate when the supply of carbohydrates from photosynthesis exceeds the sink demand for carbohydrates (Allen, 1994). Arp (1991) pointed out that strong downregulation was observed when plants were grown in smaller pots that would have resulted in a much smaller root sink for carbohydrates than occurs in nature. According to Conroy (1992), the greatest absolute increases in productivity with CO_2 enrichment will occur when soil N and P availabilities are high. More recently Žiska *et al.* (1996) showed for rice that growth response to elevated CO_2 was negligible when soil N was very deficient and increased with greater supply of N to the soil. These observations denoted that increases in N supply increase the ratio of carbohydrate sinks to the photosynthetic source, which than result in increased growth responses to elevated CO_2 (Rogers *et al.*, 1996). A positive correlation between responsiveness to elevated CO_2 with that of increase in RGR was reported by Poorter *et al.* (1996). That means faster growing species and those with larger sink strength were more responsive.

The increase in both grain yield and harvest index resulted from inadvertent selection of plant types were better adapted for higher CO_2 of the last 20th century compared with the low CO_2 in earlier centuries. Hall and Allen (1993) demonstrated that plants with higher harvest index would be expected to have greater sinks for carbohydrates in relation to their photosynthetic capacity and thus be more responsive to elevated CO_2. Ahmed *et al.* (1993) demonstrated that cowpea (*Vigna unguiculata*) with heat tolerance genes produced more shoot biomass and greater pod yield with elevated CO_2 than genetically similar cultivar that does not have heat tolerance gene. These studies also indicate that selections for higher harvest index through enhancing pod set might produce cowpea cultivars whose grain yield are more responsive to elevated CO_2.

The older cultivars of wheat showed greater responsiveness to elevated CO_2 under high soil nitrogen due to simulation of tillering by high nitrogen (Manderscheid and Weigel, 1997; Rogers *et al.*, 1996). Neither of these studies gave clear indication of whether cultivars of wheat with greater tillering tendency would be more responsive than modern cultivars to elevated CO_2 in terms of grain yield under optimum field conditions and this is a critical issue for the plant breeding program.

Ziska *et al.* (1997) confirmed that grain yield responses to elevated CO_2 were associated with increases in number of fertile tillers in rice; but the grain yield responses (+15 % and +27 %) were much smaller than the responses of total biomass (+31 % and +40 %). This indicates that grain yield may have been sink limited and that responsiveness to elevated CO_2 in rice might be enhanced by breeding to increased reproductive sink strength.

Very few reports of studies determining the specific traits that enhance responsiveness of grain yield to elevated CO_2. Comparative studies of responsiveness to elevated CO_2 under optimum field conditions with genetic lines that differ mainly in harvest index and also exhibit positive correlations between grain yield and harvest index under ambient CO_2. There is a critical need for field studies of this type because current selection strategies may be appropriate or in the opposite direction from those needed for breeding cultivars for future environment that have elevated CO_2.

The studies which are focused on the effects of elevated CO_2 on gene expression are very few. In controlled environments, a decrease of rbcs and rbcl transcript levels at elevated CO_2 was found in tomato (Van Oosten *et al.*, 1994) and in nitrogen deficient wheat cultivars (Nie *et al.*, 1993). Indirect evidences suggest that regulation of the expression of photosynthetic genes via increased soluble carbohydrate concentration may underlie acclimation to growth in elevated CO_2. Decreased expression of several photosynthetic genes has occurred when sugar concentration have been increased. Sheen (1990) demonstrated that addition of soluble carbohydrates to a maize protoplast transient expression system inhibited specifically the transcriptional activity of seven photosynthetic gene promoters including that for rbcs. According to Drake *et al.* (1999) decrease in Rubisco is commonly correlated with an increase in leaf non-structural carbohydrates. Sucrose and starch increased 60% and 160%, respectively, in elevated CO_2. The low concentration of glucose required for this repression suggests that other sugars, in particular sucrose and fructose, may be effective in the cell via metabolism to glucose. Based on glucose signaling in yeasts, a hypothetical scheme, whereby hexokinase (HXK) associated with a glucose channel in the plasmalemma or tonoplast would release an effector in response to glucose has been proposed.

Webber *et al.* (1994) provided an attractive explanation of acclimation of photosynthetic capacity, hypothesizing that increased soluble carbohydrate concentration causes the repression of the genes coding for Rubisco and other photosynthetic proteins in leaves in elevated CO_2. Furthermore, correspondence between the increased soluble carbohydrate content and decreased gene transcript levels was often lacking, suggesting other external and internal signals may be more dominant in the field. Although repression of photosynthetic genes, by accumulation of soluble carbohydrates in laboratory grown plants, remains an attractive mechanistic explanation of acclimation of photosynthetic capacity in elevated CO_2, the stage of development modify this effect in nature. It was shown that growth at an elevated CO_2 of approximately 50% above current levels does result both in increased leaf soluble carbohydrate concentrations and decreased mRNA transcripts for some genes coding for the photosynthetic apparatus in particular rbcs and rbcl.

Grey *et al.* (2000) reported identification of *Arabidiopsis* gene HIC (for high carbon dioxide) that encodes a negative regulator of stomatal development that responds to CO_2. This gene encodes a putative 3-keto acyl coenzyme A synthase (KCS), an enzyme involved in the synthesis of very long chain fatty acids. Mutant HIC plants exhibit upto 42% increase in stomatal density in response to a doubling CO_2. These results identify a gene involved in the signal transduction pathway responsible for controlling stomatal numbers and showed that HIC plants carry a mutation in a 3-keto acyl coenzyme A synthase (KCS) gene that results in disruption in the signal transduction pathway responsible for the control of stomatal patterning in response to elevated CO_2. CO_2 responsive genes have been identified in cynobacteria (Kapalan *et al.*, 1998) and *Chlamydomonas* (Eriksson *et al.*, 1998). Certain cyanobacterial genes are transcribed only under low CO_2, in case of others, the level of transcript increases during acclimation.

The use of genetically manipulated plants may turn out to be very powerful tool in studying plant responses to elevated CO_2, as no elevated CO_2 is required because the carbon or nitrogen relation can be accelerated or downregulated by genetic manipulations (Stitt and Schulze, 1994). In addition, specific plant internal adjustment can be studied at the enzyme level. The disadvantage of this technology is that at present, only few plant species can be manipulated (tobacco, potato, *Arabidiopsis*) and that only relatively few proteins have been cloned. It is expected that this will change in near future.

REFERENCES

Ahmed F.E., Hall A.E., and Madore M.A. (1993). Interactive effects of high temperature and elevated carbon di oxide concentration on cowpea (*Vigna unguiculata* L. Walp). *Plant, Cell and Environment,* 16: 835-842.

Akita S. (1980). Studies on the difference in photosynthesis and photorespiration among crops.II.The differential response of photosynthesis, photorespiration and dry matter production to carbon dioxide concentration among species. Bull. National Instt. Agric. Sci, Series D, 31: 59-94 (in Japanese with Eng. summary).

Allen L.H.Jr. (1994). Carbon dioxide increase: direct impacts on crops and indirect effects mediated through anticipated climatic changes. In: Boote, K.J., Bennett,J.M., Sinclair, T.R.and Paulson, G.M. (eds) *Physiology and Determination of Crop Yield.* Crop Sci. Soc. of America, Madison, Wisconsin, pp. 425-459.

Amthor J.S. (1994). Respiration and carbon assimilation use. In: Boot,K.J. et al (eds) *Physiology and Determination of crop yield.* American Soc. of Agronomy, Madison, Wisconsin, pp. 221-250.

Amthor J.S., Mitchell R.J., Runion G.B., Rogers H.H., Prior S.A. and Wood C.W. (1994). Energy content, construction cost and phytomass accumulation of *Glycine max(L)* Merr and *Sorghum bicolor(L)* Moench grown in elevated CO_2 in the field. *New Phytologist* 128: 443-450.

Andreeva T.F., Strogonova L.E., Voevudskaya S.Yu, Maevskaya S.N. and Cherkanova N.N. (1989). Effect of enhanced CO_2 concentration on photosynthesis, carbohydrate and nitrogen metabolism and growth process in mustard plants. Fiziol. Rast. 36: 40-48. [In Russian].

Anten N.P.R., Hirose T., Onoda Y., Kinugasa T., Kim H.Y., Okada M. and Kobayashi K. (2003). Elevated CO_2 and nitrogen availability have interactive effects on canopy carbon gain in rice. *New Phytologist* 161: 459-471.

Arp W.J. (1991). Effects of source-sink relations on photosynthetic acclimation to elevated CO_2. *Plant, Cell and Environment,* 14: 869-875.

Baker J.T., Allen L.H.Jr., Boote K.J., Jones P. and Jones J.W. (1990). Developmental responses of rice to photoperiod and carbon dioxide concentration. *Agriculture and Forest Meteorology* 50: 201-210.

Baker J.T., Allen L.H.Jr., and Boote K.J. (1992). Temperature effects on rice at elevated CO_2 concentration. *J. Expt. Botany,* 43: 959-964.

Baker J.T. and Allen L.H.Jr. (1993). Effects of CO_2 and temperature on rice: a summary of five growing seasons, *J. Agric. Meteorology,* 48: 575-582.

Baker J.T., Allen L.H.Jr., Boote K.J. and Pickering, N.B. (1997). Rice responses to drought under carbon dioxide enrichment. II.Photosynthesis and evapotranspiration. *Global Change Biology,* 3: 129-138.

Baker J.T. (2004). Yield responses of southern US rice cultivars to CO_2 and temperature. *Agric. and Forest Meteorology,* 122: 129-137.

Batts G.R., Ellis R.H., Morison J.I.L., Nkemka P.N., Gregory P.J. and Hadley P. (1998). Yield and partitioning in crops of contrasting cultivars of winter wheat in response to CO_2 and temperature in field studies using temperature gradient tunnels. *J. Agric. Sci.,* Cambridge, 130: 17-27.

Bazzaz F.A. and Garbutt K. (1988). The response of annuals in competitive neighbourhoods: effects of elevated CO_2, *Ecology*, 69: 937-946.

Bloom A.J., Smart D.R., Nguyen D.T. and Searles P.S. (2002). Nitrogen assimilation and growth of wheat under elevated carbon dioxide. *Proc. National Acad. Sci. USA*, 99: 1730-1735.

Bowes G. (1993). Facing the inevitable: plants and increasing atmospheric CO_2. *Ann. Rev. Plant Physiol. & Plant Mol. Biol.* 44: 309-332.

Brown R.A. and Rosenberg N.J. (1997). Sensitivity of crop yield and water use to change in a range of climatic factors and CO_2 concentrations: a simulation study applying EPIC to the central USA. *Agricultural and Forest Meteorology,* 83: 171-203.

Bunce J.A. and Sicher R.C. (2003). Daily irradiance and feedback inhibition of photosynthesis at elevated carbon dioxide concentration in *Brassica oleracea. Photosynthetica,* 41: 481-488.

Chakravarty N.V.K. and Gautam R.D. (2002). For warming mustard ephid. Misson mode project on development of weather based forewarning system for crop pest and diseases. IARI, New Delhi.

Cole C.V., Cerri C., Minami K., Mosier A., Rosenberg N., Sauerbeck D. *et al.*, (1996). Agricultural options for mitigation of greenhouse gas emissions. In: Watson, R., Zinyowera, M.C. and Moss, R (eds) *Climate Change 1995. Impacts,Adaptations and Mitigation of Climate Change*: Scientific and Technical Analyses. IPCC Working Group II, Cambridge University Press, Cambridge, UK, pp. 745-771.

Collette E. (1995). Will plants profit from high CO_2 ?. *Science.* 268: 654-655.

Conroy J.P. (1992). Influence of elevated atmospheric CO_2 concentrations on plant nutrition. *Aust. J. Botany,* 40: 445-456.

Cousins A.B., Adam N.R., Wall G.W., Kimball B.A. and Pinter P.J. (2001). Reduced photorespiration and increased energy use efficiency in young CO_2-enriched sorghum leaves. *New Phytol.* 150: 275-284.

De Costa W.A.J.M., Weerakoon W.M.W., Herath H.M.L.K. and Abeywardena R.M.I. (2003). Response of growth and yield of rice (*Oryza sativa*) to elevated atmospheric carbon dioxide in the subhumid zone of Sri Lanka. *J. Agron. and Crop Sci.* 189: 83-95.

Delgado E., Mitchell R.A.C., Parry M.A.J., Driscoll S.P., Mitchell V.J. and Lawlor D.W. (1994). Interacting effects of CO_2 concentration, temperature and nitrogen supply on the photosynthesis and composition of winter wheat leaves. *Plant, Cell and Environment* 17: 1205-1213.

Delucia E.H., Sasek T.W. and Strain B.R. (1985). Photosynthetic inhibition after long term exposure to elevated levels of atmospheric carbon dioxide. *Photosynthesis Research* 7: 175-184.

Drake B.G., Leadley P.W., Arp W.J., Nassiry D. and Curtis P.S. (1989). An open top chamber for field studies of elevated atmospheric CO_2 concentration on salt marsh vegetation. *Functional Ecology,* 3: 363-371.

Drake B.G., Azcon-Bieto J., Berry J., Dijkstra P., Farrar J., Gifford R.M., Gonzalez-Meler M.A., Koch G., Lambers H., Siedow J. and Wullschlegar S. (1999). Does elevated CO_2 concentration inhibit mitochondrial respiration in green plants? *Plant Cell and Environment* 22: 649-657.

Drake B.G., Gonzalez-Meler M.A. and Long S.P. (1997). More efficient plants: a consequenceof risingatmospheric CO_2? *Ann. Rev. Plant Physiol. and Mol. Biol.* 48: 609-639.

Ellis R.H., Craufurd P.Q., Summerfield R.J. and Roberts E.H. (1995). Linear relationship between carbon dioxide concentration and rate of development towards flowering in sorghum, cowpea and soybean. *Annals of Botany* 75: 193-198.

Esau K. (19778). Developmental features of the primary phloem of *Phaselous vulgaris* L. *Ann. Bot.* 42: 1-13.

Estiarte M., Penuelas J., Kimball B.A., Idso S.B., LaMorte R.L., Pinter P.J.Jr., Wall G.W. and Garcia R.L. (1994). Elevated CO_2 effects on stomatal density of wheat and sour orange trees. *J. Expt. Botany* 45: 1665-1668.

Eswaran H., Van Den Berg and Reich P. (1993). Organic carbon in soils of the world. *Soil Sci. Soc. America J.* 57: 192-194.

FAO (1995). *Production Year Book* 49: 70-71.

FAOSTAT (1997). Statistical Databases/Agriculture/crops/primary.

Foyer C.H. and Harbinson J. (1994). Oxygen metabolism and the regulation of photosynthetic electron transport. In; C.Foyer, P.Mullineaux, (eds) Causes of photooxidative stress and amelioration of defense systems in plants. CRC Press, Boca Raton, Fl, pp 1-43.

Frick J., Nielsen S.S. and Mitchell C.A. (1994). Yield and seed oil content response of dwarf, rapid-cycling *Brassica* to nitrogen treatments, planting density and carbon dioxide enrichment. *J. Amer. Soc. Hort.* 119: 1137-1143.

Garcia R.L., Long S.P., Wall G.W., Osborne C.P., Kimball B.A., Nie G.Y., Pinter P.J.Jr., Lamorte R.L. and Wechsung F. (1998). Photosynthesis and conductance of spring wheat leaves: field response to continuous free air atmospheric CO_2 enrichment. *Plant, Cell and Environment,* 21: 659-669.

Gesch R.W., Kang I.-H., Gallo-Megher M., Vu J.C.V., Boote K.J., Allen L.H.Jr. and Bowes G. (2003). Rubisco expression in rice leaves is related to genotypic variation of photosynthesis under elevated growth CO_2 and temperature. *Plant, Cell and Environment,* 26: 1941-1950.

Gifford R.M. (1995). Whole plant respiration and photosynthesis of wheat under increased CO_2 concentration and temperature: long term vs short term distinctions for modelling. *Global Change Biology,* 1: 385-396.

Goudriaan J. and Unsworth M.H. (1990). Implications of increasing carbon dioxide and climate change for agricultural productivity and water resources. In: Kimball B.A., Rosenberg N.R. and Allen L.H.Jr (eds) *Impact of Carbon Dioxide, Trace Gases and Climate Change on Global Agriculture,* Special Publication Number 53, American Soc. of Agronomy, Madison, Wisconsin, pp. 291-297.

Graham E.A. and Nobel P.S. (1996). Long term effects of a doubled atmospheric CO_2 concentration on the CAM species *Agave deserti. J. Expt. Botany,* 47: 61-69.

Habash D.Z., Paul M.J., Parry M.A.J., Keys A.J. and Lawlor D.W. (1995). Increased capacity for photosynthesis in wheat grown at elevated CO_2-the relationship between electron-transport and carbon metabolism. *Planta,* 197: 482-489.

Hall A.E. and Allen H.Jr. (1993). Designing cultivars for the climatic conditions of the next century. In: Buxton D.R., Shibles R., Forsberg R.A., Blad B.L., Asay K.H., Paulson G.M. and Wilson R.F. (eds) *International Crop Science Soc. of America*, Madison, Wisconsin. pp. 291-297.

Hamid A., Haque M.M., Khanam M., Hussain M.A., Karim M.A., Haliq Q.A., Biswas D.K., Gomasta A.R., Chaudhary A.M. and Uprety D.C. (2003). Photosynthesis, growth and productivity of rice under elevated CO_2. *Indian J. Plant Physiol.* 131, N.S., 253-258.

Hardy R.W.F. and Havelka V.D. (1975). Photosynthate as a major limiting N_2 fixation by field grown legumes with emphasis on soyabeans In Nutman PS (ed). International Biological Programme, 7. *Symbiotic nitrogen fixation in plants. Cambridge University Press,Cambridge.*

Heagle A.S., Body D.E. and Heck W.W. (1973). An open top chamber to assess the impact of air pollution on plants. *J. Environment Qual.* 2: 365-368.

Hendry G.R. and Kimball B.A. (1994). The FACE program. *Agriculture and Forest Meteorology* 70: 3-14.

Hocking P.J. and Meyer C.P. (1991). Effects of CO_2 enrichment and nitrogen stress on growth and partitioning of dry matter and nitrogen in wheat and maize. *Aust. J. Plant Physiol.* 18: 339-356.

Homma K., Nakagawa H., Horie T., Ohinishi H., Kim H.Y. and Ohnishi M. (1999). Energy budget and transpiration characterstics of rice grown under elevated CO_2 and high temperature conditions as determined by remotely sensed canopy temperatures. *Japanese J. Crop Sci.* 68: 137-145. (in Japanese with English abstract).

Horie T. (1993). Pridicting the effect of climate variation and elevated CO_2 on rice yield in Japan. *J. Agric. Meteorology,* 48: 567-574.

Horie T., Baker J.T., Nakagawa H., Matsui T. and Kim H.Y. (2000). Crop Ecosystem Responses to Climatic Change: Rice. *Climate Change and Global Productivity.* K.R. Reddy and H.F. Hodges (eds). pp. 81-106. Cambridge University Press, Cambridge, UK.

Horie T., Nakagawa H., Nakano J., Hamotani K. and Kim H.Y. (1995). Temperature gradient chambers for research on global environmental change. III. A system designed for rice in Kyoto, Japan *Plant, Cell and Environment,* 18: 1064-1069.

Houghton J.T., MeiraFilho L.G., Callander B.A., Harris N., Kattenberg A. and Maskell K. (eds) (1996). Climate Change 1995: *The Science of Climate Change* IPCC, Cambridge Univ. Press, Cambridge, UK, 572 pp.

Hunsaker D.J., Kimball B.A. Pinter P.J.Jr., Wall G.W. and LaMorte R.L. (1997). Soil water balance and wheat evatranspiration as affected by elevated CO_2 and variable soil nitrogen. In: *Annual Research Report 1997. US Water Conservation Laboratory,* USDA, ARS, Phoenix, Arizona, pp. 79-82.

Hurrell J.W. (1998). Relationships among recent atmospheric circulation changes, global warming and satellite temperatures. *Science Progress,* 81: 205-224.

Idso S.B. (1995). CO_2 and Biosphere: *The Incredible Legacy of the Industrial Revolution.* Spl. Publi. University of Minnesota, St. paul.

Imai K., Colman D.F. and Yanagisawa T. (1985). Increase of atmospheric partial pressure of carbon dioxide and growth and yield of rice (*Oryza sativa* L.). *Japanese J. Crop Science,* 54: 413-418.

Imai K. and Murata Y. (1978). Effect of carbon dioxide concentration on growth and dry matter production of crop plants. IV. After-effects of carbon dioxide treatments on apparent photosynthesis, dark respiration and dry matter production. *Japanese J. Crop Science,* 47: 330-335.

Jones P., Allen L.H.Jr., Jones J.W., Campbell W.J. and Boote K.J. (1984). Soybean canopy growth, photosynthesis and transpiration responses to whole season carbon dioxide enrichment. *Agron. J.* 76: 633-637.

Joshi M., Boyer J. and Kramer P. (1965). Growth, carbon dioxide exchange, transpiration and transpiration ratio of pineapple. *Botanical Gazette,* 126: 174-179.

Keeling C.D., Whorf M., Wahlen T.P. and Van der Plicht J. (1995). Interannual extremes in the rate of rise of atmospheric carbon dioxide since 1980. *Nature,* 375: 666-670.

Kim H.Y., Horie T., Nakagawa H. and Wada K. (1996). Effects of elevated CO_2 concentration and high temperature on growth and yield of rice. II. The effect on yield and its component of Akihikari rice. *Japanese J. Crop Sci.* 65: 644-651.

Kimball B.A., Kobayashi K. and Bindi M. (2002). Responses of agricultural crops to free air CO_2 enrichment. *Advances in Agronomy,* 77: 293-368.

Kimball B.A. (1985). Adaptation of vegetation and management practices to a higher carbon dioxide world. In: Strain B.R. and Cure J.D. (eds.) *Direct Effects of Increasing Carbon dioxide on Vegetation.* US Department of Energy, Carbon Dioxide Research Division, DOE/ ER-0238, Washington, DC, pp. 185-204.

Kimball B.A., Pinter P.J.Jr., Wall G.W., Garcia R.L., LaMorte R.L., Jac P., Frumau K.F.A. and Vugts H.F. (1997). Comparisons of responses of vegetation to elevated CO_2 in free air and open top chamber facilities. In: Allen L.H.Jr., Kirkham M.B., Olszyk D.M. and Whitman D.M. (eds.) *Advances in CO_2 Effects research.* Special Publication No. 61, American Soc. Agronomy, Madison, Wisconsin, pp. 113-130.

Kimball B.A. (1983). Carbon dioxide and agricultural yield: an assemblage and analysis of 430 prior observations. *Agronomy J.* 75: 779-788.

Kobayashi K., Okada M. and Kim H.Y. (1999). The free air CO_2 enrichment (FACE) with rice in Japan. In: Horie et al. (eds.) *World Food Security and Crop Production Technologies for Tomorrow.* Graduate School of Agriculture, Kyoto University, Kyoto, Japan. pp. 213-215.

Lal R., Kimble J.M., Follet R.F. and Cole V. (1998). *Potential of US Cropland for carbon sequestration and Greenhouse effect mitigation.* USDA-NRCS, Washington, DC. Ann Arbor Press, Chelsea, Michigan, 128 pp.

Lawlor D.W. and Mitchell R.A.C. (1991). The effects of increasing CO_2 on crop photosynthesis and productivity: a review of field studies. *Plant, Cell and Environment,* 14: 803-818.

Lawlor D.W. and Uprety D.C. (1993). Effects of water stress on photosynthesis of crops and Biochemical mechanism. *Photosynthesis and Crop productivity.* Kluwer, Academic Publishers, Dordrecht, Germany. pp. 419-450.

Lawlor D.W., Delgado E., Habash D.Z., Driscoll S.P., Mitchell V.J., Mitchell R.A.C. and Parry M.A.J. (1995). Photosynthetic acclimation of winter wheat to elevated CO_2 and temperature. In: Mathis P. (ed.) *Photosynthesis: From Light to Biosphere,* Vol. V. Kluwer Academic Publishers, Dordrecht, The Netherlands, pp. 989-992.

Leakey A.D.B., Bernacchi C.J., Dohleman F.G., Ort D.R. and Long S.P. (2004). Will photosynthesis of maize (*Zea mays*) in the US corn belt increase in future $[CO_2]$ rich atmosphere? An analysis of diurnal courses of CO_2 uptake under free-air concentration enrichment (FACE). *Global Change Biology,* 10: 951-962.

Leemans R. and Solomon A.M. (1993). Modeling the potential change in yield and distribution of the earth's crops under a warmed climate. *Climate Research,* 3: 79-96.

Lin W., Ziska L.H., Namuco O.S. and Bai K. (1997). The interaction of high temperature and elevated CO_2 of photosynthetic acclimation of single leaves of rice in situ. *Physiologia Plantarum,* 99: 178-184.

Long S.P. (1991). Modification of the response of photosynthetic productivity to rising temperature by atmospheric CO_2 concentrations: has its importance been underestimated? *Plant Cell and Environment,* 14: 729-739.

Long S.P. (1999). Environmental responses. In: Sage R.F. and Monson R.K. (eds) C_4 *Plant Biology*. Academic Press, San Diego, pp. 215-250.

Long S.P., Ainsworth E.A., Rogers A. and Ort D.R. (2004). Rising atmospheric carbon dioxide: Plants FACE the future. *Ann. Rev. Plant. Biol.* 55, 591-628.

Manderscheid R. and Wiegel H.J. (1997). Photosynthetic and growth responses of old and modern spring wheat cultivars to atmospheric CO_2 enrichment. *Agriculture, Ecosystems and Environment,* 64: 65-73.

Marina M.J., Mikkelsen T.N. and Rikke B.J. (2002). CO_2 exploitation and genetic diversity in winter varieties of oilseed rape (*Brassica napus*); varieties of tomorrow. *Euphytica,* 128: 75-86.

Maxon Smith J.W. (1977). Selections for response to CO_2 enrichment in glasshouse lettuce. *Horticultural Research,* 17: 15-22.

Mayeaux H.S., Johnson H.B., Polley H.W. and Malone S.R. (1997). Yield of wheat across a subambient carbon dioxide gradient. *Global Change Biology,* 3: 269-278.

McKee I.F. and Woodward F.I. (1994). CO_2 enrichment responses of wheat: interactions with temperature, nitrate and phosphate. *New Phytologist,* 127: 447-453.

Mishra R.S., Abdin M.Z. and Uprety D.C. (1999). Interactive effect of elevated CO_2 and moisture stress on photosynthesis, water relations and growth of *Brassica* species. *J. Crop Sci. and Agron.* 182: 223-230.

Mitchell R.A.C., Mitchell V.J., Driscoll S.P., Franklin J. and Lawlor D.W. (1993). Effects of increased CO_2 concentration and temperature on growth and yield of winter wheat at 2 levels of nitrogen application. *Plant Cell and Environment,* 16: 521-529.

Mitchell R.A.C., Lawlor D.W., Mitchell V.J., Gibbard C.L., White E.M. and Porter J.R. (1995). Effects of elevated CO_2 concentration and increased temperature on winter wheat-test of ARCWHEAT1 simulation model. *Plant Cell and Environment,* 18: 736-748.

Mitchell R.A.C., Black C.R., Burkart S., Burke J.I., Donelly A., deTemmerman L., Fangmeir A., Mulholland B.J., Theobald J.C. and Oijen M. van (1999). Photosynthetic responses in spring wheat grown under elevated CO_2 concentrations and stress conditions in the European multiple-site experiment ESPAC-wheat. *European J. Agron.* 10: 205-214.

Morrison J.I.L. and Gifford R.M. (1983). Stomatal activity to carbon dioxide and humidity. A comparison of two C_3 and two C_4 grass species. Plant Physiol. 71: 789-796.

Morrison J.I.L. and Gifford R.M. (1984). Plant growth and water use with limited water supply in high CO_2 concentrations. II. Plant dry weight, partitioning and water use efficiency. *Aust. J. Plant Physiol.* 11: 375-384.

Morison J.I.L. and Lawlor D.W. (1999). Interactions between increasing CO_2 concentration and temperature on plant growth. *Plant, Cell and Environment,* 22: 659-682.

Moya T.B., Ziska L.H., Weldon C., Quilang J.E.P. and Jones P. (1997). Microclimate in open top chambers: implication for predicting climate change effects on rice production. *Transaction of the American Soc. Agric. Engineers,* 40: 739-747.

Nakagawa H., Horie T., Nakano J., Kim H.Y., Wada K. and Kobayashi M. (1993). Effect of elevated CO_2 and high temperature on growth and development on rice. *J. Agric. Meteorology,* 48: 799-802.

Nakagawa H., Horie T., Kim H.Y., Ohnishi H. and Homma K. (1997). Rice responses to elevated CO_2 concentrations and high temperatures. *J. Agric. Meteorology,* 52: 797-800.

Nakayama F.S. and Kimball B.A (1988). Soil carbon dioxide distribution and flux within open top chamber. *Agronomy, J.* 80: 394-398.

Natr I., Driscoll S. and Lawlor D.W. (1996). The effect of increased CO_2 concentration on the dark respiration rate of etiolated wheat seedlings. *Cereal Research Communications,* 24(1): 53-59.

Nie G.Y., Hendrix D.L., Webber A.N., Kimball B.A. and Long S.P. (1995). Increased accumulation of carbohydrates and decreased photosynthetic transcript levels in wheat grown at an elevated CO_2 concentration in the fields. *Plant Physiol.* 108: 975-981

Nobel P.S. (2000). Crop ecosystem responses to climatic change: Crassulacean acid metabolism crops. Reddy K.R. and Hodges H.F. (eds.) *Climate change and Global Crop productivity.* pp. 315-331. Cambridge University Press.

Pearson P.N. and Palmer M.R. (2000). Atmospheric carbon dioxide concentration over the past 60 million years. *Nature,* 406: 695-699.

Poorter H. (1993). Inter specific variation in the growth response of plants to an elevated ambient CO_2 concentration. *Vegetatio,* 104/105: 77-97.

Poorter H., Roumet C. and Campbell B.D. (1996). Inter specific variation in the growth responses of plants to elevated CO_2: a search for functional types. In: Korner C. and Bazzaz F.A. (eds.) *Carbon Dioxide, Populations and Communities.* Academic press, San Diego, pp. 375-412.

Poorter H., Gifford R.M., Kriedemann P.E. and ChinWong S. (1992). A quantitative analysis of dark respiration and carbon content as factors in the growth response of plants to elevated CO_2. *Aust. J. Botany,* 40: 501-513.

Rabha B.K. and Uprety D.C. (1998). Interactive effect of elevated CO_2 and moisture stress on *Brassica juncea. Photosynthetica,* 35: 597-602.

Rao M.V., Beverly A.H., and Ormrod D.P. (1995). Amelioration of ozone induced oxidative damage in wheat plants grown under high carbon dioxide. *Plant Physiol* 109: 421-432.

Rawson H.M. (1995). Yield responses of two wheat genotypes to carbon dioxide and temperature in field studies using temperature gradient tunnels. *Aust. J. Plant physiol.* 22: 23-32.

Reddy G.V.P., Tossavainen P., Nerg A.M. and Holopainen J.K. (2004). Elevated atmospheric CO_2 affects the chemical quality of *Brassica* plants and the growth rate of the specialist, *Plutella xylostella,* but not the generalist, *Spodoptera littoralis. J. Agric. Food Chem.* 52: 4185-4191.

Reekie E.G., Mac Dougall G., Wong I. and Hicklenton P.R. (1998). Effect of sink size on growth response to elevated atmospheric CO_2 within the genus *Brassica. Can. J. Bot.* 76: 829-835.

Rogers A., Fischer B.U., Bryant J., Frehner M., Blum H., Raines C.A. and Long S.P. (1998). Acclimation of photosynthesis to elevated CO_2 under low N nutrition is effected by the capacity for the assimilation utilization. Perennial ryegrass under free-air CO_2 enrichment (FACE). *Plant Physiol.* 118: 683-689.

Rogers H.H., Sionit N., Cure J.D., Smith J.M. and Bingham G.E. (1984). Influence of elevated carbon dioxide on water relations of soybean. *Plant Physiol.* 74: 233-238.

Rogers G.S., Milham P.J., Thibaud M.C. and Conroy J.P. (1996). Interactions between rising CO_2 concentration and nitrogen supply in cotton. I. Growth and leaf nitrogen concentration. *Aust. J. Plant physiol.* 23: 119-125.

Rogers G.S., Gras P.W., Batey I.L., ilham P.J., Payne L. and Conroy J.P. (1998). The influence of atmospheric CO_2 concentration on the protein, starch and mixing properties of wheat flour. *Aust. J. Plant Physiol.* 25: 387-393.

Rogers H.H., Bingham G.E., Cure J.D., Smith J.M. and Surano K.A. (1983). Responses of selected plant species to elevated carbon dioxide in the field. *J. Environ. Quality* 12: 569-574.

Rosegrant M.W., Paisner M.S., Meijer S. *et. al.* (2001) *Global food projections* to 2020: *emerging trends and alternative futures,* International Food Policy Research Institute.

Rosenzweig C., and Parry M.L. (1994). Potential impact of climate change on world food supply. *Nature* 367, 133-138.

Samarakoon A.B. and Gifford R.M. (1996). Elevated CO_2 effects on water use and growth of maize in wet and drying. *Aust. J. Plant Physiol.* 23: 53-62.

Samarakoon A.B., Muller W.J. and Gifford R.M. (1995). Transpiration and leaf area under elevated CO_2-effects of soil water status and genotype in wheat. *Aust. J. Plant Physiol.* 22: 33-44.

Sheen J. (1990). Metabolic repression of transcription in higher plants. Plant Cell 2: 1027-1038.

Shewry P.R., Broadhead J., Harwood J., Williams M., Grimwade B., Freedmore R., Napier J., Fido R., Lawlor D.W. and Tatum A.S. (1994). Environmental and developmental regulation of gluten protein synthesis. In: *Wheat kernel proteins. Molecular and Environmental aspects.* Proc. Meeting Univ. Tuscany, Viterbo, Italy, pp 273-275.

Singh U. and Padilla J.L. (1995). Simulating rice response to climate change. In: Rosenzwseig, C. *et al.* (eds.) *Climate Change and Agriculture; Analysis of Potential International Impacts.* American Soc. Agronomy, Madison, Wisconsin, pp 99-122.

Singh B., Elmaayar M., Andre P., Bryant C.R. and Thouez J.P. (1998). Impacts of a GHG-induced climate change on crop yields: effects of acceleration in maturation, moisture stress and optimal temperature. *Climatic Change,* 38: 51-86.

Stitt M. and Krapp A. (1999). The interaction between elevated carbon dioxide and nitrogen nutrition: the physiological and molecular background. *Plant Cell and Environment.* 22: 583-621.

Theobald J.C., Mitchell R.A.C., Parry M.A.J. and Lawlor D.W. (1998). Estimating the excess investment in Ribulose-1,5-bisphosphate carboxylase/oxygenase in leaves of spring wheat grown under elevated CO_2. *Plant Physiol.* 118: 945-955.

Uprety D.C. (2004). Effect of rising atmospheric CO_2 on the photosynthesis and productivity of crop plants. *National Fellow Project consolidated report* (2000-2004), *Indian Council of Agricultural Research (ICAR), New Delhi, India.*

Uprety D.C. and Tomar V.K. (1993). Photosynthesis and drought resistance of *Brassica carinata* and its parent species. *Photosynthetica.* 29: 321-327.

Uprety D.C. (1998). Carbon dioxide enrichment technology: Open top chambers a new tool for global climatic research. *J. Sci. Industr. Rese.* 57: 266-270.

Uprety D.C. and Mahalaxmi V. (2000). Effect of elevated CO_2 and nitrogen on some physiological characters in *Brassica juncea. J. Crop Sci. and Agron.* 184: 271-276.

Uprety D.C., Abrol Y.P., Bansal A.K. and Mishra R.S. (1996). Comparative study on the effect of elevated CO_2 in mungbean and maize cultivars. *Proc. Indian Nat. Sci. Academy(B)* 62: 29-36.

Uprety D.C., Dwivedi N. and Mohan R. (1998). Characterization of CO_2 responsiveness in *Brassica oxycamp* interspecific hybrid. *J. Crop Sci. and Agron.* 180: 7-13.

Uprety D.C., Dwivedi N., Mohan R and Paswan G. (2001). Effect of elevated CO_2 concentration on leaf structures of *Brassica juncea* under water stress. *Biol. Planta.* 44: 149-152.

Uprety D.C., Dwivedi N, Jain V., Mohan R. (2002). Effect of elevated CO_2 on the stomatal parameters of rice cultivars. *Photosynthetica.* 40: 315-319.

Uprety D.C., Dwivedi N., Jain V., Mohan R., Saxena D.C., Jolly S. and Paswan G. (2003). Responses of rice varieties to elevated CO_2. *Biol. Plantarum,* 46: 35-39.

Uprety D.C., Garg S.C., Tiwari M.K. and Mitra A.P (2000). Crop responses to elevated CO_2: Technology and Research (Indian studies). *Global Environmental Research*, 3: 155-167.

Uprety D.C., Kumari S. Dwivedi N, and Mohan R. (2000). Effect of elevated CO_2 on the growth and yield of rice variety 'Pusa 834'. *Indian J. Plant Physiol* 5(N.S.): 105-107.

Uprety D.C., Mishra R.S. and Abrol Y.P. (1995). Effect of elevated CO_2 and moisture stress on photosynthesis and water relations in *Brassica* species. *J. Crop Sci. and Agron.* 175: 231-237.

Uprety D.C., Dwivedi N., Saxena D.C., Raj A. and Paswan G. (2004). Impact of enhanced CO_2 concentration on crop growth. *Agriculture, Forestry and Natural ecosystem* pp 19-22.

Uprety D.C., Dwivedi N. and Mohan R. (1997). Interactive effect of rising atmospheric CO_2 and drought on nutrient constituents of *Brassica juncea* seeds. *Science and Culture* 63: 291-292.

Uprety D.C. and Rabha B.K. (1999). Interactive effect of elevated CO_2 and moisture stress on the carbon nitrogen dynamics in *Brassica juncea* plants *Biol. Planta.* 42: 133-136.

Van Oosten J.J. and Besford R.T. (1995). Some relationship between the gas exchange, biochemistry and molecular biology of photosynthesis during leaf development of tomato plants after transfer to different carbon dioxide concentrations. *Plant Cell and Environment,* 18: 1253-1266.

Van Oosten J.J., Wilkins D., and Besford R.T. (1994). Regulation of the expression of photosynthetic nuclear genes by CO_2 is mimicked by regulation of carbohydrates: a mechanism for the acclimation of photosynthesis to high CO_2. *Plant, Cell and Environment,* 17: 913-923.

Vu J.C.V., Allen L.H.Jr., Boote K.J. and Bowes G. (1997). Effects of elevated CO_2 and temperature on photosynthesis and rubisco in rice and soybean. *Plant, Cell and Environment,* 20: 68-76.

Wall G.W., Adam N.R., Brooks T.J., Kimball B.A., Ademson F.J., Pinter P.J.Jr. and LaMorte R.I. (1997). Acclimation of photosynthetic apparatus of spring wheat grown under free air CO_2 enrichment (FACE) and two soil nitrogen regimes. In *Annual research report 1997,US Water Conservation Lab*, USDA ARS, Phoenix, Arizona, pp 61-64.

Watson R.T., Zinyowera M.C. and Moss R.H. (1996). Climate change 1995: *Impacts,Adaptations and Mitigation.Summary for policy makers.*Contribution of working group II to the second assessment report. IPCC, Cambridge University Press. Cambridge, U.K., 22 pp.

Webber A.N., Nie G.Y. and Long S.P. (1994). Acclimation of photosynthetic proteins to rising atmospheric CO_2. *Photosynthesis Research* 39: 413-426.

Weerakoon W.M.W., Ingram K.T. and Moss D.N. (2000). Atmospheric carbon dioxide and fertilizer nitrogen effects on radiation interception by rice. *Plant and Soil,* 220: 99-106.

Wheeler T.R., Batts G.R., Ellis R.H., Hadley P. and Morison J.I.L. (1996). Growth and yield of winter wheat (*Triticum aestivum*) crop in response to CO_2 and temperature. *J. Agr. Sci. Cambridge,* 127: 35-48.

Williams M., Shewry P.R., Lawlor D. WandHarwood J.L. (1995). The effects of elevated temperature and atmospheric carbon dioxide concentration on the quality of grain lipids in wheat (*Triticum aestivumL.*) grown at two levels of nitrogen application. *Plant, Cell and Environment,* 18: 999-1009.

Wittwer S.H. (1986). Worldwide status and history of CO_2 enrichment:an overview. In: Enoch H.Z. and Kimball B.A. (eds.) *Carbon dioxide Enrichment of Greenhouse Crops*, Vol.1. CRC Press, Boca Raton, Florida, pp. 3-15.

Wong S.C. (1979). Elevated atmospheric partial pressure of CO_2 and plant growth. *Oecologia,* 44: 68-74.

Wood C.W., Torbert H.A., Rogers H.H., Runion G.B. and Prior S.A. (1994). Free air CO_2 enrichment effects on soil carbon and nitrogen. *Agric. For. Meteorol.* 70: 103-116.

Young K.J. and Long S.P. (2000). Crop ecosystem responses to climatic change: maize and sorghum. Climate Change and Global Crop Productivity. Reddy, K.R. and Hodges, H.F. (eds.) CAB International 2000. pp. 107-131.

Zhu J., Bartholomew D.P. and Goldstein G. (1997). Effect of elevated carbon dioxide on the growth and physiological responses of pineapple, a species with crassulacean acid metabolism. *J. American Soc. Hort. Sci.* 122: 233-237.

Ziska L.H., Weerakoon W., Namuco O.S. and Pamplona R. (1996). The influence of nitrogen on the elevated CO_2 response in field-grown rice. *Aust. J. Plant Physiol.* 23: 45-52.

Ziska L.H., Hogan K.P., Smith A.P. and Drake B.G. (1991). Growth and photosynthetic response of nine tropical species with long term exposure to elevated carbon dioxide. *Oecologia,* 86: 383-389.

Ziska L.H., Namuco O., Moya T. and Quilang J. (1997). Growth and yield response of field-grown tropical rice to increasing carbon dioxide and air temperature. *Agronomy J.* 89: 45-53.

Ziska L.H., Sicher R.C. and Bunce J.A. (1999). The impact of elevated carbon dioxide on the growth and gas exchange of three C_4 species differing in CO_2 leak rates. *Physiologia Plantarum,* 105: 74-80.

Chapter **22**

Bioinformatics in Biodiversity Management

Pramod Tandon and Pallavi Bhattacharjee
Bioinformatics Centre, North-Eastern Hill University, Shillong - 793 022, Meghalaya

Biodiversity is very important for human survival and is the most precious and fragile resource in the world. Yet, in an increasingly modern and technological world, people often forget how fundamental biodiversity is to daily life and are unaware of the impact of its loss. We are eliminating populations and species faster than we can discover new ones. Nations can prosper only if their biodiversity is preserved. While the Earth's biotic inventory is far from complete, our records and data from specimens, fieldwork and other research have enormous value for building our knowledge of biodiversity on our planet and for applied uses in solving real-world problems. There are two areas in biology where enormous amounts of information are generated. One is in molecular biology, which deals with the base sequences in DNA and amino acid sequences in proteins and the other is the enormity of biodiversity related data. Information Technology is being used to tackle these problems with procedures, which come under the label of Bioinformatics. Bioinformatics harnesses the power of computational and information technologies to organise and analyse biological data and deliver them to users throughout the world. In the present day world, the Internet and World Wide Web (www) have proved to be powerful tools for linking and utilising the data of biodiversity and at the same time, help devicing methods to maintain and exploit it in a sustainable manner.

INTRODUCTION

Biodiversity or Biological Diversity is the term used for the variability among living organisms from all sources. These include diversity within species, between species and of ecosystems. Diversity is the key for ensuring the continuance of life on earth. It is also a fundamental requirement for the adaptation, survival and continued evolution of species. The concept of biodiversity represents how life is organised and how it interacts on our planet. These interactions can take place on scales ranging from the smallest at the gene/chromosome level to organisms, ecosystems and even to entire landscapes. The entire gamut of biodiversity is often referred to as *'ecosystem complexity'* (Sala *et al.*, 1999).

Bioinformatics is a discipline that generates computational tools, databases and methods to organise and analyse biological data. These data are derived from research collections, experiments, remote sensing, modelling, database searches and instrumentation – and delivered to users throughout the world. It comprises the study of DNA structure and functions, gene and protein expression; protein production, structure and functions; and genetic regulatory systems and clinical applications.

Biological research has generated a huge quantity of data, which requires high-throughput and large-scale technologies to manipulate this data. This trend is growing exponentially with a shift in emphasis from individual biomolecules to analysis of how they interact in complex networks, which control the developmental and physiological processes of whole biological systems. The ultimate aim of this research is to relate its results directly or indirectly to human health. This transition has increased the importance of bioinformatics, and raises key challenges which make it imperative that computer scientists work closely with biologists to refine existing bioinformatics tools and develop new ones.

REQUIREMENT OF BIODIVERSITY CONSERVATION

Biodiversity conservation is of prime importance to the Earth's health and would ultimately determine our fate and existence. However, our ignorance of most of the world's biodiversity, particularly at the level of individual species, poses a major threat. Under the current scenario, about 25% of the higher plant species are expected to be lost in the next few decades and another 25% by the end of the 21st century (Tandon, 2004b, Singh and Khurana, 2002). There is a heavy loss of species, populations, and ecosystems due to the loss of habitat and destruction. Alterations in ecosystem composition, such as the loss or decline of a species, can lead to a loss of biodiversity. The threats to biodiversity posed by the destruction of the natural environment by urbanisation, global warming and other factors have given rise to a high level of public awareness in this regard (Tandon and Kumaria, 2005).

Although biodiversity conservation is not a new field in itself, it has realigned itself and brought the existing sciences of systematics, ecology and evolution closer. It has also given a platform for a unified cry for the conservation of our ecosystems. Conservation is not an end in itself, but a means of ensuring that plant and animal genetic resources are available for use by present and future generations. Efforts for conservation and management of our natural resources must derive from a set of clear objectives, mechanisms for action and commitment from all stakeholders. Apart from this, halting the process of degradation and species loss requires specialised solutions and an understanding of ecological processes. Protecting biodiversity does not merely involve setting aside chunks of area as reserves. Instead, all ecological processes that maintains the area's biodiversity, involving complex interactions between several species of plants and animals, need to be conserved (Terborgh, 1999). "Ecologically destructive economic activities" are inefficient, not merely because of the resulting resource misallocation, but also because of the excessive scale of activity levels. In order to bring about sustainable resource conservation and management, it is essential to adopt several different approaches for managing our forests and biodiversity.

ROLE OF BIOINFORMATICS IN BIODIVERSITY CONSERVATION

Bioinformatics is a modern science that applies database and computer modelling technologies for the study of biodiversity and habitats. It has been realised that the first step towards biodiversity conservation would be to establish an elaborate catalogue that would describe and classify all the surviving species of the world and simultaneously, take appropriate steps for conservation. This would undoubtedly be an overwhelming task and would need sustained and dedicated effort, as revealed by the researchers and biotechnicians of the Human Genome Project. This is where the role of informatics has gained importance. In the current scenario, bioinformatics is placed at the core of contemporary biology, in fields ranging from structural biology to genomics to biomedical imaging. Ready access to data and analytical tools has fundamentally changed an investigator's approach to problems. Today, Internet presents state-of-the-art tools for linking and utilising the extraordinary assets of natural history institutions.

Biodiversity informatics helps build tools to provide a standard format for the collection of data, offers an infrastructure for linking collections and supports predictive models for habitat expansion or destruction. A number of software packages relating to biodiversity and molecular systemics are available on the Internet. These are either downloadable shareware or are web-based applications. It is important to note that information management for biodiversity conservation is somewhat different from traditional scientific disciplines, which use mechanistic, laboratory and field studies. The vision is to create information systems for the inventory, analysis, reporting and visualisation of the vast natural resources and biological life on Earth. Currently, one of the main technologies for the inventory of biodiversity is the Geographic Information Systems, which is software for digitising geographic information and producing maps. [*http://www.usgs.gov*]

One of the major problems in the management of biodiversity information is its sheer magnitude and variety. The data is available in different formats. Moreover, biological data can be politically and commercially sensitive, leading to conflicts of interests. Because of all these complexities, it is feared that biodiversity information may not be very amenable to automatic correlation, analysis, synthesis and presentation (Schnase, 2000). Integration and analysis of such diverse data requires novel data mining tools that are either under development or still at the conceptual level. Coordinated effort is required nationally to collate, store, analyse and interpret all data related to biodiversity. Bioinformatics can play a determinable role in biodiversity conservation.

COMPUTERISATION OF BIODIVERSITY DATA

Till recently, biodiversity data were held as card indexes and taxonomic monographs, as well as physical specimens in museums and herbaria. Data storage and retrieval are vital for the research community. This was true even before gene-sequencing efforts put data storage and its analysis in the centrefold of active taxonomy research. Recorded information on specific taxonomic groups was only available to a few experts affiliated to various premium institutions of the world. In addition, information on biodiversity was only available after detailed search of a number of disjoined databases or could only be collected from paper-based libraries. Databases tailored exclusively for biodiversity conservation studies, based on field observations and scientists' notebooks, in addition to museum data, are perhaps the first step in this direction to manage the existing biodiversity information.

Estimates say that we know only 10% of the species living on earth (Lane *et al.*, 2000). In India the Ministry of Environment and Forests (MoEF) has developed a catalogue comprising 47,000 species of plants and 80,000 species of animals (MoEF, 1999). A number of species still remain undiscovered, as roughly 30% of India's geographical area still remains to be fully explored. An inventory and a comprehensive informatics system has to be developed which will help in the efforts towards the conservation of biodiversity (Gundu *et al.*, 2005).

The gradual involvement of information technology to biological science has, in effect, given birth to the bioinformatics revolution, which is finally enabling researchers to communicate effectively with each other and derive the benefits of this modern science. Computerisation of large amounts of biodiversity data has developed databases, which are now available in the public domain through the Internet and are easily accessible to one and all (Sugden and Pennisi, 2000). However, just getting biodiversity related information on-line is not the only issue here, but studying this data and devising ways to protect our precious biodiversity is the ultimate aim. Specific examples of research directions and applications to the management and conservation of biodiversity are:

- Large distributed biological databases
- Biologically relevant database
- Species distribution modelling
- Biodiversity modelling
- Ecological process modelling
- Reserve system planning
- Error estimation
- Visualisation/map making
- Interactivity

(Source: Venkatesh and Bansal, 2005)

BIODIVERSITY DATABASES

Global communication has expanded remarkably since the inception of the internet. Technical advances have made the distribution of data from major centres to remote parts of the world possible, if those data are in digital form. Biodiversity is found around the world; however, it is not distributed evenly across the face of the planet. Then information about biodiversity (natural history collections, library materials, databases) likewise, is not distributed evenly around the globe. Three-quarters or more of the data about biodiversity are stored in the developed world. However, most of the data that may be needed can't be transferred because, either they are not digitised, or the capacity to handle digital information is lacking, or both.

Information on different aspects of biodiversity is expanding rapidly. When genome sequencing projects started to pour data for the scientific community, computer-based databases were established for processing, storing and sharing this data all around the world. However, as Maurer *et al.,*. (2000) rightly observed, large sophisticated databases do not just happen. They have to be planned carefully with interoperability in mind. Recent browser technology, such as extensible markup language (XML), allows a user to search several databases. Benefits of this database approach are manifold, some of them being: reduction of redundancy and inconsistency of information, fast and simultaneous sharing of data for a large number of users, standards can be maintained and last, but not the least, integrity and security restrictions of data information can be enforced.

SOME OF THE GLOBAL BIODIVERSITY DATABASES

Species 2000 is a "Federation" of database organisations working closely with users, taxonomists and sponsoring agencies. It is an ambitious undertaking, which aims to enumerate all known species of plants, animals, fungi and microbes on earth, in a comprehensive resource that ties together a variety of smaller online taxonomic indexes.
[*http://www.sp2000.org/*]

Species Analyst provides access to a variety of natural history databases through a Web interface. Data includes museum collection information such as date, specimen, latitude and longitude of the collection site and the collector's name, etc. This can be downloaded in a variety of formats. The site also generates a global distribution map of the collection sites for a given query and can link users to online tools for further analysis and modelling of the data at the San Diego Supercomputer Centre.
[*http://speciesanalyst.net/*]

Global Biodiversity Information Facility is responsible for the digitisation and global dissemination of primary biodiversity data, so that people from all countries can benefit from this information. This establishes a standard method for data exchange about specimens for researchers in different countries.
[*http://www.gbif.org/*]

Tree of life (ToL) is a collaborative effort of biologists from around the world. In more than 3000 World Wide Web pages, the project provides information about the diversity of organisms on earth, their evolutionary history (phylogeny) and characteristics. Each page contains information about a particular group of organisms. ToL pages are linked to one another hierarchically, in the form of the evolutionary tree of life. Starting with the root of all life on earth and moving out along diverging branches to individual species, the structure of the ToL project thus, illustrates the genetic connections between all living things.
[*http://www.tolweb.org/tree/phylogeny.html*]

Integrated Taxonomic Information System is a cooperative effort of U.S., Canadian and Mexican agencies, building "an online, scientifically credible list of biological names" for North American taxa.
[*http://www.itis.usda.gov/*]

NatureServe is an "online encyclopedia of life" sponsored by the Association for Biodiversity Information. It provides data on nomenclature, conservation, geographic distribution and the life history of more than 50,000 U.S. and Canadian plant and animal species and ecological communities. Users can search by common or scientific species name, species association and other criteria.
[*http://www.natureserve.org/*]

In addition to the above, the following sites provide links to biodiversity information:

1. http://nature.ac.uk
2. http://www.nscalliance.org/bioinformatics/index.asp
3. http://www.biodiv.org/programmes/default.shtml

SOME OF THE INDIAN BIODIVERSITY DATABASES

Biotechnology Information System (BTISNet)

It is a major Satellite and Terrestrial network on biotechnology in the country, set up by the Department of Biotechnology (DBT), Government of India. It has established and networked various bioinformatics centres in the country, that are actively involved in research and development in the field of modern biology and bioinformatics. BTISnet has now become the single largest information resource for all references to biotechnology related literature, scientific data, patent information, policy matters and related issues in India.
[*http://www.btisnet.nic.in/*]

Environmental Information System (ENVIS)

It is a decentralised system with a network of distributed subject-oriented Centres, ensuring integration of national efforts in environmental information collection, collation, storage, retrieval and dissemination to all concerned (decision makers,

policy planners, scientists, engineers and research workers, etc.), all over the country.
[*http://www.envis.nic.in/*]

Agricultural Databases and Information on Sacred Groves

M.S. Swaminathan Research Foundation is involved in genetically characterising coastal bioresources, in particular mangroves, identifying, isolating and characterising novel genetic combinations from mangroves.
[*http://www.mssrf.org*]

The ICRISAT programme

'International Crops Research Institute for the Semi-Arid Tropics' is an international organisation devoted to science based agricultural development. Its work involves helping countries apply science to increase crop productivity and food security, reduce poverty and protect the environment. ICRISAT focusses on the farming systems of the semi-arid tropical areas of the developing world, where erratic rainfall, low soil fertility and extreme poverty are formidable constraints to agricultural development. ICRISAT conserves the seeds of 113,000 lines or types of crops and related information that are important for diet.
[*http://www.icrisat.org*]

Agricultural Research Information Network (ARISNET)

It is a network of agricultural research, extension and educational institutions in India over a Satellite based Computer Communication Network of NIC called NICNET. It provides global access of information to scientists and facilitates them to build up their links with the national and international research community and promotes voluntarily exchange of information between different institutions and sharing of their resources.
[*http://www.dacnet.nic.in/arisnet/*]

Biodiversity Characterisation using Remote Sensing/Geographic Information Systems

This is a multi-institutional programme on bioprospection of biological wealth, jointly supported by Department of Space (DOS) and DBT. The study includes integrating spatial data, like land use and land cover data, disturbance regimes and biological richness maps with non-spatial data like taxonomic and genetic information and creating landscape level information linked with a comprehensive plant species database.
[*http://www.gisdevelopment.net/magazine/gisdev/2002/nov/mabc.shtml*]

BODHI

It is developed at the Supercomputer Education and Research Centre, Indian Institute of Science, Bangalore. It is a database system, which stores plant biodiversity information. The unique feature of BODHI is that it seamlessly integrates diverse types of data, including taxonomic characteristics, spatial distributions and genetic sequences, thereby spanning the entire range from molecular to organism-level information.
[*http://eprints.iisc.ernet.in/archive/00001033/*]

CDROMs on Marine Prawns, Marine Crabs, Mangroves, Lignicolous Fungi and corals of India

National Institute of Oceanography has created database that collates information on marine biology and biotechnology. The published literature about the biota in marine and estuarine environment of India is made available in an electronic format.
[*http://www.nio.org*]

Endemic Trees of the Western Ghats

This database presents the geographic database of the endemic evergreen and semi-evergreen tree species of the Western Ghats, with maps, species synonyms, superimposition of forest types and species distribution areas with pictures and graphics.
[*http://www.etfrn.org/etfrn/newsletter/nl22_pub.html#datta*]

Ethnobotany

National Botanical Research Institute (NBRI) is involved in extensive and intensive ethnobotanical surveys for the collection, identification and documentation of plants and plant products of ethnobotanical value, used by the tribals of Uttar Pradesh, Uttaranchal, Madhya Pradesh, Himachal Pradesh and Andaman and Nicobar Island.
[*http://www.nbri-lko.org/randdarea/biodiversity/ethnobotany.htm*]

Sampada

It is a natural science collection computerisation initiative by the Information Division of the National Chemical Laboratory, Pune. It is an application for the storage and retrieval of data from biological collections. This includes information about museum specimens, field information and notes pertaining to biodiveristy.
[*http://www.ncbi.org.in/sampada*]

Digitally Compiled Flora

Jawaharlal Nehru Centre for Advanced Scientific Research has a collation of data from over 100 Regional Flora, Monographs and other literature. [*http://www.jncasr.ac.in/bdu/*]

Medicinal plants database

Foundation for Revitalisation of Local Health Traditions's (FRLHT) contains botanical database of medicinal plants and provides correlation between over 0.1 million botanical and vernacular names of around 6000 plant species (herbs, shrubs, trees, climbers, orchids, grasses, tubers and even lichens), with 5000 images and over 460 literature references to corresponding systems of Indian medicine viz., Ayurveda, Siddha, Unani, Swa. Rigpa (Tibetan), folk medicine, homeopathy and allopathy.
[*http://www.frlht-india.org/html/crg.htm*]

National Register of Green Grassroots Innovations and Traditional Knowledge

National Innovation Foundation helps grassroot innovators and traditional knowledge holders build their links with institutional scientists, technologists and designers so as to add value in their technologies. It provides support to help convert innovations into enterprises, protects their intellectual property rights and creates a culture of creativity and innovation in society.
[*http://www.nifindia.org/5th_announcement.htm*]

National Wildlife Database and Zoo Database

Wildlife Institute of India maintains a National Wildlife database that deals with all the protected areas and biodiversity of wild life. It maintains an information system on zoo and wildlife.
[*http://www.envfor.nic.in/wii/wii.html*]

Birds of India

Centre for Biodiversity Informatics at the National Chemical Laboratory, Pune, maintains a checklist of Indian waterbirds and wetland dependant birds.
[*http://www.ncbi.org.in/wbni/final_checklist.html*]

Plants of India and Legume Database of South Asia

NBRI has created a specialised database for species diversity on Leguminosae. The premier purpose of ILDIS is to provide a global species diversity information service of over 19000 taxa of the plant family Leguminosae.
[*http://www.nbri-lko.org/randdarea/bioinformatics/bioinformatics.htm*]

Sahyadri

It is a database of the Western Ghats flora and fauna and also a Biodiversity Bibliography developed and maintained by the Centre for Ecological Sciences, Indian Institute of Science.
[*http://wgbis.ces.iisc.ernet.in/biodiversity/*]

Helminth Parasite Spectrum of North-East India

The Bioinformatics Centre at North-Eastern Hill University, Shillong, maintains a database of parasites of NE India. Software has the facility to view scientific classification details, host, habitat, place of occurrence and images of parasites. This database will be available for public access via the Internet in the near future.

MOLECULAR DATABASES

Applied research in biodiversity is critically dependent on molecular and genetic databases. Such databases, which can be used for research on biodiversity, are being developed and maintained by various organisations at the global level. The three major organisations working in this field are:

- National Centre for Biotechnology Information (NCBI) in USA
- European Molecular Biology Laboratory (EMBL) in Germany
- DNA Database of Japan (DDBJ) in Japan

The above three databases collaborate with each other through data exchange and information on the internet and by regularly holding meetings with the International DNA Data Banks Advisory Meeting and the International DNA Data Banks Collaborative Meeting. In addition to the above three, there are numerous organisations that are actively involved in the development and maintenance of biotechnology databases in the world.

NCBI is an international resource for molecular biology information. NCBI creates public databases, conducts research in computational biology, develops software tools for analysing genome data and disseminates biomedical information — all for the better understanding of molecular processes affecting human health and diseases.
[*http://www.ncbi.nlm.nih.gov/*]

EMBL maintains the EMBL Nucleotide Sequence Database (also known as EMBL-Bank) in an international collaboration with GenBank (under NCBI, USA) and the DNA Database of Japan (DDBJ).
[*http://www.ebi.ac.uk/embl/*]

DDBJ is functioning as one of the International DNA databases in collaboration with EBI (European Bioinformatics Institute; responsible for the EMBL database) in Europe and NCBI (responsible for GenBank database) in USA. DDBJ is the sole DNA data bank in Japan, which is officially certified to collect DNA sequences from researchers and to issue the internationally recognised accession number to data submitters. DDBJ also provides numerous tools for the retrieval and analysis of data.
[*http://www.ddbj.nig.ac.jp/*]

Some of the most commonly used genetic databases of the world are:

PubMed (NCBI)

PubMed, available via the NCBI Entrez retrieval system, is located at the National Institute of Health (NIH), USA. PubMed is designed to provide access to citations from biomedical literature and also provides access and links to the other Entrez molecular biology resources. In addition, PubMed provides a Batch Citation Matcher, which allows users to match their citations to PubMed citations using bibliographic information such as journal, volume, issue, page number and year.
[*http://www.ncbi.nlm.nih.gov/Entrez/*]

GenBank (NCBI)

GenBank is the NIH genetic sequence database. It is an annotated collection of all publicly available DNA sequences. GenBank is part of the International Nucleotide Sequence Database Collaboration, which is comprised of the DDBJ, EMBL and GenBank of NCBI. Each GenBank entry includes a concise description of the sequence, the scientific name and taxonomy of the source organism and a table of features that identifies coding regions and other sites of biological significance, such as transcription units, sites of mutations or modifications and repeats. Protein translations for the coding regions are included in the feature table. Bibliographic references are included, along with a link to the Medline unique identifier for all published sequences. Sequence data are submitted to the GenBank by individual scientists from around the world, as well as by large centres involved in the Human Genome Project.
[*http://www.ncbi.nlm.nih.gov/Entrez/*]

OMIM (NCBI)

OMIM (Online Mendelian Inheritance in Man) database is a catalogue of human genes and genetic disorders, developed for the WWW by NCBI. The database is intended for use by physicians and other professionals concerned with genetic disorders, by genetics researchers and by advanced students in science and medicine. While the OMIM database is open to the public, users seeking information about a personal medical or genetic condition are urged to consult a qualified physician for diagnosis and for answers to personal questions. [*http://www.ncbi.nlm.nih.gov/Omim/*]

In addition to the above databases, numerous search tools are also available for biotechnological data, some of which are:

Sequence Retrieval System (SRS)

SRS is a Network Browser for databanks in Molecular Biology. It allows one to search multiple databases simultaneously, by entering a single text-based query. The user chooses databases of interest from a large list of databases, categorised by subject such as sequence, sequence-related, metabolic pathway, transcription factor, three-dimensional structure, mapping, mutation, etc. SRS provides a homogenous interface to more than 80 biological databases. [*http://www.ebi.ac.uk/services*]

Entrez (Ncbi)

Entrez is a search and retrieval system which integrates scientific literature, DNA and protein sequence databases, 3D protein structure and protein domain data, population study datasets, expression data, assemblies of complete genomes and taxonomic information into a tightly interlinked system. [*http://www.ncbi.nlm.nih.gov/Entrez/*]

Fasta

FASTA is a sequence comparison software that uses the method of Pearson and Lipman. The basic FASTA algorithm assumes a query sequence and a database over the same alphabet. It searches a DNA sequence in a DNA database or a protein sequence in a protein database. Practically, FASTA is a family of programmes, also allowing queries of DNA vs. a protein database, or vice versa. In these variants, there is a further distinction, regarding the location of gaps: one may assume that gaps occur only in the codon frames corresponding to amino-acid insertion; alternatively, one can assume the gap location to be arbitrary, accounting for insertion/deletion of nucleotides. This search tool is preferred for searching nucleotides. [*http://www.ebi.ac.uk/services/*]

Blast 2.0

It is a Basic Local Alignment Search Tool that provides a method for rapid searching of nucleotide and protein databases. This search tool is better for proteins than for nucleotides. BLAST information guide is designed to assist new and veteran users in employing NCBI tools, such as BLAST and PSI-BLAST in their research. Sequence alignments provide a powerful way to compare novel

sequences with previously characterised genes. Both functional and evolutionary information can be inferred from well-designed queries and alignments. Since the BLAST algorithm detects local as well as global alignments, regions of similarity, embedded in otherwise unrelated proteins, could be detected. Both types of similarity may provide important clues to the functioning of uncharacterised proteins.
[*http://www.ebi.ac.uk/services/*]

TOOLS AND SOFTWARE FOR INTEGRATING ONLINE BIODIVERSITY DATA

Alice Software

This website produces easy to use software for the creation, management and publication of biodiversity data. One who is interested in the cladistics, collation, management or publication of information about biological organisms or wishes to publish species lists or annotated checklists etc, and share with others can use this software.
[*http://www.alicesoftware.com*]

Spice for Species 2000

Its approach for creating a catalogue of life has been defined in the Species 2000 Programme (refer above to the list of 'some of the global biodiversity databases'). It makes use of autonomous Global Species Databases (GSDs), created and maintained by appropriate experts. These provide the central array in a federated architecture. This project investigates how uniform authority and wide taxonomic coverage can be obtained through accurate mapping of quality GSDs from many organisations. It investigates how the federated system can be designed to overcome the inherent problems of extreme heterogeneity, scalability and stability.
[*http://www.systematics.reading.ac.uk/spice/*]

Biodiversity Pro Software

This software is used for calculations based on single samples; also known as univariate diversity measures, it has the following features.
[*http://www.nhm.ac.uk/zoology/bdpro/*]

- *Alpha Calculations*: Abundance Plot-Dominance, Rank, Abundance Model-Log-Series, Broken Stick, Rarefaction, Diversity Indices-Shannon, Alpha, Caswell, Berger-Parker, Simpson, Hill, Margalef, McIntosh.
- *Beta Calculations*: SHE Analysis, Species Richness, Species Distribution
- *Multivariate Calculation*: Principal Components, Correspondence Analysis, Cluster Analysis Non-Metric MDS

- *Comparisons*: Descriptive Statistics, Kulczynski, Mann-Whitney, Rank Correlation, Correlation, Variance-Covariance. ANOSIM
- *Tools*: Send Data, Transform Data, Standardise Data, Add-Ins Options

Platypus Software

It is a comprehensive relational database programme for faunal-based taxonomy, managing taxonomic, geographic, ecological, bibliographic and graphic information, all of which are organised around a graphically displayed checklist, etc.
[*http://www.environment.gov.au/abrs/platypus.html*]

Linnaeus II 2.x

Expert Centre for Taxonomic Identification's (ETI) Linnaeus II interactive software facilitates biodiversity documentation and species identification. It supports the creation of taxonomic databases, optimizes the construction of easy-to-use identification keys, expediates the display and comparison of distribution patterns and promotes the use of taxonomic data for biodiversity studies.
[*http://www.eti.uva.nl/products/linnaeus.html*]

LITCHI Project

This project is belong to a collaborative group of researchers at three U.K. universities, who are seeking to write "taxonomically intelligent" software for tying together various databases on species diversity.
[*http://www.litchi.biol.soton.ac.uk/*]

CODA Software

Conservation Options and Decisions Analysis (CODA) assists in the design of networks of nature reserves or protected areas. It has been used for major reserve planning studies, as a teaching resource and for research in conservation planning methods. It provides a framework and a set of tools for regional conservation planning. Using CODA, one can build and compare alternative conservation proposals in a simple and flexible way.
[*http://www.members.ozemail.com.au/~mbedward/coda/coda.html*]

MultiFlora Project

This project is used for "information extraction" techniques to place descriptive data "locked" in natural-language texts into a more structured electronic database, to facilitate automated analysis.
[*http://www.cs.man.ac.uk/ai/Software/MultiFlora/*]

Phylip Software

PHYLIP (the PHYLogeny Inference Package) is a package of programmes for inferring phylogenies (evolutionary trees). Methods that are available in the package include parsimony, distance matrix and likelihood methods, including bootstrapping and consensus trees. Data types that can be handled include molecular sequences, gene frequencies, restriction sites, distance matrices and 0/1 discrete characters.
[*http://evolution.genetics.washington.edu/phylip.html*]

Biota Software

Biota helps us manage specimen-based biodiversity and collected data by providing an easy-to-use graphical interface to a fully relational database structure. For ecologists, conservation biologists, reserve managers and biogeographers, for taxonomists, systematists and collections managers, Biota offers rigorous tools for recording data and images for specimen determination, as well as for revision and evolutionary studies.
[*http://viceroy.eeb.uconn.edu/Biota*]

The Specify Collections Management System

The SPECIFY software provides a standard software system for diverse disciplines of biodiversity to manage and collect information and to improve the efficiency and effectiveness of development.
[*http://www.usobi.org/*]

The Biodiversity Species Workshop

Biodiversity Species Workshop is a place for the experimental analysis and mapping of the distribution data of your own species.
[*http://biodi.sdsc.edu/bsw_home.html*]

World Map

World Map is a Software for exploring geographical patterns in diversity, rarity and conservation priorities from large biological datasets.
[*http://www.nhm.ac.uk/science/projects/worldmap/*]

BIOINFORMATICS IMPLICATIONS OF THE INTERNATIONAL BIODIVERSITY CONVENTIONS

Over the past 23 years, nations have agreed to a number of international treaties intended to ensure the on-going sustainability of the biota of this planet. Five of these are completely or closely related to biodiversity:

- The Convention on Biological Diversity (CBD)
- The Convention on International Trade in Endangered Species (CITES)
- The Convention on Migratory Species (CMS or the "Bonn" Convention)
- The Ramsar Convention on Wetlands of International Importance
- The World Heritage Convention

India is one of the countries which has confirmed its commitment to the principles of the above treaties. These treaties, in conjunction with others, aimed at the stabilisation of the atmosphere (the Framework Convention of Climatic Change and Montreal Protocol), or at specific aspects of environmental degradation (Convention to Combat Desertification), may mark the beginning of a new era of international co-operation, wherein nations act in consort to ensure the sustainability of the biosphere. It is still a beginning. The next few decades will tell us if the enthusiasm and sense of urgency, which surrounded the Earth Summit in Rio in 1992, will itself be sustainable. If the hope and impetus engendered by the Earth Summit is to bear fruit, this new wave of major treaties must be fully and efficiently implemented, and must increase participation so as to be truly global.

Firstly, although not all countries have joined, participation is extensive and mostly includes a broad spectrum of industrialised, emerging and developing nations. Secondly, all of them require, in fact, significant information reporting and exchange.

Moreover, these treaties were formed in the new "Information Age", an age which has in its command information technology and infrastructure with a capacity to organise, communicate and exchange information rapidly and easily. The result is that "Biodiversity Information", formerly the domain of the notebook of the "naturalist", or the specimen in the museum, botanical garden or zoo, is now openly available to the full international scientific community, as well as the general public. One of the factors which will determine whether the treaties can, in the long run, achieve their purpose, is the ability of party nations, separately and collectively to take advantage of the appropriate tools of modern information technology ("informatics"). Moving towards this partnership will cause direct consequences to the way in which biological information is gathered, organised, maintained and disseminated within countries, that is, national "bioinformatics".

Bioinformatics implications in principle

Integration into National Strategies

The treaties insist or imply that consideration of biodiversity information must become integral to all national strategy developments and decision-making, with regard to the utilisation of biological resources. [*Implications:* National information systems and information flow must crosscut traditional discipline boundaries and bridges between the scientific and socio-economic sectors.]

External Impacts

The treaties require knowledge of the resources at risk, beyond national borders, and the potential consequences of actions and decisions on ecosystems. [*Implications:* Information must be exchanged and shared on a regional basis to complement national information. In addition, decision support systems, which are capable of forecasting consequences, are needed.]

Reporting Requirements

The treaties require nations to report, through a coordinating secretariat to a Conference of Parties, on actions taken and on summary statistics or indicators reflecting the level of implementation of the treaty. [*Implications:* Bioinformatics systems must be available to consistently summarise national actions, progress and the status of biodiversity.]

It is abundantly clear that to meet the above demands effectively and efficiently, countries must develop a national biodiversity information system, which must be closely linked with national economic and social information systems.

Bioinformatics implications in practice

Decisions should be taken well in time so as to avert adverse consequences, or when proposed, the measures can have the best effect. The process of biodiversity data collection, integration and conversion into "information products" suitable for decision makers can be very time-consuming (even if assisted by modern computer systems). Just as it is necessary to have "water management infrastructure" in place to avert floods, before they occur, there must be an "information management infrastructure" in place, before particular instances of decision making become critical. This means:

- having available in advance, essential "core" datasets likely to be needed for a range of decision making purposes
- having the information systems in place with the processing capabilities to be able to quickly produce the specific information products required
- information exchange agreements and facilities already established.

This basic information management infrastructure must be developed to support a range or classes of biodiversity issues, in anticipation of likely decision making scenarios and requirements. The information systems capabilities would normally be required by specific expert institutions, which are the key custodians of the essential core datasets.
[*http://www.iupac.org/symposia/proceedings/phuket97/crain.html*]

BIODIVERSITY INFORMATION SYSTEM (BIS)

The DBT and DOS have joined hands to undertake studies on biodiversity characterisation. The main objectives of this project are to:

- identify the biorich areas
- evaluate the forest types for their value
- provide geographical information on the economically important species for bioprospecting.

The outcome of this project will assist forest managers, decision makers and the leading national institutes involved in genetic diversity and bioprospecting. BIS targets to provide user–level information regarding the hotspots of biodiversity.

However, it was felt that in addition to characterising information on biodiversity, it should also cater to decision-making, and monitoring and management of various resources. It is for this reason that the BIS was conceptualised as a 5-component unit.

- BioSPATIAL (Biodiversity characterisation at the landscape level)
- BIOSPEC (Bioprospecting and molecular taxonomy programme)
- FRIS (Forest resource information system)
- PhytoSIS (Species information system)
- Biocon SDSS (Spatial decision support system for biodiversity conservation prioritisation)

(Source: Roy, 2005)

Networking of Biodiversity Databases

For hundreds of years, explorers and scientists have collected plant and animal specimens from every region and habitat of the earth. Millions of these specimens are preserved with the great natural historical collections of the world, along with detailed notes and logs describing locations and collecting events. The collected data include records spanning many decades, showing where specific species of plants and animals lived, how they lived and how they responded to changes in the environment. This vast global archive of collection data is a priceless source of information on the earth's biosystems; however, most of this data are catalogued through incompatible, roll-your-own data-management applications, developed for individual museums. Scientists therefore, must study each collection in isolation, without the aid of advanced modelling and data-management techniques, for identifying patterns across multiple collections. In some cases, an individual collection may have only one or two specimens of a specific plant or animal—not enough for any meaningful predictions regarding the species. Other museums, however, may have hundreds of examples of this same plant or animal. Combining several collections into one study leads to more accurate conclusions and, in some cases, it leads to new conclusions that, otherwise would have been unobservable or statistically insignificant.

At present, there is inconsistency in biodiversity and its information that is distributed globally (Lane *et al.*, 2000). Biodiversity is mostly distributed in the developing world, whereas the information on biodiversity exists in the developed countries. For example, it is more likely that information pertaining to a plant in an African region may be found in some herbarium in Europe. Due to this, information is not readily available in the region where the biodiversity is mainly distributed and this in turn, is severely affecting initiatives for proper policy making by governmental and non-governmental agencies towards biodiversity conservation (Gundu *et al.*, 2005). By linking the world's natural history collections into a single virtual database, scientists can build a better and deeper understanding of questions such as:

- Where on the earth are you most likely to find a certain species (for example, a species of hawk or an endangered wildflower)?
- What climatic conditions are necessary for this species to thrive?
- How will the range of this species be affected by changes in the environment, such as changes due to global warming?
- What will happen if other competing species are introduced into the environment through human error or intervention?

Bioinformatics helps in accomplishing the task of moving natural history collection data from the gilded age to the information age by developing programmes and techniques that apply to all forms of habitat modelling, regardless of whether the supporting data originates from natural history collections.

India possesses a wealth of knowledge associated with biodiversity, be it the orally held knowledge of folk healers or herders, or the traditional knowledge codified in Ayurvedic, Sidha or Yunani texts. Information also exists in the form of specimens, grey literature, such as unpublished reports of the District Flora Project or Forest Working Plans and books, monographs and scientific papers. Potential exists in building upon the country's biodiversity resources and its associated knowledge; upon promoting biodiversity-based enterprises in the modern and traditional sectors; upon developing biotechnology industries at the cutting edge of new technologies, as well as encouraging local-level value-addition to biodiversity resources.

The Biological Diversity Act visualises the establishment of a National Biodiversity Authority (NBA), State Biodiversity Boards (SBB) and Biodiversity Management Committees (BMC) at the level of all local bodies, namely, Gram, Taluk and Zilla Panchayats, as well as Municipalities and Corporations. The NBA working with SBBs and BMCs will have the responsibility of handling all issues related to conservation and the sustainable use of biodiversity and bioresources of our country. This would need a well-organised in place information system on India's biodiversity resources and associated modern as well as traditional, codified as well as oral knowledge, so as to do justice to achieve their objectives. Such a system will have to deal with a whole range of spatial scales from local to national, as well as link properly with global databases. It will also have to address issues such as linking to information on the large holdings of biological specimens of Indian origin located in herbaria and museums abroad.

FUTURE APPROACHES

India's strength in Information Technology and Biotechnology is in an excellent position to turn the above array of significant challenges into welcome opportunities. This calls for the networking of the country's existing biodiversity databases so as to take advantage of synergies and to link all of these to activities leading to value-addition. As a part of this process, the existing biodiversity databases will need to be considerably augmented and strengthened and new ones created. We will have to come up with novel ways of bringing on board, the substantial knowledge base of the country's barefoot ecologists and grass-roots innovators. We also need to devise a country-wide decentralised system of monitoring biodiversity. Such a decentralised system could serve to enhance the quality of education by engaging teachers and students in the first hand understanding of biodiversity and its associated knowledge and in creating, using and managing electronic databases, including those employing Indian languages. Computer-aided storage and retrieval systems of plant and animal genetic resources will form an important tool for developing technology packages for conservation and ensuring the exchange of information (Tandon, 2004a).

A good beginning has been made in organising a part of this information in the form of electronic databases. There are, however, many lacunae and we may summarise the situation as follows:

- Databases are in heterogeneous formats.
- Few are on the web, while many are available offline.
- Some of these are well structured, others are largely project/species specific and/or unstructured.
- These databases exist independently.
- There is no framework to link the scattered data so as to facilitate exchange of data amongst the different databases.
- There is no meta-data.
- The gap between data managers and data producers is widening.

Well-planned institutional arrangements and legal provisions at the national level, as well as in terms of links with international agencies to address these manifold concerns regarding biodiversity conservation are required. DBT with its National Bioresources Development Board, the MoEF and Governmental agencies such as Ministries of AYUSH (Ayurveda, Unani, Siddha, and Homeopathy), Health and Commerce, Council of Scientific and Industrial Research, Indian Council for Agricultural Research and DOS will play a significant role. The challenge before us is to set standards and make technological choices that would facilitate the networking of databases and add real value to the information being brought together, and at the same time, maintain the autonomy of the various databases, so as to ensure the security of the data and protect all legitimate intellectual property rights. This would obviously have to be worked out as a group exercise by all concerned institutions and individuals, to involving an inventory of the on-going Indian efforts and a review of the various pertinent standards. These surveys would have to address issues of data characterisation and classification, validation and authentication, organisation and structuring, storage, archival, warehousing, retrieval, dissemination, sharing and interoperability, access, security, visualisation, analysis and value addition, use of multiple languages and capacity building needs. [*http://wgbis.ces.iisc.ernet.in/biodiversity/Network*]

Conservation, sustainable utilisation and management of biodiversity should become an important agenda, as it will be the key to the survival and economic well-being of man kind in the 21st century (Tandon, 2003). Sciences dealing with biodiversity management involve analysis, maintenance, cataloguing and conservation of all different life forms. Bioinformatics, being an applied discipline, it utilises computational tools to conceptualise biology, and would be capable of assisting in biodiversity conservation and its sustainable utilisation. Bioinformatics would provide methods for collecting and storing genes, detecting and eliminating diseases in gene bank collections, identifying useful genes, improving techniques for long-term storage, distributing germplasm to users in a safer and more efficient manner. Optimised use of advanced information technology tools for the management of biological resources is the need of the hour. Government agencies, institutions and NGOs will require working as a group to network the biodiversity databases of the country. Advanced computation and data archives would produce quantitative improvements in the analysis of biodiversity, through more accurate and efficient solutions.

ACKNOWLEDGEMENTS

The authors thank the Department of Biotechnology, Government of India for providing the facilities at the Bioinformatics Centre, North-Eastern Hill University, Shillong, India.

REFERENCES

Gundu R.K., Bose B., Swamynathan S., Sreenu V.B., Pavan N., Acharya S. and Nagarajaram H.A. (2005). Frontiers in Bioinformatics Research: The Biodiversity Issues. In: Biodiversity: Status and Prospects (Tandon P, Sharma M and Swarup R eds), Narosa Publishing House, India p 197-201.

Lane M.A., Edwards J.L. and Nielsen E.S. (2000). Biodiversity Informatics: The challenge of rapid development, large databases and complex data. In: Proceedings of the 26th International Conference on Very Large Databases, Cairo, Egypt.

Maurer S.M., Firestone R.B. and Scriver C.R. (2000). Science's neglected legacy. Nature 405 (6783): 117-120.

MoEF (1999). National policy and macro-level action strategy on biodiversity. Ministry of Environment and Forests, Government of India, New Delhi.

Roy P.S. (2005). Biodiversity conservation: Perspective from space. In: Biodiversity: Status and Prospects (Tandon P, Sharma M and Swarup R eds), Narosa Publishing House, India p 170-196.

Sala O.E., Chapin F.S., Gardner R.H., Lauenroth W.K., Mooney H.A. and Ramakrishnan P.S. (1999). Global change, biodiversity and ecological complexity. In: The Terrestrial Biosphere and Global Change: Implications for Natural and Managed Ecosystems (Walker B, Steffen W, Canadell J and Ingram J eds), Cambridge: Cambridge Univ. Press.

Schnase J.L. (2000). Research directions in biodiversity informatics. In: Proceedings of the 26th International Conference on Very Large Databases, Cairo, Egypt.

Singh J.S. and Khurana E. (2002). Paradigms of biodiversity: An overview (Biodiversity, Ecology and Conservation-Reviews and Tracts). Proceedings of the Indian National Science Academy. Part B, 68(3): 273-296.

Sugden A. and Pennisi E. (2000). Diversity digitised. Science 289(5488): 2305.

Tandon P. (2003). Biodiversity – A scientific approach: Agenda for the 21st century. National Academy Science Letters. 26(5&6): 111-118.

Tandon P. (2004a). Conservation and sustainable development of plant resources of North-East India. Man and Society 1(1): 49-59.

Tandon P. (2004b). Role of biotechnology in the conservation of plant genetic resources in the 21st century – an Indian perspective. In: Platinum Jubilee Lectures – 87th and 88th Session of Indian Science Congress Association (Banerjee SP and Mukherjee SP eds), Auto Print and Publicity House, Kolkata, India, p 40-67.

Tandon P. and Kumaria S. (2005). Prospects of plant conservation biotechnology in India with special reference to the northeastern region. In: Biodiversity: Status and Prospects (Tandon P, Sharma M and Swarup R eds), Narosa Publishing House, India p 79-92.

Terborgh J. (1999). Requiem for Nature. Washington, DC, Island Press, USA p 234.

Venkatesh S. and Bansal M. (2005). Role of Informatics in biodiversity conservation: Overview of available resources. In: Biodiversity: Status and Prospects (Tandon P, Sharma M and Swarup R eds), Narosa Publishing House, India p 134-145.

Chapter 23

Conservation of Splendid Orchids of North-East India

Stadwelson Dohling, Meera Chettri Das, Suman Kumaria and Pramod Tandon

Plant Biotechnology Laboratory, Department of Botany, North-Eastern Hill University, Shillong - 793 022, Meghalaya

Orchids are amongst the most advanced flowering plants, highly specialised in many ways. They are distributed all over the world except the Antarctica. The North-East region of India has the highest representation of orchid genera. Some of these genera are not found in any other part of the world. Unfortunately, the orchid diversity of North-East India and the country as a whole is being threatened for reasons, such as increased biotic influences, socio-economic development and uncontrolled commercial exploitation of forest wealth. Habitat destruction, which is occurring at an alarming rate due to deforestation and other unplanned human activities, has led to a considerable depletion of orchids in nature. Therefore, concerted efforts have to be made to conserve the declining population of orchids in the region. Conservation can be accomplished, both by *in situ* and *ex situ* methods. *In situ* conservation involves the protection of genetic resources in their natural environment, through the protection of the environment itself, while the maintenance and propagation of species in botanical gardens and *in vitro* techniques comes under *ex situ* conservation. Biotechnological approaches towards conservation should be complementary to conventional methods. For conservation of plants, seeds are, in general, preferred for propagation as they maintain maximum genetic diversity and the *in vitro* seed germination techniques have ensured germination in the hard-to-germinate taxa. Micropropagation using explants, besides seeds, can be done for the commercialisation of orchids, in order to maintain identical genotypes. Artificial seed technology, not only helps in easy handling and transportation of plantlets, but can also be used for the conservation of rare, endangered and desirable genotypes. Artificial seed production has been tried in only a few non-orchid angiosperms through encapsulation of different plant parts. In orchids, Sharma *et al.*, (1992) for the first time reported the regeneration of complete plantlets of *Dendrobium wardianum* from synthetic seeds. For the conservation of orchid germplasm, cryopreservation can also be an effective tool and till date, very less work has been done on orchids.

INTRODUCTION

Orchids belong to the family, Orchidaceae that forms the largest family of monocotyledonous plants, constituting about 7% of angiosperms and nearly 40% of monocots. Orchidaceae includes 800 genera and between 25,000 to 30,000 species spread all over the world (Chowdhery, 2001). They are distinct in types of vegetative structure and morphology. They exhibit an amazing diversity in terms of size, shape, structure, colour and fragrance of flowers. Orchids are distributed all over the world, except the Antarctica and a few isolated islands. They are found in all kinds of habitats, except the aquatic systems. Depending on their growth habitat, they may either be terrestrial, epiphytic or saprophytic. Most of the orchids are epiphytic in nature, while the rest are terrestrial. In some of the genera, e.g., *Cymbidium* and *Liparis*, both the epiphytic as well as the terrestrial mode of life is seen. Morphologically, orchids exhibit lesser diversity and consist of stems or pseudobulbs, which bear many or few flowers. The highest range of diversity is seen in floral morphology, where the dorsal petals of the flower have been modified to form the 'labellum' and the column which is formed as a result of the fusion of the sex organs.

The wide-ranging altitudinal variations, from the foothills to the high Himalayan mountains and deep river valleys with high rainfall and high humidity, distinctive soil conditions, etc., have all played a significant role in the development of the highly rich orchid diversity in the North-Eastern region (Chowdhery, 1998; Hynniewta, 2000). Depending on the presence of orchids, the region has been classified into the following orchid habitats: (1) Tropical forests (between 100-1000m), (2) Sub-tropical forests (between 1000-2000m), (3) Temperate forests (between 2000-3500m), and (4) Alpine forests (3500-5000m). Around 750-800 species of orchids are found in North-East India. Arunachal Pradesh harbours the maximum number of orchid species, followed by Sikkim and Meghalaya. Tripura has the lowest number of orchid species in the region. The North-East region also has the highest concentration of monotypic orchid genera. Some of the monotypic orchid genera in North-East India are *Anthogonium, Arundina, Acrochaene, Bulleyia, Cremastra, Cleisocentron, Dickasonia, Diglyphosa, Eriodes, Herpysma, Jejosephia, Mischobulbum, Myrmechis, Neogyna, Ornithochilus, Risleya, Renanthera, Stereosandra, Tipularia, Thecostele, Trachoma, Vandopsis,* etc. Many of these genera do not occur anywhere else in India. Another significant feature of the orchid flora of the region is the presence of a large number of saprophytic orchid species belonging to the genera *Aphyllorchis, Cymbidium, Eulophia, Gastrodia* etc. North-East India is reported to have a large number of endemic, rare and threatened species of orchids (Nayar and Sastry, 1997-1998, 1999; Ahmedullah, 2000). Some of these are listed in Table 23.1.

Table 23.1: Some rare, threatened and endangered orchids of North-East India

Acanthephippium sylhetense Lindl. (R, T)
Aerides longicornu Hook.f. (eR)
Anoectochilus grandiflora Lindl. (R)
A. sikkimensis Kng and Prantl. (R)
Bulbophyllum hookeri (Duthie) J.J.Sm. (En,R)
B. leptanthum Hook.f. (R)
B. moniliforme Par. and Reichb.f. (En,R)
B. obrienianum Rolfe (eR)
B. triste Reichb.f. (R)
Bulleyia yunnanensis Schltr. (eR,T)
Calanthe alpina Hook.f. ex Lindl. (En,R)
C. chloroleuca Lindl. (En)
C. herbacea Lindl.(R)
C. odora Griff. (R)
Ceratostylis himalaica Hook.f. (R)
C. subulata Blume. (En,R)
Coelogyne arunachalensis Choudhery and Pal.(eR)
C. nitida (Wall. ex D. Don) Lindl. (R)
C. treutleri Hook.f. (eR)
C. cristata Lindl. (R)
Cymbidium cochleare Lindl. (R,T)
C. eburneum Lndl. (R,T)
C. devonianum Paxt. (R,T)
C. giganeum Wall. ex Lindl. (En,R)
C. simonsianum King & Prantl. (eR)
C. tigrinum Par. ex Hook. (eR,T)
Dendrobium aqueum Lindl. (eR) (R)
D. chrysotoxum Lindl. (R,T)
D. densiflorum Wall. ex Lindl. (eR,T)
D. folconeri Hook. (R,T)
D. wardianum Warner (eR,T)
D. ochreatum Lndl. (eR)
Diplomeris pulchella D. Don. (R)
Epigeneium rotundatum (Lindl.) Summerh. (R,T)
Eria crassicaulis Hook.f. (R,T)
E. fragrans Reichb.f. (R,T)
E. spicata (D. Don) H. Mazz. (R)
Eulophia candida (Lindl.) Hook.f. (En,R)
E. mannii (Reichb.f.) Hook.f. (R)
Galeola cathcartii Hook.f. (R)
Habenaria cumminsiana King & Prantl. (R)
H. khasiana Hook.f. (R)
Herminium haridasanii A. N. Rao. (R)
Liparis pulchella Hook.f. (En,R)
L. acuminata Hook.f. (R)
Listera divaricata Panigr & P. Taylor. (eR)
Malaxis aphylla (King & Prantl.) Tang and Wang. (R)
Nervilia scottii (Reichb.f.) Schltr. (En,R)
Oberonia clarkei Hook.f. (R)
Paphiopedilum fairrieanum (Lindl.) Stein (eR,En,T)
P. insigne (Wall. ex Lindl.) Stein. (eR, En,T)
P. spicerianum (Reichb.f.) Pfitz. (eR, En,T)
P. venustum (Sims.) Pfitz. (eR,En,T)
P. villosum (Lindl.) Stein. (eR,En,T)
Phaius tancarvilliae (Banks ex L'Herit) Bl. (En,R)
Phalaenopsis mastersii King and Prantl. (R)
Pleione maculata (Lindl.) Lindl. (eR)
P. praecox (J. E. Sm.) D. Don. (R)
Renanthera imschootiana Rolfe. (eR,En,T)
Taeniophyllum khasianum Joseph and Yogan.
Tainia khasiana Hook.f. (R)
Thunia marshalliana Rolfe. (R)
Vanda coerulea Griff. ex Lindl. (eR,T)
V. parviflora Lindl.(R)
V. pumila Hook.f. (eR,T)
Vanilla pilfera Holtt. (R,T)
Zeuxine longifolia (Benth.) Hook.f. (R)

R, rare; eR, extremely rare; T, threatened; En, endangered

Orchids are of considerable economic importance because of their ornamental and medicinal value. They have been used variously, e.g., *Dendrobium acinaciforme* as love charms, *Vanda tessellata, Eria musicol*a for magic and species, and *Cymbidium, Habenaria commelinifolia*, etc. as food. 'Vanillin', an odourous constituent obtained from *Vanilla planifolia*, is the most important commercial product that is used for the flavouring of chocolates, beverages, confectionery cakes, custards, pudding and ice creams, besides being used as a deodorant in soaps, perfumes and powders. In India, orchids have been used in various indigenous systems of medicines since the Vedic period (Vij *et al.*, 1997). Some of the medicinally important orchids are *Acampe papillosa, Bulbophyllum neilgherrense, Calanthe triplicate, Coelogyne punctulata, Cymbidium aloefolium, C. ensifolium,*

Dendrobium nobile, D. ovatum, Eria pannea, Eulophia spp., *Habenaria* sp., *Malaxis acuminate, Pholidota articulate, Rhynchostylis retusa, Vanda testacea, Zeuxine strateumatica*, etc. But it is their ornamental value for which orchids are most important, economically.

Commercially, orchids are grown in a large number of countries. The world export and import trade of cut flower and orchid plants exceeded $150 million in the year 2000. Of this, $ 128 million was in cut orchid flowers and about $ 23 million in orchid plants, according to the United Nations Comstats (Source: http://www.floracultureintl.com/archive/article/293.asp). Asia dominates world trade, where as countries like Taiwan, Thailand, Singapore and Malaysia are the main exporters of orchids, followed by Brazil, Philippines, Madagascar, New Zealand, Vietnam and China. Japan, which imports the largest number of orchids is the leader, followed by Italy, France, Germany, USA, UK and the Netherlands. Unfortunately, orchids in North-East India have not been fully exploited. Though different research institutes are making efforts to grow and refine various varieties, they are not sufficient to pay us rich dividends. Floriculture industry in India is still in its infancy, though some private companies have started work on the mass multiplication of native orchids.

DEPLETION OF ORCHIDS

Unfortunately, the orchid diversity in North-East India and the country as a whole, is being threatened for various reasons, such as the increased biotic influences, socio-economic development and uncontrolled commercial exploitation of forest wealth. Almost all the epiphytes, because of their habitat specificity and slow growing nature, fail to withstand habitat destruction pressures, and all of them figure prominently in the list of endangered plants (Rajeevan and Shobhana, 1993). Although the decline in the population of orchids has been attributed to ruthless commercial exploitation, by the Convention on International Trade in Endangered species (CITIES) of Wild Flora and Fauna, it is observed that habitat destruction is the major factor involved.

CONSERVATION OF ORCHIDS

The modern tools of Biotechnology can be utilised for the propagation and conservation of plant genetic resources. In general, these could be accomplished both by *in situ* and *ex situ* methods. These techniques were initially introduced on plant species having agricultural and horticultural importance, but are now rapidly being applied for the collection, propagation, preservation and evolution of rare and endangered plant germplasm. *In situ* conservation, which is an ideal and dynamic approach that allows plants to interact and co-evolve, involves the protection of genetic resources in the natural environment through the protection of the environment itself (Tandon, 2004b). It is the most suitable option to ensure the natural growth, proliferation and perpetuation of different species. The government, in order to promote the cause of *in* situ conservation, has already declared many areas in North-Eastern India as national parks, wildlife sanctuaries and biosphere reserves. However, they are costly to maintain and are highly susceptible to natural calamities. *Ex situ* conservation programmes have played an important role in acclimatisation, rehabilitation, multiplication and judicious exploitation of various species. Biotechnological approaches towards conservation are complementary to conventional methods. These can directly assist plant conservation programmes through molecular marker technology, molecular diagnostics, *in vitro* technologies and cryopreservation (Tandon, 2004b; Tandon and Kumaria, 1998).

For conservation of plants, the seeds are, in general, preferred for propagation as they maintain maximum genetic diversity. The *in vitro* seed germination techniques have ensured germination in the hard to germinate taxa and have offered possibilities to study the morphogenetic changes occurring in course of seedling development. The mass propagation of orchids through asymbiotc seed germination is a tool for the conservation of the declining orchid population in nature. The orchid seeds are very minute in nature and are very difficult to germinate because they do not possess endosperm and the embryo is immature (Zeigler *et al.*, 1967). Because of their particular fungal requirement, less than 5% of the orchid seeds germinate in nature. A number of rare species of orchids have been germinated *in vitro* through seeds (Table 23.2). During the last few years, tissue culture techniques using seeds and appropriate explants have been developed for the large-scale of propagation of orchids. Different workers have suggested a number of media with modifications for orchid seed germination (Olivia and Arditti, 1984; Pierik *et al.*, 1988; Yam and Weatherhead, 1988; Yam *et al.*, 1989; Kumaria and Tandon, 1991; Kondo *et al.*, 1997; Gangaprasad *et al.*, 1999). However, it has been observed that in certain genera for, e.g., *Bletilla striata, Dendrobium chrysotoxum, Dactylorhiza, Habenaria, Ophrys,* seed germination could be initiated only in distilled water, but for further development there is a requirement of certain nutrients (Stoutamire, 1974). This finding suggested that the orchid embryos have simpler nutritional requirements at the initial stages of development, which, of course, may vary from species to species. The stage of development of the orchid embryos contributes significantly to their potentiality to develop (Pathak *et al.*, 2001). Micropropagation of orchids using explants from germinating seeds *in vitro*, has been also achieved (Christenson, 1988).

Apart from seeds, micropropagation using other explants can be done for the commercialisation of orchids, in order to maintain identical genotypes. Mericloning of orchids was pioneered by Morel (1960, 1964) and since then, it has been used for the commercial multiplication of a number of orchids. The different explants which have been used for orchid micropropagation include roots and rhizomes (Arditti and Ernst, 1993; Zhou, 1998; Chang and Chang, 2000), pseudobulb segments (Vij *et al.*, 2000), shoot meristems (Kumaria and Tandon, 1994; Laishram and Devi, 1999), stem explants (Van *et al.*, 1999), nodal explants (Teng *et al.*, 1997), axillary buds (Laishram and Devi, 1999), foliar explants (Nayak *et al.*, 1997; Teng *et al.*, 1997), leaf explants (Sinha and Hedge, 1999; Laishram and Devi, 1999; Payati and Murty,1999), foliar peels (Vij and Kaur, 1992) and protocorms (Kumaria *et al.*, 1990; Corrie and Tandon, 1993;

Sheelavanthmath and Murty, 2001). A large number of plants have been generated from shoot tips of actively growing plants, as well as from stoloniferous stem explants (Latha, 1999 a, b). The young inflorescence can also be used to achieve direct or protocorm-like bodies (plbs) and mediated development of plantlets (Vij and Sharma, 1996). Shimura and Koda (2004) have reported the regeneration of plantlets from the plbs of *Cypripedium macranthos* var. *rebunense* on hormone-free medium. The use of flower stalks for the initiation of shoot buds and plbs in liquid medium has been reported (Park *et al.*, 1996). Calli regenerated somatic embryogenesis and regeneration of orchids has also been reported (Ichihashi and Hiraiwa, 1996; Ishii *et al.*, 1998). In addition, meristem culture of orchids has made it possible to mass multiply a desirable clone in a short time, with the elimination of viruses. The use of liquid media has been found to enhance the *in vitro* multiplication rate in most of the orchids (Park *et al.*, 1996; George and Ravishanker, 1997).

Table 23.2. List of some orchids propagated through *in vitro* seed germination

Name	*Status*	*References*
Coelogyne punctulata	Endemic, overcollection	Sharma and Tandon (1986)
Cattleya dowiana	Rare	Marlow and Butcher (1987)
Dendrobium spectatissimum	Rare	Marlow and Butcher (1987)
Spiranthes parsksii	Less than 2000 in the wild	Christenson (1988)
Disa uniflora	Rare	Ronse (1990)
Hetaeria cristata	Propagation for reintroduction and horticulture	Yam and Weatherhead (1990)
Dendrobium fimbriatum var. *oculatum*	Overcollection, threatened	Kumaria and Tandon (1991)
Spiranthes magnicamporum	Rare, propagation for horticulture	Anderson (1991)
Cypripedium calceolus	Rare	Malmgren (1992)
Cypripedium debile	Rare	Hoshi *et al.*,. (1994)
Habenaria radiate	Rare	Nagayoshi *et al.*,. (1996)
Cypripedium reginae	Rare	Faletra *et al.*,. (1997)
Dendrobium lindleyi	Rare	Kaur and Sharma (1997)
Platanthera praeclara	Rare	From and Read (1997)

The physiology and development of orchids in culture are greatly influenced by several physico-chemical factors, like light, pH and temperature. Many workers have reported the effect of light, both qualitative and quantitative including photoperiod, on orchid seed germination and growth (Arditti and Ernst, 1993). For some terrestrials, even a pre-treatment with light is inhibitory for germination. Stoutamire (1974) considered this feature to be an adaptation for the terrestrial orchids, growing in well-drained woodlands and open grassy areas, to tolerate drought. The optimal temperature for seed germination of most orchid species is reported to be between 20°C and 25°C (Grillo *et al.*, 1985). The requirement of chilling temperature has also been reported in several species (Stoutamire, 1974). A pre-chilled treatment of 4°C for 3 months was required for the seeds of *Cypripedium macranthos* var. *rebunense* to germinate (Shimura and Koda, 2004). The specific pH value between 5.0 to 6.0 has been recommended for orchid seed germination (Arditti and Ernst, 1993). Besides, growth regulators like auxins (a-naphthaleneacetic acid, indole-3-acetic acid, 2, 4-dichlorophenoxyacetic acid, indole-3-butyric acid, etc.), cytokinins (6-benzylaminopurine, kinetin, etc.) and gibberellins have been incorporated in the media to promote the growth of orchids (Arditti and Ernst, 1984,1993; Sharma and Tandon, 1986). The use of thidiazuron in direct shoot regeneration, from foliar explants of an epiphytic orchid *Acampe praemorsa* (Nayak *et al.*, 1997) and callus cultures of *Cymbidium ensifolium* var. *misericors* (Chang and Chang, 1998), has been reported to be beneficial. Certain organic growth supplements, such as peptone, yeast extract, casein hydrosylate, coconut water, etc. have been continuously used for enriching the culture media for seed germination (Arditti and Ernst, 1984).

Artificial seed technology is an exciting and rapidly growing area of research in the delivery of propagules. This technology not only helps in easy handling and transportation of plantlets, but also can be used for the conservation of rare, endangered and desirable genotypes (Kumaria and Tandon, 2001). As propagation of many ornamental plant species is labour intensive, therefore integration of simple artificial seed system would dramatically reduce labour requirements, thus lowering production costs (Gray and Compton, 1993). Artificial seed production has been tried in only a few non-orchid angiosperms through encapsulation of different plant seeds (Patel *et al.*, 2000), flower buds (Mitra and Chaturvedi, 1972), axillary buds (Soneji *et al.*, 2002), shoot tips (Wang *et al.*, 2002), nodal explants (Rout *et al.*, 2001) and roots (Micheli *et al.*, 1996; Piccioni *et al.*, 1997). Orchids, which are the most precious ornamental plants, are one of the few plants that can be propagated *in vitro*, both by using seeds and tissue culture techniques. Sharma *et al.*,. (1992) for the first time reported the regeneration of complete plantlets of *Dendrobium wardianum* from synthetic seeds. Subsequently, complete plantlets of *Cymbidium giganteum*, an endangered orchid, were obtained by the germination of artificial seeds (Corrie and Tandon, 1993). There are also reports of the production of orchid plantlets through the germination of encapsulated protocorm-like bodies of *Dendrobium*, *Oncidium*, *Cattleya* etc. (Saiprasad and Polisetty, 2003). Besides, several other workers have also reported the production of orchid plantlets through the germination of synthetic seeds (Khor *et al.*, 1998; Dutta *et al.*, 1999). Martin (2003) has successfully accomplished clonal propagation, encapsulation and reintroduction of an endangered orchid *Ipsea malabarica*. The reintroduction of the *in vitro* - raised plantlets into their natural habitat was done to conserve this orchid species *in situ*.

Cryopreservation of plant genetic resources, which provides stable long-term storage in liquid nitrogen (LN) at –196° C, is the most promising approach to conserve valuable germplasm. At the temperature of LN, almost all the metabolic activities of cells are at a standstill and they can be preserved in such a state for extended periods (Tandon, 2004a; Tandon and Kumaria, 2005). In the past, mostly vegetatively propagated plants were cryopreserved, but with rapid progress in plant transformation, cryopreservation is widely used in the preservation of experimental materials of primary transformed tissues, and secondary cultures, etc. (Kendell *et al.*, 1993). A survey of literature in the recent past shows that cryopreservation has been applied successfully to a large number of agricultural crops, vegetables, and fruit trees, etc. For conservation of orchid germplasm, cryopreservation can also prove to be an important tool. However, limited work on the cryopreservation of orchids has been done, till date. Some of the orchids that have been successfully cryopreserved are the shoot primordial of *Vanda pumila* (Na and Konda, 1996), *Catteleya loddigesii, Catteleya walkeriana, Dendrobium* (Kondo *et al.*, 2001), zygotic embryos of *Bletilla striata* (Ishikawa *et al.*, 1997), seeds of *Doritis pulcherrima* (Thammasiri, 2000) and protocorm-like bodies of *Cymbidium* (Kondo *et al.*, 2001).

The *in vitro* raised plantlets of orchids are exposed to conditions of minimal stress and optimal growth conditions. They can be hardened or acclimatised for establishment. Hardening involves the transfer of plants from laboratory condition to the external environment. As orchids are delicate plants, they find it difficult to withstand sudden changes of temperature and humidity, during *in vitro* to *in vivo* transfer. This is due to poorly developed cuticles, stomatal apparatus, photosynthesis ability and conducting tissues of the *in vitro* raised plantlets (Vij, 1995). Therefore, it is essential that a gradual reduction of humidity and temperature during acclimatisation, should take place, so as to reduce the mortality rate of plantlets during their transfer from laboratory to land. Different potting mixtures have been suggested for the healthy growth and development of orchids (Kumaria and Tandon, 1994; Sharma and Kaur, 1998). According to Kanjilal *et al.,*. (1999), the rooting of *Dendrobium moschatum* shoots could be induced by incorporating different concentrations of IAA in the potting mixture; however, NAA in the mixture proved to be more beneficial for the purpose. Decruse *et al.,*. (2003) provided an efficient *in vitro* multiplication protocol for the ecorestoration and horticultural exploitation of *Vanda spathulata*, a threatened floriferous orchid of the Indian Subcontinent.

To ensure the natural growth and proliferation of orchids in nature, *in situ* conservation is the best option. There is a strong need to conserve orchids in their natural habitat through Botanical Gardens, Sanctuaries and National Parks, which automatically provide *in situ* conservation to the vast number of orchid within them. Some of the states have also established orchid sanctuaries, e.g., Sikkim at Singtam and Deorali, and Arunachal Pradesh at Sessa and Dirang. According to the report by Singh (2001), the Botanical Survey of India (BSI), the apex organisation under the Ministry of Environment and Forests, Government of India, has mandated survey, documentation and conservation of the plant resources of the country for cultivation, multiplication and *ex situ* conservation of orchids. It has established a National Orchidaria at Shillong, Yercaud and Howrah. In order to disseminate various data and focus attention on the threatened orchid species, BSI has already compiled Red Data sheets on a number threatened orchid species of the country (Nayar and Sastry, 1997-1998, 1999). Besides, the Indian Council of Forestry Research and Education; Indian Council of Agricultural Research; National Botanical Research Institute; Tropical Botanic Garden and Research Institute; Botany Department of Punjab University; North-Eastern Hill University, Shillong; Orchid Research and Development Centre of Tipi, Arunachal Pradesh; National Research Centre on Orchids, Pakyong, Sikkim;

etc. are some of the institutions actively engaged in research and conservation of orchids in India.

REFERENCES

Ahmedullah M. (2000). Endemism in the Indian flora. In: Flora of India- Introductory Volume, Part II (Singh NP ed), Botanical Society of India, Calcutta pp 246-265.

Anderson A.B. (1991). Symbiotic and asymbiotic germination and growth of *Spiranthes magnicamporum* (Orchidaceae). Lindleyana 6: 183-186.

Arditti J. and Ernst R. (1984). Physiology of germinating orchid seeds — A manual. In: Orchid Biology, Reviews, and Perspectives Vol III (Arditti J ed), Cornell University Press, New York pp 172-222.

Arditti J. and Ernst R. (1993). Micropropagation of orchids. John Wiley and Sons Inc, New York.

Chang C. and Chang W.C. (1998). Plant regeneration from callus culture of *Cymbidium ensifolium* var. *misericors*. Plant Cell Rep 17: 251-255.

Chang C. and Chang W.C. (2000). Micropropagation of *Cymbidium ensifolium* var. *misericors* through callus derived rhizomes. *In Vitro* Cell Dev Biol Plant 36: 517-520.

Chowdhery H.J. (1998). Orchid flora of Arunachal Pradesh. Bishen Singh Mahendra Pal Singh, Dehradun, India.

Chowdhery H.J. (2001). Orchid diversity in North-East India. J Orchid Soc India 15(1-2): 1-17.

Christenson E.A. (1988). Conservation of *Spiranthes parksii*: A beginning. Orchid Review 96: 148-149.

Corrie S. and Tandon P. (1993). Propagation of *Cymbidium giganteum* Wall. through high frequency conversion of encapsulated protocorms under *in vivo* and *in vitro* conditions. Indian J Exp Biol 31: 61-64.

Decruse S.W., Gangaprasad A., Seeni S. and Menon S. (2003). Micropropagation and ecorestoration of *Vanda spathulata*, an exquisite orchid. Plant Cell Tiss Org Cult 72: 199-202.

Dutta K.B., Kanjilal B. and De Sarker D. (1999). Artificial seed technology: Development of a protocol in *Geodorum densiflorum* (Lam.) Schltr.—An endangered orchid. Curr Sci 76: 1142-1145.

Faletra P., Dovholuk A., King T. and Sokolski K. (1997). Saving *Cypripedium reginae*. Orchids 66(2): 139-143.

From M. and Read P.E. (1997). *Platanthera praeclara*: A threatened temperate prairie orchid species. In Vitro Cell Dev Biol Plant 33: 17A.

Gangaprasad A.S., William Decruse S., Seeni S. and Menon V.S. (1999). Micropropagation and restoration of the endangered malaber daffodil orchid *Ipsea malabarica*. Lindleyana 14(1): 47-56.

George P.S. and Ravishanker G.A. (1997). *In vitro* multiplication of *Vanilla planifolia* using axillary bud explants. Plant Cell Rep 16: 490-494.

Gray D. and Compton M. (1993). Use of somatic embryos as synthetic seeds for reducing micropropagation costs. Hortscience 28: 515.

Grillo M.E., Pena S. and Perez D. (1985). Mass *in vitro* germination of some species of orchids in the National Botanical Garden (Cuba). Rev Jard Bot Nac 6(2): 95-100.

Hoshi Y., Kondo K. and Hamatani S. (1994). *In vitro* seed germination of four Asiatic taxa of *Cypripedium* and notes on the nodal micropropagation of American *Cypripedium montanum*. Lindleyana 9: 93-97.

Hynniewta T.M. (2000). Orchids of Nagaland. Botanical Society of India, Calcutta.

Ichihashi S. and Hiraiwa H. (1996). Effect of solidifier, coconut water and carbohydrate source on the growth of embryogenic callus in *Phalaenopsis* and allied genera. J Orchid Soc India 10(1-2): 81-88.

Ishii Y., Takamura M., Goi M. and Tanaka M. (1998). Callus induction and somatic embryogenesis of *Phalaenopsis*. Plant Cell Rep 17: 466-470.

Ishikawa K.K., Harata M., Mii M., Sakai A., Yoshimatsu K. and Shimomura K. (1997). Cryopreservation of zygotic embryos of a Japanese terrestrial orchid (*Bletilla striata*) by vitrification. Plant Cell Rep 16: 754-757.

Kaur S. and Sharma C.M. (1997). Selection of the best medium for the *in vitro* propagation of *Dendrobium lindleyi* Steud. Advances in Plant Sciences 10: 1-5.

Kanjilal B.D., Sarker D., Mitra J. and Dutta K.B. (1999). Stem disc culture: Development of a rapid mass propagation method for *Dendrobium moschatum* (Buch Ham) Swartz. An endangered orchid. Curr Sci 77(4): 497-499.

Kendall E.J., Kartha K.K., Qureshi J.A. and Chermak P. (1993). Cryopreservation of immature spring wheat zygotic embryos using abscisic acid pretreatment. Plant Cell Rep 12: 89-94.

Khor E., Ng W.F. and Loh C.S. (1998). Two-coat systems for encapsulation of *Spathoglottis plicata* (Orchidaceae) seeds and protocorms. Biotechnol Bioeng 59(5): 635-639.

Kondo K., Tanaka C., Shimada T. and Ohtani M. (1997). Developmental morphology of seed micropropagation of *Orchis aristata* Fischer (Orchidaceae) in axenic culture. Ann Tsukuba Bot Gard 16: 41-48.

Kondo K., Tatarenko I.V., Verghese S.B., Iwai Y. and Matsumoto K. (2001). Orchid cryopreservation by using induced shoot primordia and protocorms—like bodies of tissue culture. In: Orchids: Science and Commerce (Pathak P, Shekhar RN, Sharma M and Sood A eds), Bishen Singh Mahendra Pal Singh, New Connaught Place, Dehra Dun, India pp 397-412.

Kumaria S. and Tandon P. (1991). Asymbiotic gemination of *Dendrobium fimbriatum* var. *oculatum* Hk.f. seeds on different media. Proc Indian Natn Sci Acad B57(3-4): 277-279.

Kumaria S. and Tandon P. (1994). Clonal propagation and establishment of *Dendrobium fimbriatum* var. *oculatum* Hk.f. In: Advance in Plant Tissue Culture in India (Tandon P ed), Pragati Prakashan, India pp 218-231.

Kumaria S. and Tandon P. (2001). Orchids: The world's most wondrous plant. In: Orchids: Science and Commerce (Pathak P, Shekhar RN, Sharma M and Sood A eds), Bishen Singh Mahendra Pal Singh, New Connaught Place, Dehra Dun, India pp 17-28.

Kumaria S., Chrungoo N.K. and Tandon P. (1990). Activities of some oxidative enzymes in axenic cultures of protocorms of *Cymbidium giganteum* Wall, as influenced by different growth regulators. J Orchid Soc India 4(1-2): 37-44.

Laishram J.M. and Devi S. (1999). Micropropagation of *Renanthera imschootiana* Rolfe. through shoot tips, axillary bud and leaf segment culture. J Orchid Soc India 13(1-2): 1-4.

Latha P.G. (1999a). *In vitro* culture of the regionally endangered medicinal orchid *Nervilea aragoana* Gaud. J Orchid Soc India 13(1-2): 59-64.

Latha P.G. (1999b). Micropropagation and *in vitro* flower induction in *Habenaria crenifera* Lindl. J Orchid Soc India 13(1-2): 47-53.

Malmgren S. (1992). Large-scale asymbiotic propagation of *Cypripedium calceolus* — Plant physiology from the surgeon's point of view. Botanic Gardens Micropropagation News 1: 59-63.

Marlow S.A. and Butcher D. (1987). Propagation of two rare orchid species from seeds. Orchid Review 95: 331-334.

Martin K.P. (2003). Clonal propagation, encapsulation and reintroduction of *Ipsea malabarica* (Reichb.f.) J. D. Hook., an endangered orchid. *In Vitro* Cell Dev Biol Plant 39: 322-326(5).

Micheli M., Piccioni E. and Standardi A. (1996). Conversion of synthetic seeds of M.26 clonal apple root stock on different substrata. *In Vitro* 32(3) 2: 92A-93A.

Mitra G.C. and Chaturvedi H.C. (1972). Embryoids and complete plants from unpollinated ovaries and from ovules of *in vitro* grown encapsulated flower buds of *Citrus* spp Bull Torrey Bot Club 99: 184-189.

Morel G.M. (1960). Producing virus free *cymbidiums*. Amer Orchid Soc Bull 29: 495-497.

Morel G.M. (1964). A new means of clonal propagation of orchids. Amer Orchid Soc Bull 31: 473-477.

Na H. and Kondo K. (1996). Cryopreservation of tissue cultured shoot primordial from shoot apices of cultured protocorms in *Vanda pumila*, following ABA preculture and desiccation. Plant Sci 118: 195-201.

Nagayoshi T., Hatanaka T. and Suzuki T. (1996). Seed propagation of *Habenaria radiata*. Nature and Human Activities 1: 67-81.

Nayak N.R., Patnaik S. and Rath S.P. (1997). Direct shoot regeneration from foliar explants of an epiphytic orchid *Acampe praemorsa* (Roxb.) Blatter and McCann. Plant Cell Rep 16: 583-586.

Nayar M.P. and Sastry A.R.K. (1997-1998), (1999). Red Data Book of Indian Plants. Vols 1-3. Botanical Society of India, Calcutta.

Olivia A.P. and Arditti J. (1984). Seed germination of North-American Orchid. II. Native California and related species of *Aplectrum, Cypridium* and *Spiranthes*. Bot Gaz 145(4): 495-501.

Park Y.S., Kakati S., Kend A. and Okabe M. (1996). Efficient propagation of protocorms-like bodies of *Phalaenopsis* in liquid medium. Plant Cell Tiss Org Cult 45: 79-81.

Patel A.V., Pusch I., Wagner G.M. and Vorlop K.D. (2000). A novel encapsulation technique for the production of artificial seeds. Plant Cell Rep 19(9): 868-874.

Pathak P., Mahant K.C. and Gupta A. (2001). *In vitro* propagation as an aid to conservation and commercialisation of Indian orchids: Seed culture. In: Orchids: Science and Commerce (Pathak P, Shekhar RN, Sharma M and Sood A eds), Bishen Singh Mahendra Pal Singh, New Connaught Place, Dehra Dun, India pp 319-362.

Payati A.N. and Murty H.N. (1999). *In vitro* propagation of *Acampe praemorsa* (Roxb.) Blatter and McCann from leaf explants. J Orchid Soc India 13(1-2): 33-35.

Piccioni E., Capuano G. and Standardi A. (1997). Conversion of M26 apple root stock synthetic seeds on perlite. *In Vitro* 33, (3)2: 51A.

Pierik R.L.M., Sprenkels P.A., Van der Harst B. and Van der Meys Q.G. (1988). Seed germination and further development of plantlets of *Paphiopedilum ciliolare* Pfitz. *In Vitro* Sci Hortic 34: 139-156.

Rajeevan P.K. and Shobhana A. (1993). Performance of certain epiphytic species in Central Kerala. J Orchid Soc India 7(1-2): 31-35.

Ronse A. (1990). *In vitro* culture at the National Botanic Garden of Belgium. Botanic Gardens Micropropagation News 1: 14-15.

Rout G.R., Das G., Samantary S. and Das P. (2001). Micropropagation of *Plumbago zeylanica* L by encapsulated nodal explants. J Hort Sci Biotechnol 76: 24-29.

Saiprasad G.V.S. and Polisetty R. (2003). Propagation of three orchid genera using encapsulated protocorm-like bodies. *In Vitro* Cell Dev Biol Plant 39: 42-48(7).

Sharma A., Tandon P. and Kumar A. (1992). Regeneration of *Dendrobium wardianum* Warner (Orchidaceae) from synthetic seeds. Indian J Exp Biol 30: 747-748.

Sharma C.M. and Kaur S. (1998). Effect of different media on the regeneration and establishment of plantlets of *Arundina grammifolia* (D Don) Hochr International festival of Orchid, Abs p 99.

Sharma S.K. and Tandon P. (1986). Influence of growth regulators on asymbiotic germination and early seedling development of *Coelogyne punctulata* Lindl. In: Biology, Conservation and Culture of Orchids (Vij SP ed), Affiliated East-West Press Private Limited, New Delhi pp 441-451.

Sheelavanthmath S.S. and Murty H.N. (2001). *In vitro* propagation of a terrestrial orchid *Habenaria marginata* Coleb. J Orchid Soc India 5(1-2): 85-88.

Shimura H. and Koda Y. (2004). Micropropagation of *Cyprepidium macranthos* var. rebunense through protocorm-like bodies derived from mature seeds. Plant Cell Tiss Org Cult 78: 273-276.

Singh D.K. (2001). Orchid Diversity in India: An overview. In: Orchids: Science and Commerce (Pathak P, Shekhar RN, Sharma M and Sood A eds), Bishen Singh Mahendra Pal Singh, New Connaught Place, Dehra Dun, India pp 35-65.

Sinha S.K. and Hedge S.N. (1999). Regeneration of plantlets from *in vitro* leaf culture of *Renades arunoday* hybrid (*Aerides rosea* Loddigesex Paxt X *Renanthera imschootiana* Rolfe). J Orchid Soc India 13(1-2): 19-24.

Soneji J.R., Rao P.S. and Mhatre M. (2002). Germination of synthetic seeds of pineapple (*Annanus*). Plant Cell Rep 20: 891-894.

Stoutamire W. (1974). Terrestrial orchid seedlings. In: The Orchids-Scientific Studies (Withner CL ed) Wiley-Interscience, New York pp 101-128.

Tandon P. (2004a). Role of biotechnology in the conservation of plant genetic resources in the 21st century—an Indian perspective. In: Platinum Jubilee Lectures- 87th and 88th session of the Indian Science Congress Association (Banerjee SP and Mukherjee SP eds), Auto Print and Publicity House, Kolkata, India pp 40-67.

Tandon P. (2004b). Conservation and sustainable development of plant resources of North East India. Man and Society 1(1): 49-59.

Tandon P. and Kumaria S (1998). Threats to plant diversity in the high altitude areas of North-East India and conservation of rare and endangered plants using biotechnological approaches. In: Science at High Altitude. Proceedings of the National Symposium (Raha S, Ray PK and Sinha B eds), Allied Publishers Limited, India pp 140-147.

Tandon P. and Kumaria S. (2005). Prospects of plant conservation biotechnology in India with special reference to the Northeastern region. In: Biodiversity: Status and Prospects (Tandon P, Sharma M and Swarup R eds), Narosa Publshing House, New Delhi, India pp 79-91.

Teng W.L., Nicholson L. and Teng M.C. (1997). Micropropagation of *Spathoglottis plicata.* Plant Cell Rep 16: 831-835.

Thammasiri K. (2000). Cryopreservation of the seeds of Thai orchid (*Doritis pulcherrima* Lindl.) by vitrification. Cryo Letters 21(4): 237-244.

Van I.B., Hang P.T.N., Ann H.L.T. and Tran T.V.K. (1999). High frequency shoot regeneration from *Rynchostylis gigantean* (Orchidaceae) using thin cell layers. Plant Growth Regulation 28(3): 179-185.

Vij S.P. (1995). Genetic resources of orchid. In: Advances in Horticulture Vol. 12- Ornamental plants I (Chadha KL and Bhattacharjee SK eds), Malhotra Publishing House, New Delhi, India pp 153-181.

Vij S.P. and Kaur P. (1992). Regeneration potential of foliar peels of *Rynchostylis retusa.* J Orchid Soc India 6(1-2): 71-74.

Vij S.P. and Sharma V. (1996). Regeneration competence of *Vanda cristata* perianth segment: A study *in vitro.* J Orchid Soc India 10(1-2): 25-29.

Vij S.P., Pathak P. and Kher A. (1997). Regeneration response of *Rynchostylis gigantea* inflorescence segments: A study *in vitro.* J Orchid Soc India 11(1-2): 75-78.

Vij S.P., Kher A. and Pathak P. (2000). Micropropagation of *Bulbophyllum careyanum* (Hook.) Spreng using pseudobulb segments. J Orchd Soc India 14(1-2): 47-55.

Wang Q.C., Batterman O., Li P., Joseph B. and Gafny R. (2002). Cryopreservation of *in vitro* grown shoot tips of royer citrange (*Poncirus trifoliate* (L.) Raf x *Citrus sinensis* (L.) Osbect) by encapsulation-dehydration. Plant Cell Rep 20: 901-906.

Yam T.M. and Weatherhead M.A. (1988). Germination and seedling development of some Hong Kong Orchids. Lindeyana 3(3): 156-160.

Yam T.M., Arditti J. and Weatherhead M.A. (1989). The use of darkening agents in seed germination and tissue culture media for orchids: A review. J Orchid Soc India 3(1-2): 35-39.

Yam T.M. and Weatherhead M.A. (1990). Early growth of *Hataeria cristata* seedlings and plantlet initiation from rhizome nodes. Lindleyana 5: 199-203.

Zeigler T.W., Sheehan T.J. and Poole R.T. (1967). Influence of various media and photoperiod on growth and amino acid content of orchid seedlings. Amer Orchid Soc Bull March: 195-201.

Zhou T.S. (1998). Propagation of *Phalaenopsis* orchids. Hortscience 33(4): 595.

Index